AF833439

Class-D Audio Power Amplifier Design

Class-D Audio Power Amplifier Design is a comprehensive and detailed guide to Class-D amplifiers. Including historical perspectives covering the evolution of Class-D to the present day, this book provides a detailed discussion of each Class-D functional block.

With design guidance to optimize performance, the reader will learn from a series of real-life examples and design challenges to equip them in designing reliable, high-performance Class-D audio amplifiers. Featuring numerous diagrams and schematics to aid learning, chapters also cover a wide range of topics, such as:

- Power supply considerations
- EMI mitigation
- Multi-channel designs
- Gate driver circuits
- L-C demodulation filters.

Class-D Audio Power Amplifier Design is the essential reference guide for students, researchers, and professionals engaged with audio electronics.

Robert N. Buono has over 40 years' experience in analog and power electronic circuit design. He is an expert in the design of Class-D audio power amplifiers and switched-mode power supplies and is a named inventor on 16 patents. He is presently a Principal Engineer at Crestron Electronics, where he has designed successful Class-D amplifier products for the residential and install markets.

"*Class-D Audio Power Amplifier Design* is a thorough, in-depth treatment brimming with practical design techniques from a top-tier design engineer. With full chapters on EMI mitigation, PCB layout, and audio performance measurement, amongst others, this book provides us with a wealth of tips and techniques that takes years to accumulate."

Dennis Fink, *IC Engineering Manager, THAT Corporation, Milford MA, USA*

Class-D Audio Power Amplifier Design

Robert N. Buono

Routledge
Taylor & Francis Group
LONDON AND NEW YORK

Designed cover image: BONDART / Getty Images

First published 2025
by Routledge
4 Park Square, Milton Park, Abingdon, Oxon OX14 4RN

and by Routledge
605 Third Avenue, New York, NY 10158

Routledge is an imprint of the Taylor & Francis Group, an informa business

British Library Cataloguing-in-Publication Data
A catalogue record for this book is available from the British Library

Library of Congress Cataloging-in-Publication Data
Names: Buono, Robert N., author.
Title: Class-D audio power amplifier design / Robert N. Buono.
Description: New York, NY : Routledge, 2025. |
Includes bibliographical references and index.
Identifiers: LCCN 2024035942 (print) | LCCN 2024035943 (ebook) |
ISBN 9781032356693 (hbk) | ISBN 9781032356327 (pbk) |
ISBN 9781003327943 (ebk)
Subjects: LCSH: Audio amplifiers–Design and construction. |
Switching power supplies–Design and construction. |
Power amplifiers–Design and construction. |
Sound–Recording and reproducing–Equipment and supplies–Design and construction.
Classification: LCC TK7871.58.A9 B86 2025 (print) | LCC TK7871.58.A9 (ebook) |
DDC 621.39/5–dc23/eng/20241228
LC record available at https://lccn.loc.gov/2024035942
LC ebook record available at https://lccn.loc.gov/2024035943

ISBN: 978-1-032-35669-3 (hbk)
ISBN: 978-1-032-35632-7 (pbk)
ISBN: 978-1-003-32794-3 (ebk)

DOI: 10.4324/9781003327943

Typeset in Bembo
by Newgen Publishing UK

Contents

About the Author xvii

Acknowledgements xviii

1 **Introduction** 1

Audio Amplification 1

Sources of Information on Class-D Audio Amplifier Design 4

Using the References 4

Comparison with Class-A, B, AB, G, H 5

Key Parameters of Class-D Audio Power Amplifiers 17

Evolution of Semiconductor Power Switches Used in Class-D Amplifiers 18

Class-D Amplifier History 20

Organization of the Book 21

Goal 22

Equations 22

Overview of the Reference Design 22

Audio Performance of the Reference Design 23

Mono or Stereo Operation 23

Page-by-Page Summary of the Reference Design Schematic 25

Bibliography 29

Appendix 1A: Reference Class-D (Fixed-Frequency) Schematic 30
Input/Preamp 30
Error Amplifier 31
PWM Comparator 32
Triangle Ramp Generator 33
Gate Driver (Half-Bridge Power Stage) 34
L-C Demodulation Filter 35
Housekeeping Circuits 36
Protection and Fault-Sensing Circuits 37
AC-to-DC Power 38
Low-Voltage Block Diagram 39
Low Voltage 1 40
Low Voltage 2 41
Appendix 1B: Bill of Materials (Class-D Reference Design) 42

2 Input Circuits 49
Input Circuitry Differences between Class-D and Class-AB 49
Requirements for Op-Amps Used in Analog Preamp Circuitry for Class-D 50
Critical Specifications 50
Specifications of Secondary Importance 51
Review of Critical Op-Amp Specifications 51
Review of Input Circuitry in the Reference Schematic 54
General Schematic Notes 54
Input Overvoltage Tolerance (Electrical Overstress) 55
Supply Rail Bypassing 58
Common-Mode Input Voltage Range 58
Input Coupling 59
Phasing 60
Bibliography 60

3	**Comparator and PWM Circuits**	62
	Comparator Characteristics for Class-D Usage	62
	Supply Voltage Range	62
	Propagation Delay	63
	Input Voltage Range	63
	Input Overdrive Tolerance	64
	Power Dissipation	65
	Clipping Detection	68
	Class-D Amplifier Output Voltage vs PWM Duty Cycle	69
	Alternative Design Approaches for the PWM Comparator Circuitry	72
	Bibliography	72
4	**Ramp Generator Circuits**	74
	Triangular Ramp versus Sawtooth Waveshape	74
	Detailed Discussion of the Reference Schematic Ramp Generator Circuitry	76
	Current Mirror Operation	76
	Current Switching / Steering, in More Detail	77
	Ramp Buffer Circuits	80
	Bibliography	82
5	**Level-Shifter Circuits**	83
	Desired Features and Characteristics of Level-Shifter Circuits for Class-D Amplifier Design	83
	Bibliography	87
6	**Gate Driver Circuits**	88
	Gate Charge	88
	Gate Drive Power Calculations	92
	Effects of Reducing or Eliminating the External Gate Resistor	95
	UVLO	97

Different Types of Drivers (Dual Half-Bridge, Independent Dual, PWM, Isolated, or Referenced to Negative Rail) 99
dv/dt CM Ratings 100
VS Transient Voltage Undershoot 101
Built-in Protection Features 101
Important Specifications for Gate Drivers 102
Propagation and Turn-on Delays 102
Quiescent Power 103
Input Voltage Thresholds and Voltage Range 105
Input Voltage Threshold Hysteresis 105
Frequency Range and Pd 107
Drive Current 107
Isolation and Level-Shift Technologies 107
SiC and GaN Driver Types 108
Higher-Frequency Operation 110
Discrete Gate Driver Circuits 110
A Word about Gate Drive Transformers 112
Bibliography 113

7 Bias Supply for Upper Gate Driver 114
Objectives and Challenges of Bias Supply Design 114
Bootstrap Circuit Design 115
Bootstrap Circuits with Internal Switch 115
Completely Floating Gate Drive Supplies 118
Advantages Explained 121
Creating Multiple Floating Bias Supplies 122
Bibliography 123

8	**Half-Bridge and Full-Bridge Switching Power Stages**	125
	Modulation Types for Full-Bridge Arrangement	125
	Characteristics of Half-Bridge and Full-Bridge Circuits	125
	Power Supply Rail Connections	126
	Power Switch Operation, in Detail	128
	Full-Bridge Switching Stages	138
	Bibliography	147
9	**L-C Demodulation Filters**	148
	L-C Filter Characteristics and Equations	148
	Residual Output Ripple Voltage at the Class-D Switching Frequency	153
	Benefits and Disadvantages of Zobel Networks	154
	Zobel Resistor Power Dissipation	154
	Non-Linearities	156
	Inductor Types	156
	Detailed Design of the Class-D Inductor	158
	Magnet Wire Size (AWG)	163
	Core Selection	163
	Checking Flux Density and Saturation	165
	Core Loss Calculation	165
	Notch Filters	171
	Capacitor Types	173
	ESR	173
	Film Capacitors	175
	Filterless Class-D Amplifiers	176
	Bibliography	181

10	**Sources of Distortion and Noise**	183
	Overview	183
	Power Switching Stage Distortion	184
	Other Sources of Noise and Distortion	191
	Passive Components	191
	Comparator and Ramp Circuitry	195
	Gate Driver	196
	Switching Power Stage	196
	Summary	196
	Bibliography	197
11	**Class-D Power Stage Losses, a Deep Dive**	199
	Overview	199
	Hard-Switched Mode	203
	Body Diode Reverse Recovery (a Detailed Discussion)	208
	Addition of External Schottky Diodes (Drain-to-Source)	213
	Mosfet Capacitive Switching Losses	214
	Gate Charge (Qg) Capacitive Power Loss	216
	Bibliography	217
12	**Feedback and Compensation**	219
	Overview	219
	A General Discussion of Negative Feedback	220
	Open-Loop Designs	220
	Stability Criteria	221
	Simplifying Simulation	222
	Feedback before the L-C Filter	232
	Feedback Loop Measurement Techniques	238
	Practical Feedback Loop Measurement Tips	239
	Bibliography	240

13	**Housekeeping and UVLO Circuits**	242
	The Roles of Housekeeping Circuitry and UVLO	242
	Detailed Circuit Description of the Reference Design	245
	+2.5V Reference	250
	Reference Biasing Option for Designs with Higher Rail Voltage	251
	Bibliography	253
14	**Protection Circuits (Overcurrent, Overtemperature, DC-Offset)**	255
	Class-D Amplifier Fault Conditions	255
	DC Output Condition (DC Fault)	256
	Disconnect Response	258
	Detecting DC Fault	262
	Overcurrent (OC) Detection and Response	270
	Class-D Amplifier Output Current versus Inductor Current	270
	Current Sensing	273
	Types of OCP Response	285
	Overtemperature (OT) Detection and Response	286
	Thermal Example (Class-AB vs Class-D)	286
	Bibliography	294
15	**EMI Mitigation**	296
	Reasons Why Class-D Amplifiers Generate EMI	296
	Current Loops (B-Fields)	296
	Voltage Nodes (E-Fields)	298
	Cables	299
	CM and DM	300
	Spread-Spectrum Techniques	304
	Radiated and Conducted Susceptibility/Immunity Tests	304
	Bypass Noise to Chassis	305

	Electrostatic Discharge (ESD)	309
	Bibliography	311
16	**PCB Layout and Thermal Considerations**	312
	Problems a Poor Layout Can Cause	312
	Develop a PCB Floorplan (Placement)	313
	Best Practices	313
	Switching Current Containment	317
	Functional Blocks	320
	Key Component Placement	321
	Current Loops	321
	Common Impedance	321
	Shielding of the Switch Node	325
	L-C Demodulation Filter Layout	327
	PCB Trace Sizing and Current-Carrying Considerations	327
	PCB Layer Count and Stackup	328
	Other Routing Considerations	329
	Thermal Considerations	330
	Thermal Design 1 (Using Metal Heatsink)	330
	Thermal Design 2	334
	Heatsinking of Through-Hole Components	335
	Thermal Considerations for Other Components	335
	Thermal Sensing	336
	Example Layout	337
	Features	337
	Bibliography	338
17	**Power Supply Considerations**	340
	Block Diagrams of ACDC and DCDC Power Distribution	340
	ACDC Converters	342

	Specifying Original Equipment Manufacturer (OEM) ACDC Converters for Audio Amplifier Applications	343
	Supply Rail Pumping	346
	Power Rail Pumping: Half-Bridge versus Full-Bridge Class-D Topologies	350
	Output Connection Phasing to Reduce Power Rail Pumping	351
	ACDC Power Supplies Optimized for Class-D Audio	352
	Housekeeping ACDC	353
	Low-Voltage Power Distribution	354
	DCDCs; Detailed Circuit Discussion	355
	–7VA DCDC	358
	LDOs	361
	Bibliography	362
	Appendix: Current Demand for Reference Design DCDCs and LDOs	362
18	**70V and 100V Constant-Voltage Amplifiers (High Impedance)**	365
	Definition and Nomenclature	365
	Constant-Voltage Amplifiers Employing Output Step-up Transformers	368
	Direct Drive	370
	Output Stage Configurations for Hi-Z and Full-Range Loads using Direct Drive	371
	Overcurrent Protection for Direct-Drive Hi-Z/Lo-Z Outputs	374
	Selectable or Variable Supply Rail Voltages for Hi-Z/Lo-Z Amplifiers	375
	Bibliography	375
19	**Multi-Channel Designs**	377
	Interactions between Channels in a Multi-Channel Amplifier Design	377
	Power Supply Induced Crosstalk	378
	Via Common High-Voltage DC Power Supply Rails	378
	Capacitive Coupling	380
	Electromagnetic Coupling	380

Coupling Due to Leakage Fields from Inductors 381
I × R Induced Voltage Disturbances (Common Impedance Coupling) 382
Crosstalk in a Multi-Gate Component 384
Performance Considerations in Multi-Channel Designs 384
Shared Circuit Allocation 385
Power Supplies 385
Ramp Generator 386
Bibliography 387

20 Audio Performance Measurements and Instrumentation 388
Overview 388
Characterizing Class-D Outputs in the Time Domain 390
Noise 390
Fast Fourier Transform (FFT) 391
THD+N 391
THD+N versus Power 391
THD+N versus Frequency 392
FFT Analysis 393
InterModulation Distortion Measurements 393
SMPTE IMD 394
Twin-Tone IMD 395
Dynamic or Transient IMD Measurements 396
A Discussion about SR (Slew Rate) 396
Frequency Response 399
Damping Factor 399
Crosstalk Measurements 401
Causes of Crosstalk 402
Setting up Crosstalk Measurements 402
Bibliography 404

21 Alternate Modulation Schemes for Class-D Amplifiers 406
Modulation Schemes 406
Fixed-Frequency (Clocked) Modulators 406
Self-Oscillating Modulators 407
Multi-Level Switching Schemes 411
Self-Oscillating Class-D Amplifier Topologies 411
Aspects of Self-Oscillating Modulators 411
Types of Self-Oscillating Modulators 414
Bibliography 420

22 Digital Applications, Interfaces, and Digital Signal Processing 424
Digital versus Analog Circuitry 424
Field-Programmable Gate Arrays (FPGAs) 426
DSP 427
Digital-Input Class-D Amplifiers 427
"Fully Digital" Class-D Closed-Loop Amplifiers 429
Bibliography 429

23 Integrated Circuit (IC) Approaches and Advances in Components and Materials 431
Integrated Circuit (IC) Solutions 431
Texas Instruments (TI) 433
Infineon 435
Analog Devices 436
Monolithic Power Systems (MPS) 436
NXP 436
ST Microelectronics 436
Cirrus Logic 437
ICEpower 437

Class-D Modules 437
Advances in Components and Materials 437
Power Switches 438
Gate Drivers Optimized for GaN 440
Comparators 441
Op-Amps 442
Current Sensors 442
Inductor Core Materials 443
Magnet Wire 443
Capacitor Technologies 443
PCB Materials and Construction 444
Thermal Management Materials: Gap Pads 444
Integrated Class-D (IC) Solutions 444
Bibliography 445

Index **447**

About the Author

I have been a lifelong audio enthusiast. I purchased my first pair of Advent Loudspeakers in 1974, shortly after turning 16 and earning enough money to buy them. These were quickly followed by a second pair, after reading an article reviewing the Double Advent System in *The Absolute Sound* magazine (Issue 1, Spring 1973). I was forever after hooked on high-quality, high-power sound reproduction. After attaining my Master's degree, my first assignments as an Electrical Engineer (EE) were designing ACDC and DCDC converters for IREX Medical Systems and LeCroy Corporation. As is common for young EEs just starting out, the assignments which followed were more of the same, but I did not mind this, as switchmode power conversion was gaining popularity (over less efficient linear power conversion techniques) and I enjoyed learning the intricacies of these power switching stages. When I was later introduced to Class-D power amplifier circuits, I had an epiphany when I realized these stages shared the same topologies as my switchmode power supplies and had the added benefit that they could produce beautiful music waveforms instead of a steady DC voltage! I felt like I had come full circle and was in my element. To pursue Class-D design, I joined Crest Audio as a Design Engineer and later, my present employer, Crestron Electronics, where I design Class-D amplifiers, as well as ACDC and DCDC converters, as a Principal Engineer.

Robert N. Buono
May 2024

Acknowledgements

I want to thank my wonderful wife, Joyce, for her 24/7 support of everything I do. I also want to thank my family: my stepdaughters Megan and Lindsay, my daughter Gabriella, my grandchildren Joshua and Avery, and my son-in-law Jay, for making the non-technical part of my life so full and enjoyable, and for always listening to my very loud (but undistorted) music, without complaint. I was truly blessed to have a mom, Lucy, who embodied unconditional love, and a dad, Joe, who always expected our best, and sisters, Phyllis Anne and Lucille, and brother, Jack, who always deliver exactly that.

In researching this book, I was struck by the wealth of knowledge shared by so many investigators. These are Professors, Engineers, and Scientists, who, by documenting and sharing their measurements and findings, help provide a continually growing knowledge base, used by future investigators to further the state of the art in Class-D audio power amplifier design. These contributors are too numerous to mention here, but please take note of their contributions when reading the references at the end of each chapter, as well as the references cited *within* each reference.

I also want to thank Frank Amato at Crestron Electronics Inc. for reviewing the manuscript, as well as Emily Tagg and Jennifer Hicks of Taylor & Francis Group for their help in guiding me on the path to publication of this book. Thanks also to Drew Stanley for his excellent editing, and to author and lifelong friend, Joel Samberg, for his unwavering encouragement during the writing of this book.

Robert N. Buono

Chapter One

Introduction

Audio Amplification

It is difficult to exaggerate the practical and economic importance of audio power amplifiers in modern society. Every single loudspeaker must be driven by an audio amplifier. This means every smartphone, every laptop, every flat-screen TV and display, every headphone and earbud. Next, consider automotive sound systems. Here the requirements are small size and light weight, with high-efficiency operation to reduce battery drain, while simultaneously providing high output power to low impedance loads. Next, consider PA systems in every public venue, ranging in size from large sports arenas, concert halls, and cinemas, to conference rooms and lecture halls. Basically, any amplified audible communication in our society requires the involvement of a loudspeaker and an audio power amplifier to drive it. It is very easy to take this arrangement for granted because it is so ubiquitous. Given the sheer numbers and market size, it is quite surprising that there are not more texts written on the subject of audio power amplifiers, especially and particularly Class-D amplifiers. Class-D amplifiers already dominate the portable and automotive markets, because their extremely high efficiency is required in order to maximize run-time on battery power. Additionally, because of the requirement for small size and weight, power dissipation with its attendant waste heat must be minimized in portable applications. Another market sector where Class-D amplifiers have replaced less efficient, Class-AB amplifiers is the pro audio market. Here, parameters that matter include power density, measured in watts per cubic inch and watts per pound. Sound and touring companies want

DOI: 10.4324/9781003327943-1

amplifiers with tremendous output power, that are as small and lightweight as possible, so that they can be packed and loaded, trucked to different locations, and unloaded, without breaking the backs of the sound crew. The high efficiency and scalability of Class-D designs make high power density pro audio amplifiers possible. Now in addition to having already dominated the portable device, automotive, and pro audio markets, Class-D is making undeniable inroads into product areas where audio sound quality has been deemed paramount. In those areas, Class-AB amplifiers have long dominated because of their inherently lower-noise (non-switching) design topologies. However, due to continual advances in Class-D audio amplifier sonic performance, this audio amplifier topology is becoming more accepted in audiophile applications as well. This book will describe the performance characteristics of Class-D amplifiers and the sources of distortion, as well as techniques and technologies used to improve performance, which have led to acceptance of Class-D audio power amplifiers in these market sectors.

High efficiency, as noted in the introductory paragraph, is a hallmark of, and figure-of-merit in, Class-D audio power amplifier design. It is therefore important to define it here. It will be understood to be the ratio of the audio output power to the total power supplied to the amplifier. This "power supplied" includes power supplied by the DC power supply voltage rails, as well as any low-voltage power supply rails which may supply power to low-voltage signal processing circuits in the amplifier.

For example, suppose a single-channel Class-D audio amplifier supplies 100W of measured output power to an 8-ohm load (speaker). Further, suppose it is of the "half-bridge" topology (to be explained later), utilizing positive and negative 48VDC power supply rails, and a +12VDC low-voltage supply, from which are derived the low-voltage supplies required by the signal processing circuits. The measured current consumption on these rails is:

+ 48VDC @ 1.13ADC = 54.35W

− 48VDC @ 1.14ADC = 54.93W (includes extra power for gate drive bias supply)

+12VDC @ 115.83mA = 1.39W

Total "Power Supplied" = 110.67W

Efficiency = $\eta = 100W_{OUT} / 110.67W_{IN}$ = **90.36%**

In this analysis, the power losses in the low-voltage circuits (powered by +12V) are included, because high efficiency of the complete amplifier solution is desired. Though it is the efficiency of the high-power circuits that normally gets the most attention, it is important to consider all power losses when designing a high-efficiency Class-D amplifier solution.

The *purpose* of this book is to provide a sufficiently thorough treatment of Class-D audio amplifier design, so that the reader gains a working understanding of the circuit design principles involved. It is the hope of this author that the reader will also come away with a "feel" for Class-D design, knowing what design decisions need to be made, and having the knowledge to make the proper tradeoffs, in order to achieve a specific set of performance and cost goals. One important characteristic of Class-D audio amplifier design is that most design parameters are interrelated. It is critical to be aware of this as early as possible. It is hoped that the reader will come away with an understanding and appreciation of these interactions. Some examples of this interrelationship are the following: Power transistor switching deadtime adjustment affects not only power loss and efficiency, but audio performance parameters such as total harmonic distortion (THD) and clipping threshold. It will also affect electromagnetic interference (EMI) performance. Selecting DC power rail voltages will affect loop gain and bandwidth, clipping levels, and power dissipation, as well as the obvious component breakdown voltage requirements. These are just two examples (which will be explained in detail later in this book) of how design parameters affect multiple aspects of the design. Frequently, the effects of these changes are in opposition, and this is one of the main challenges of Class-D design… understanding the opposing effects and making the best design decisions for the given application. For example, increasing switching deadtime will decrease power dissipation, improving efficiency, but will result in increased THD, unless feedback loop gain can be increased to compensate for this effect. Similarly, increasing the DC supply rails will allow for more clipping headroom (in a fixed-frequency design with fixed deadtime), and increased loop gain and therefore improved THD, but will result in greater power loss and lower efficiency, and may require components with higher breakdown voltage ratings, which usually comes with higher resistance specifications (such as higher $R_{DS}ON$ for mosfets).

The content presented will be both broad and detailed. It will be presented in a way that will provide valuable information to practicing professionals, while providing enough supporting background material so that students and novices can understand the concepts as well. There are a great deal of helpful references that deal with all aspects of Class-D amplifier design. These references should be used when a deeper understanding of a particular aspect of Class-D design is needed. These references are cited at the end of each chapter for this purpose. The assumed prerequisites for the reader are a basic understanding of electronic circuits, and the ability to read and understand schematic diagrams of electronic circuits.

Sources of Information on Class-D Audio Amplifier Design

The technical information available on Class-D audio amplifier design is spread among a wide range of industry sources, including application notes, datasheets, and white papers from many different electronic component manufacturers. Many of these appear as references at the end of each chapter. This design information will be presented here in an orderly and comprehensible manner, in a single comprehensive text, starting with a reference Class-D audio amplifier design, which is then analyzed at a functional block level and then at the component level. That design then serves as a starting point from which specific topics in Class-D audio amplifier design are treated in greater detail. The reader will benefit by having a single source of Class-D design information, with valuable references in each chapter, to promote further investigation on each topic.

Using the References

As previously mentioned, there is a great deal of technical information available on Class-D audio amplifier design, distributed among a wide variety of sources. The references listed at the end of each chapter are a great starting point for learning more about a given topic and taking a deeper dive. Each of these references typically has its own list of references on the same topic. The reader is strongly encouraged to dig deeper and deeper into a topic by pursuing this thread of references, until they are satisfied that their understanding is complete.

Part of what makes Class-D design so interesting, exciting, and challenging is that it involves many different disciplines, including analog and digital circuit design and high-frequency switch-mode power circuit design, as well as low-noise circuit design and layout, and thermal design. In order to present this information from so many different areas of electronic circuit design in an orderly and comprehensible manner, a complete Class-D design will be presented initially as a reference. Then, each Class-D functional block will be discussed in detail, as well as alternate circuit topology choices available for each block. Design guidance will be given for optimizing the performance of each functional block. Component selection is critical to Class-D performance, and new components and technologies targeted to Class-D applications are constantly being developed. This book will provide guidance for component selection for Class-D, explaining which characteristics and specifications are critical for each component and which need to be optimized in order to meet the performance objectives.

With green, energy-saving initiatives continually evolving and being imposed by regulatory authorities, the high energy efficiency of Class-D makes it the topology of choice in more and more application areas. New technologies and components, such as wide bandgap (WBG) transistors (SiC and GaN), are poised to make a big impact on Class-D design, allowing significant improvements in the key performance metrics. This will serve to further increase market adoption of, and interest in, Class-D design.

Comparison with Class-A, B, AB, G, H

The distinguishing characteristic of Class-D amplifiers is that they utilize a switching power stage, employing electronic switches (typically mosfets) switching at ultrasonic (>20,000Hz) frequencies. These switching devices are thought of as being in only one of two possible idealized states: either fully off (non-conducting) or fully on (fully conducting with ideally zero on-resistance). This is in direct contrast to amplifiers using Class-A, B, AB, G, or H stages. Those classes utilize power devices which are operated in a *linear* mode, in which their conductivity is modulated over a continuous range. The main impetus behind the development of Class-D amplification topologies is to improve power conversion efficiency. There are many follow-on benefits that accrue from improvements in efficiency. These are mainly reduced size and weight, reduced waste heat and heatsink size, and reduced cost. It can be argued that the evolution of all the different amplifier classes has been driven by a quest for improved power conversion efficiency. Class-A, being the first and simplest topology, had the lowest efficiency and highest quiescent power dissipation. Classes B and AB were invented specifically to improve upon the efficiency of Class-A designs. Subsequently, Classes G and H were developed as attempts to improve upon the efficiency of Class-AB designs. Class-D has always held the promise of achieving the highest overall power conversion efficiency, because of the two (ideally) lossless states of the power devices. Jiang (2017) includes a very informative graph (Figure 7 in that document), which compares the efficiency of Class-D to Classes AB, G, and H. That graph is reproduced here as Figure 1.1.

It can be seen that Class-D maintains the highest overall efficiency from low output power to high output power. Class-H has slightly less efficiency, and Class-G shows the "cusp" effect resulting from the switching of different rail voltages as needed. Two-tier and three-tier examples of Class-G appear in the graph. If the number of tiers were extrapolated to infinity, Class-H operation would result. *Note:* In this example, Class-H employs a tracking rail voltage scheme, where the rail voltages track the output voltage level, always staying slightly higher than the required output voltage. This accomplishes the goal of low dropout voltage (headroom) across the linear pass device, achieving low power dissipation and high efficiency in a linear (non-switching) power stage. In this example, Class-G uses discrete voltage steps in the rail voltages supplying the power stage. This distinction is stated here, because in some parts of the world, these definitions of Class-G and Class-H are swapped. Class-AB has much lower

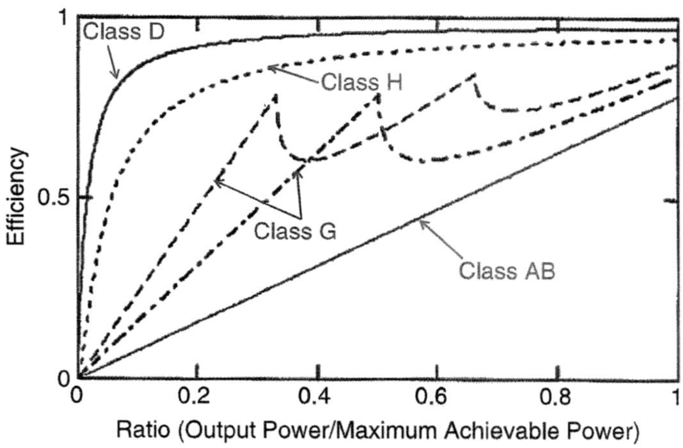

Figure 1.1 Efficiency Comparison of Class-D to Classes AB, G, and H

Source: © 2017, IEEE. Reprinted, with permission, from *IEEE Solid-State Circuits Magazine* (Volume: 9, Issue: 3, Summer 2017).

efficiency and the graph illustrates its linear characteristic, with efficiency being extremely low at low output power.

When discussing the efficiency of Class-D amplification, it is extremely important to include a discussion of quiescent power. When designing a Class-D amplifier, the designer must always be mindful of, and attempt to reduce, quiescent power. Quiescent power is the baseline power required to operate the amplifier. This includes bias power for low-level audio circuits (preamp and modulator circuits) as well as gate drive circuitry and protection circuitry. In general, the power consumption of these circuits remains constant over the output power range of the amplifier, and so represents a fixed power loss. Therefore, this quiescent power loss becomes a larger portion of the total power loss as output power decreases. This results in decreased efficiency of Class-D amplifiers at lower levels of output power. It is for this reason that quiescent power consumption must be minimized. "Exploding the efficiency myth" (2010) uses this characteristic as a counter-argument to the claims about the superiority of Class-D efficiency. However, if quiescent power consumption is minimized through careful design, the Class-D topology will result in excellent efficiency over the full output load range of the amplifier, with comparatively less circuit complexity compared to alternate approaches.

The main impediments to the adoption of Class-D amplification as the dominant class have been the technical development of power switching devices with parameters specifically applicable to Class-D, and reservations on the part of designers and users alike about the effects of high-power switching on amplifier performance, including audio performance, noise, and EMI considerations.

The figures below illustrate the basic topology of each audio power amplifier class, beginning with Class-A (see Figure 1.2).

Figure 1.3 illustrates that the single output transistor (Q1) operating in Class-A mode is active (ON) and conducting current over the entire output voltage swing. The power dissipated by Q1 is the product of I_{OUT} times the voltage across Q1 ($V_{C\text{-}E}$). Since *both* these values can be large simultaneously, this leads to high power dissipation in Class-A designs, and very low efficiency.

In Class-B (see Figure 1.4), each transistor conducts over half (or less) of the output voltage waveform, and the two transistors never conduct at the same time. That is to say, their conduction intervals never overlap. Q1 conducts for the positive half-cycle of Vout, and Q2 conducts for the negative half-cycle. The graphic in Figure 1.5 is meant to indicate a case where there is an output voltage region

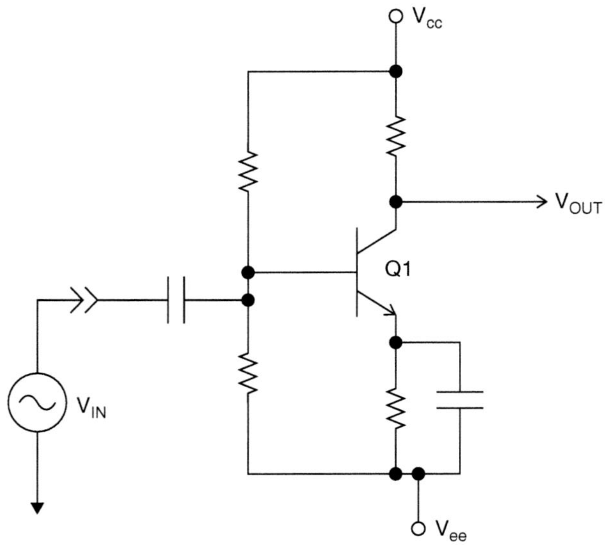

Figure 1.2 Class-A

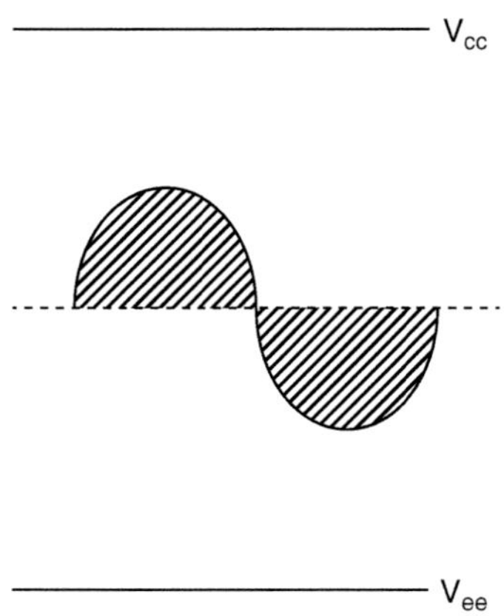

Figure 1.3 Class-A Conduction Interval

near the zero-crossing of the output waveform, during which neither transistor is ON. This will cause a specific type of distortion known as crossover distortion.

In Class-AB (see Figure 1.6) (see Cordell, 2019), there is a slight overlap in the conduction intervals of the two transistors. The graphic in Figure 1.7 is intended

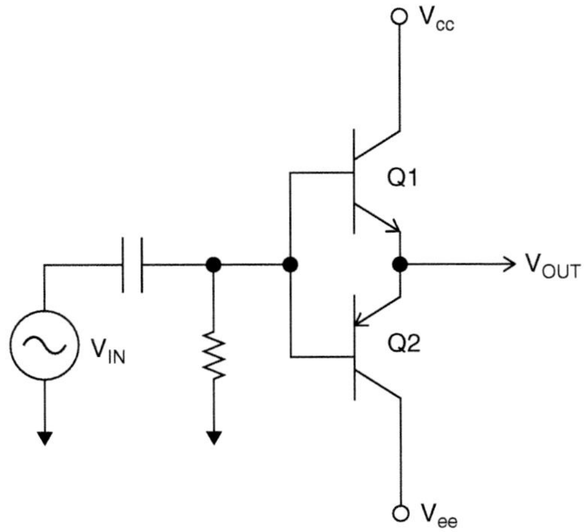

Figure 1.4 Class-B

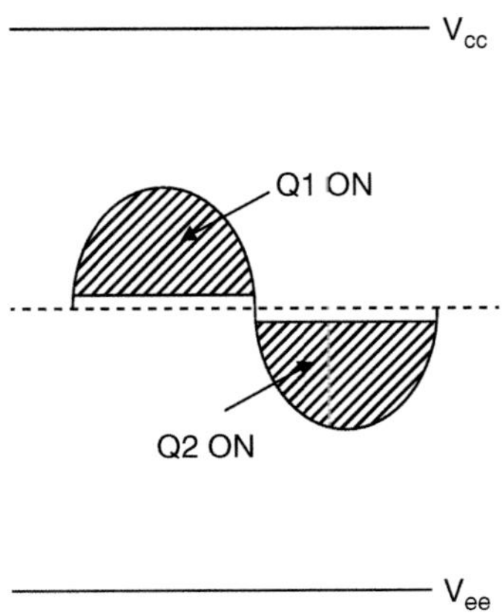

Figure 1.5 Class-B Conduction Intervals

to indicate that during the positive output voltage interval, Q1 is ON and the output current flows through Q1; however, Q2 remains slightly ON during the positive half-cycle of output voltage and conducts a small bias current. Likewise, during the negative half-cycle of output voltage, Q2 is ON and conducting the full output current; however, Q1 remains slightly ON, and conducts a small bias current during the negative half-cycle of the output voltage waveform. The fact

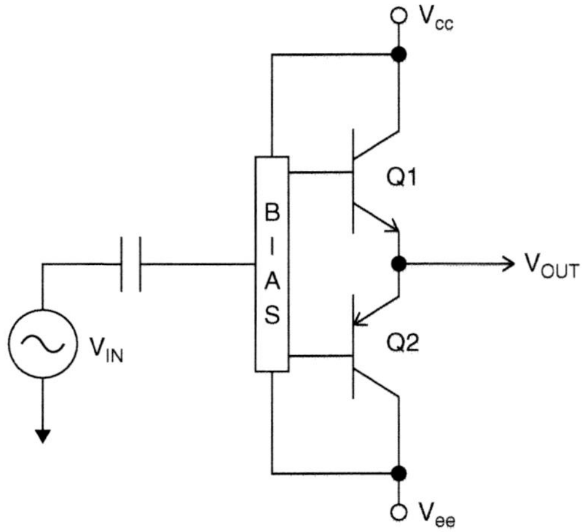

Figure 1.6 Class-AB

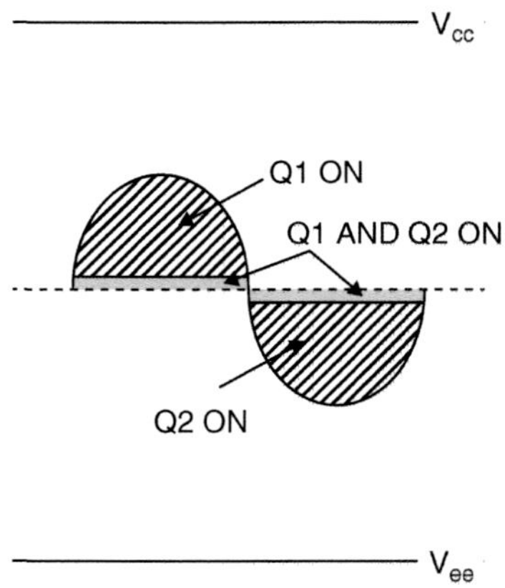

Figure 1.7 Class-AB Conduction Intervals

that the conduction intervals of both transistors overlap helps to greatly decrease crossover distortion. Cordell (2019) is an excellent source of information on Class-AB audio amplifier design.

Note: The designation of Class-G and Class-H can be confusing. These designations have been swapped in different parts of the world. In this text, the Class-G topology uses multiple fixed supply voltages, and Class-H employs variable, "tracking" power supply voltage rails.

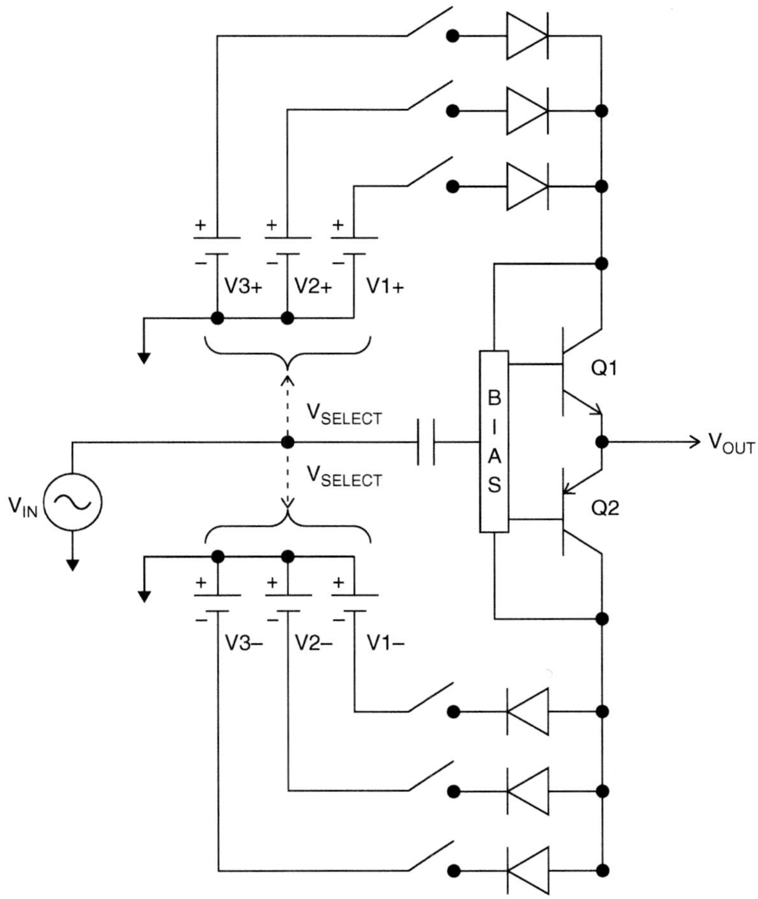

Figure 1.8 Class-G

The conduction intervals of Q1 and Q2 in Class-G (see Figures 1.8 and 1.9) are the same as for Class-AB. What is different is the way the amplifier's supply rails behave. In Class-G, there are more than one positive, and more than one negative power supply rail. In Figure 1.8, there are three "tiers" to the supply rails. At lower output voltages, the lowest supply voltage rails supply Q1 and Q2. As the output voltage approaches the level of the lowest power supply rails, sensing circuitry switches in the next higher power supply voltage rail tier. As Vout increases still further, the third power supply voltage rail tier will be switched in. This arrangement of output voltage tiers increases efficiency, because it reduces the voltage drop across the output transistors.

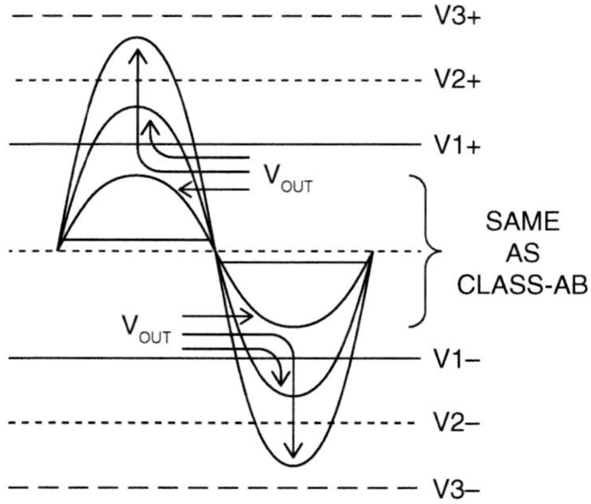

Figure 1.9 Class-G Conduction Intervals

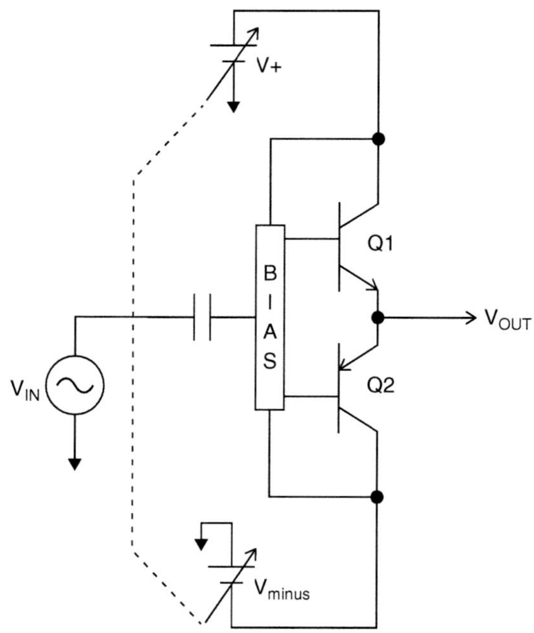

Figure 1.10 Class-H

The conduction intervals of Q1 and Q2 in Class-H (see Figures 1.10 and 1.11) are the same as for Class-AB. What is different is the way the amplifier's supply rails behave. In Class-H, the positive and negative supply rails track the output voltage, constantly maintaining a low voltage drop across the ON transistor. In

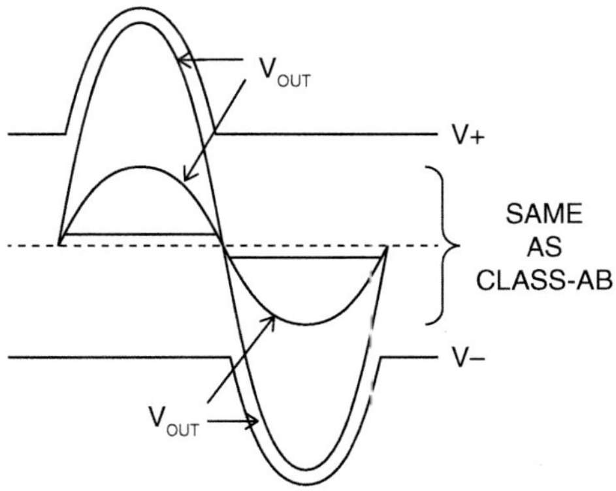

Figure 1.11 Class-H Conduction Intervals

a sense, Class-H can be thought of as Class-G with an infinite number of power supply voltage tiers.

Texas Instruments (2019) describes an interesting application of Class-G and Class-H boost technology to Class-D designs. Typically, Class-G and Class-H techniques are applied to Class-AB amplifiers to boost their efficiency. In Texas Instruments (2019), such techniques are used to increase the available output power of a Class-D design, while simultaneously increasing efficiency, to extend battery life in portable products.

The power stage of a Class-D amp (as depicted in Figure 1.12) is very different than those of the linear classes shown above. Whereas the linear topologies all involve *modulating* the ON-state, or conduction of the power transistors in a linear fashion, the Class-D amp operates the power transistors as switches. They are either fully OFF, and non-conducting, or they are fully ON, with (ideally) their lowest value of on-resistance. From a power dissipation standpoint, this arrangement produces extremely high efficiency in the power stage, because power loss is the product of the current through the power switch times the voltage across it. When the switches are OFF, the current term is zero, and therefore the power dissipation is zero. When the (ideal) switch is ON, its resistance (R) is zero, which means the voltage across it will be zero, at any level of current ($V = I \times R$), and power dissipation in the ON switch will be zero. Of course, in reality, these power switches are not ideal (though they keep improving!), so on-resistance is not zero, and there are losses associated with the switching ON and OFF of the power switches (switching losses), which will be discussed

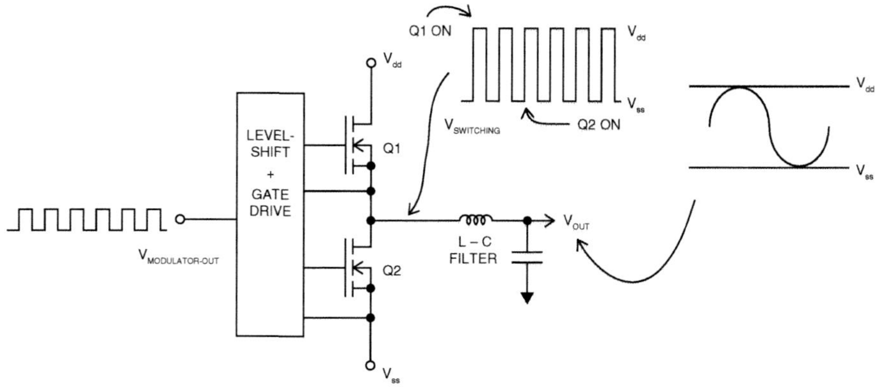

Figure 1.12 Class-D Switching Power Stage

in detail. However, to a first approximation, it can be appreciated that the efficiency of this type of switching power stage can be extremely high, especially when compared to linear power stages in which the power transistors, operating in a linear mode, experience both high current and high on-state voltage simultaneously.

The switching power stage of the Class-D amplifier produces a rail-to-rail voltage waveform, switching at very high frequency (typically greater than 250KHz), with extremely fast switching edges and variable duty cycle. One technique used to create a varying duty cycle waveform is called Pulse Width Modulation (PWM). This pulse width modulated waveform contains all the audio information, and just needs to be low-pass filtered, demodulated, or integrated to recover the audio bandwidth (20Hz–20KHz) output voltage waveform. Typically, an L-C filter is used for this purpose, although it is possible to employ filterless designs in low-power Class-D implementations, where the inductance of the speaker itself is involved in the demodulation.

Figure 1.13 describes in a little more detail the relationship between the modulator waveform, the switching waveform, and the demodulated output waveform, as well as the effect of PWM. It can be seen that the low-voltage modulator waveform is amplified by the switching power stage but has the same timing relationship. Examining the switching waveform, it can be seen that as the ON-time of Q1 increases, the time spent at the Vdd level increases, and as the ON-time of Q2 increases, the time spent at the Vss level increases. After this switching waveform

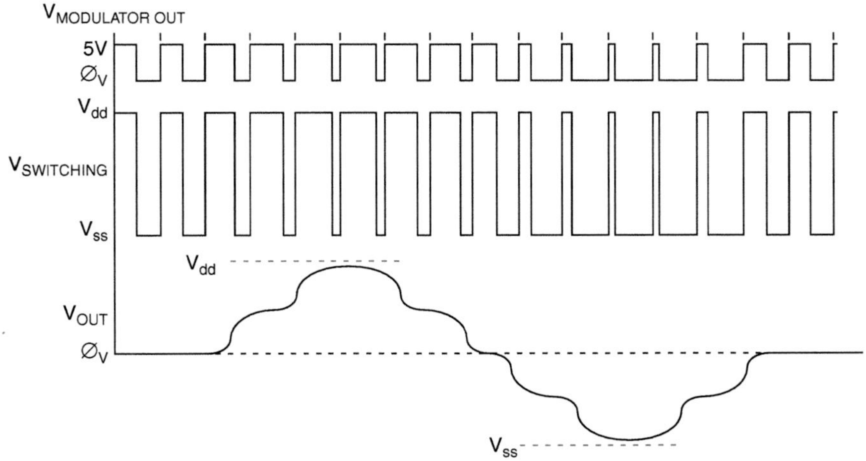

Figure 1.13 Class-D Waveforms

is low-pass filtered by the L-C filter, the audio output waveform results. Typically, this output voltage waveform has some residual voltage ripple at the switching frequency, and this must be carefully controlled to prevent EMI, especially when cables are used to connect the amplifier output to remote speakers.

One critically important aspect of the Class-D switching stage is that Q1 and Q2 must be driven alternately, and must *never* be ON at the same time, or else a short circuit across the power supply rails will result, causing damage or, at the very least, high power dissipation and reduced efficiency. This aspect of Class-D design is critical. A parameter called *deadtime* is inserted between the ON-time of the opposing power switches, to ensure that cross-conduction will not occur. The careful control of switching deadtime is one of the most critical aspects of Class-D amplifier power stage design.

The Class-D block diagram in Figure 1.14 shows a simple fixed-switching-frequency Class-D design, similar to the reference amplifier that will be presented in this text as an aid to learning. It includes the switching power stage discussed in Figure 1.12, as well as more detail in the modulator section. The MODULATOR is responsible for generating the pulse-train which will serve as *input* to the switching power stage. The modulator in the above example serves as an Analog-to-Digital (A-to-D) converter. It takes the analog output of a linear ERROR AMPLIFIER (usually configured as an integrator) and compares it to a TRIANGLE RAMP waveform, which is operating at the switching frequency.

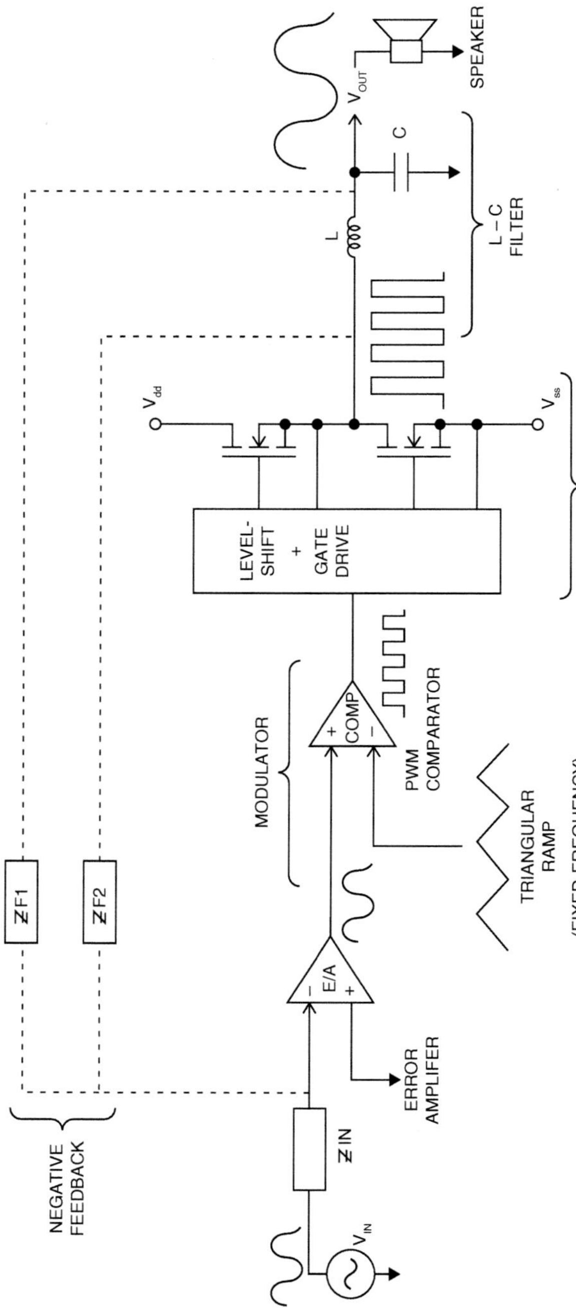

Figure 1.14 Class-D Block Diagram

The high-speed voltage COMPARATOR compares the output of the error amplifier to this RAMP waveform. During the time interval when the error amplifier output voltage exceeds the ramp voltage, the comparator output will be high, and during the interval where the ramp voltage exceeds the error amplifier output voltage, the comparator output will be low. In this block diagram, the comparator output voltage waveform is *in-phase* with the switching voltage waveform at the switching node (the junction of Q1 and Q2). Negative feedback can be added to Class-D designs to improve many performance parameters, and this will be discussed in detail. In the block diagram, two examples of feedback connections are shown. These are: feedback taken from the switching node, and feedback taken from the amplifier output. The advantages and disadvantages of this connection choice will be discussed as well. For now, it is sufficient to see that if Vout rises too high (positive), the output of the error amplifier will swing in the negative direction and decrease the ON time of the upper switch (Q1) and increase the ON time of the lower switch (Q2). This will cause the amplifier output voltage to be less positive (i.e. negative feedback).

Key Parameters of Class-D Audio Power Amplifiers

Some *key* parameters which will be addressed in discussing Class-D audio amplifier design are:

- Output Power level and Load Impedance:
  - Low or High Power, Low Impedance loads (4, 6, 8 ohms, or High Impedance "Constant-Voltage" loads)
- Topology:
  - Half-Bridge or Full-Bridge (Bridge-Tied Load (BTL) or Parallel Bridge-Tied Load (PBTL)), with single or dual power supply rails
- Efficiency
- Quiescent Power
- Switching frequency:
  - Fixed or Variable switching frequency
- Input type:
  - Analog or Digital
- Feedback type (none, pre-filter, or post-filter feedback)
- L-C Filter or Filterless.

Evolution of Semiconductor Power Switches Used in Class-D Amplifiers

It should be understood when reading any technical book of this type, written about a dynamic and quickly evolving technical subject, that it is a snapshot of the state of the art, at the time of writing. Therefore, it is recommended to imagine that each technological pronouncement in this book is preceded by the unwritten preface: "At the time of writing..." In some cases predictions may be offered about what the next technological developments will be, but this is an area fraught with uncertainty. It is perhaps better to diligently describe and explain the state of the art as it presently exists, and allow the reader, armed with the information provided, to make their own informed predictions of what the future may hold in store.

The evolution of Class-D amplifier design has, for the most part, been governed by the development and availability (including cost constraints) of power semiconductor switching devices. The earliest Class-D designs utilized vacuum tubes as the power switching elements. However, these had limitations in switching speed and on-resistance. Since vacuum tubes operate at high voltage and feature high on-state resistance, a step-down transformer was typically used to drive low impedance loudspeaker loads. If the application called for output bandwidth extending down to low frequencies (such as below 200Hz), this would require a large transformer with a large, laminated transformer core to support the volt-seconds required for low-frequency signals. This transformer requirement, it can be argued, would negate one of the main goals of Class-D design, which is reduced size and weight.

The next candidate for a power switching device was the bipolar silicon transistor. This low-voltage, low on-resistance device obviated the need for the step-down transformer, and also allowed much higher switching frequency operation, as compared to vacuum tube designs. Typically, this would allow switching frequencies up to approximately 200KHz, before switching losses (to be discussed later in detail) became difficult to handle. Some of the drawbacks of bipolar power transistor switches are their lower current gain (Beta) at higher values of collector current, and their characteristic storage time. Storage time is due to silicon bipolar power transistors being minority carrier devices. The minority carriers need to be swept out of the conducting region before the device can again reach a high-resistance off state. These characteristics made it necessary to design robust, high-current, and high-speed driver circuits, in order to properly drive these power switching transistors at high speed. Again, this increased design complexity, requiring gate drive transformers and baker clamps, and other design "tricks" to achieve high switching speed with reasonable switching losses.

The real paradigm shift in Class-D amplifier design and acceptance came with the development of the power mosfet (metal oxide semiconductor field effect transistor) power transistor switch in the mid- to late 1980s. This power device boasted voltage drive (as opposed to the current drive required for bipolar power transistors). It also had no storage time, due to being a majority carrier device, as opposed to bipolar transistors which are minority carrier devices. Mosfets also have no SOA (Safe Operating Area) restrictions.[1] With the adoption of power mosfets in Class-D designs, the switching frequency increased to 400KHz typically, and as high as 1MHz, significantly improving many performance parameters. The influence of switching frequency on Class-D performance parameters will be discussed in detail later in this book. The principal drawback of power mosfets as Class-D amplifier power switches is their intrinsic body diode, and, to a much lesser extent, their gate charge characteristics, and intrinsic capacitances. The body diode has an antiparallel connection to the mosfet's channel and allows current to flow from Source to Drain (for N-channel mosfets, which are used almost exclusively over P-channel mosfets in Class-D design) when the drain voltage is more negative than the source voltage. This is also called third-quadrant operation. This p-n body diode is a minority carrier device and has problematic reverse recovery characteristics. These include reverse recovery time (Trr) and reverse recovery charge storage (Qrr). These parameters degrade Class-D amplifier switching performance and create EMI challenges. This topic will be handled in greater detail in future chapters.

The latest development in power switching devices is the advent of the wide bandgap (WBG) devices. These are Gallium Nitride (GaN) and Silicon Carbide (SiC) transistor switches. These are majority carrier devices and boast zero reverse recovery time. The more popular WBG device type for Class-D amplifiers is GaN, because SiC is targeted for higher-voltage, higher-power applications, whereas GaN devices are offered in smaller surface-mount device (SMD) packages, better suited to low- and medium-voltage/power Class-D amplifiers. These devices will be discussed in detail in future chapters. The principal benefit of using these devices is the elimination (GaN) or reduction (SiC) of the reverse recovery penalties imposed by the minority carrier p-n junction body diode of mosfets. They also switch faster, due to decreased parasitic elements (inductances and capacitances), and have lower gate charge, and therefore allow for the use of shorter imposed switching deadtimes and produce cleaner switching waveform edges with higher dv/dt slew rates and less ringing.

Class-D Amplifier History

When discussing the history of any technology, it is always with some trepidation that a particular product or achievement be declared as "first", unless there are patents in place to prove precedence. It is therefore prudent to describe what is generally accepted as the timeline, without making definitive statements.

The first Class-D amplifier is purported to have been invented in the 1950s, by Alec Reeves, a British Scientist, who also helped in the development of the first transatlantic telephone cable, and electronic countermeasures used in World War II (Fletcher, 2020). The first commercially available class-D amps are generally believed to have appeared in the 1960s from the British company Sinclair Radionics (now Thurlby Thandar Instruments) (Fletcher, 2020; Zorpette, 2010).

In 1964, a Class-D *kit* (the X-10) (Zorpette, 2010) was marketed by Sinclair Radionics (Fletcher, 2020; Zorpette, 2010). It had a claimed power output of 10W but was reported to really produce far less than that. The X-20 (20W output power) was produced in 1966. These early designs were hampered by the reliability of the bipolar germanium power transistors that were used (https://en.wikipedia.org/wiki/Class-D_amplifier). John Ulrick designed a couple of class-D amps for Infinity Systems in the early and mid-1970s, and a decade later in 1984 (Fletcher, 2020; Zorpette, 2010) Brian Attwood designed a series of Class-D amplifiers called "digital energy conversion amplifiers" for Peavey Electronics Corp (Zorpette, 2010). Class-D designs in those days were challenging. They were hampered by the components then available. It could be argued that commercially successful Class-D amplifiers were not possible before the development of reliable mosfets. In those days, Class-D theory was promising, but reliable implementation required better components. In 1978 Sony produced a successful studio power amp called the TA-N88, using early mosfets (VFETS) (Fletcher, 2020; Zorpette, 2010; www.thevintageknob.org/sony-TA-N88B.html).

The principal problem was switching speed: Class-D amps switched at a minimum of 250KHz in those days. Mosfet performance improved through the 1980s, but it wasn't until the late 1980s and early 1990s that inexpensive and reliable mosfets became available, that were fast enough to switch at such high frequencies (Fletcher, 2020; Zorpette, 2010).

Pioneering companies at that time were Tact Audio, ICEpower, and Tripath Technology (Zorpette, 2010). All three companies introduced their first Class-D product offerings in the late 1990s. One of these products was called the Tact Millenium, which was designed by the Danish engineer Lars Risbo (Zorpette, 2010).

One of the early Class-D companies, ICEpower, was founded jointly by another Danish engineer, Karsten Nielsen, who designed the company's first amplifiers and amplifier modules, and the Danish audio company Bang & Olufsen (Zorpette, 2010). Another early Class-D company, Tripath, in San Jose, Calif., was the first to introduce a Class-D amplifier integrated circuit (IC), the TA1101, in 1996 (Zorpette, 2010). It was used in Apple's Power Mac G4 Cube. Tripath's TA2020, produced in 1998, could deliver 20W per channel into 4-ohm speakers. According to President and CEO Adya Tripathi, quoted in Zorpette (2010), "We were shipping millions of chips every quarter to companies like Samsung, Panasonic, Toshiba, Sanyo, NEC, Onkyo, and Sony."

From the 1990s to the present time, power switching components, including mosfets and GaN devices, have steadily improved, and along with that improvement, switching frequencies have increased and audio performance parameters have improved. Engineers working in this field have produced a wealth of knowledge about circuit design techniques for Class-D, which have helped others move the state of the art forward. Many IC companies have entered the Class-D arena and produce dedicated high-performance Class-D ICs (discussed in other chapters), which facilitate the design of Class-D amplifiers over a wide power range. Today, Class-D designs can easily yield audiophile performance at extremely high efficiency, at low cost. Advances in both active and passive devices, which are happening continually, hold promise for continued improvement.

Organization of the Book

This book will present an example (reference design) of a fixed-frequency Class-D amplifier. An output power level of 100W into 8 ohms was chosen as this is a typical medium power level. Of course, it must be appreciated that terms such as low, medium, and high power have different meanings in different applications. For example, a power level of 100W would be considered to be extremely low in professional audio applications where large music venues typically use amplifiers of thousands of watts per channel.

This reference design will be used as a practical *starting point* in the technical discussion of each circuit block. But it is only a starting point. After the circuit details and theory are discussed for the reference design, the discussion will then shift to alternative approaches for accomplishing the same or similar circuit functions. By this method it is hoped that a great deal of Class-D subject matter can be covered in an organized manner.

Goal

The goal of this book, as already stated, is to provide a thorough treatment of Class-D amplifier design, so that the reader comes away with an understanding of the present state of the art, as well as an intuitive "feel" for how circuit design parameters affect performance, and how these parameters interact, and what tradeoffs need to be made, and how to make these. A very interesting characteristic of Class-D design is how one circuit parameter affects many different performance parameters, often in opposite ways. When this happens, as it so often does in Class-D design, an informed tradeoff must be made. One aim of this book is to explain these tradeoffs, and how they are best made. The phrase "the conflicting goals are" appears often in this book in this context.

Equations

Equations will be presented that will aid in circuit design and component selection, and where possible, these equations will be presented in a way that will increase transparency and understanding of the circuits. Often, these equations will be solved using actual numerical values, to illustrate and confirm their utility.

Precision: Numerical values presented will be rounded off when convenient, unless a more precise value is mandatory.

Overview of the Reference Design

The reference design is a discrete, analog, fixed-frequency, Class-D audio power amplifier design, capable of providing 100W root-mean-square (RMS) into low impedance loads (4–8 ohm impedance range).

Audio Performance of the Reference Design

The purpose of the reference design is to provide a foundational circuit from which a discussion of Class-D audio amplifiers can be initiated and developed. Its first and foremost purpose is to be a *teaching tool.* With that goal in mind, the specific circuit blocks were designed in such a way as to facilitate understanding of the function of each circuit block. To that end, a discrete design was used, instead of a potentially simpler IC-based design utilizing an IC Class-D solution. A discrete design provides better visibility into the workings of each of the functional circuit blocks in a Class-D amplifier design. Likewise, circuit additions that would enhance audio performance, or efficiency, at the expense of increased complexity were avoided for the sake of clarity and ease of understanding. For these reasons, the audio performance of the reference design, though respectable, is not equal to the present state of the art for Class-D audio amplifiers. Table 1.1 lists the performance parameters of the reference design.

Mono or Stereo Operation

In order to achieve the stated goal of its having clarity as a teaching tool, the reference design schematic is drawn as a single-channel (mono) design. "Hooks"

Table 1.1 Audio Performance of the Reference Amplifier[a,b]

Specification	*(Typical)*	*Conditions*
Noise (A-weighted)	175μV	22Hz–20KHz/8 ohms
SNR (signal-to-noise ratio) (A-weighted)	104dB	22Hz–20KHz/8 ohms
THD (@ 1KHz)	0.03%	100W/8 ohms
THD (@ 1KHz) sweep	See Figure 1.15	1–100W/8 ohms
THD vs Frequency sweep	See Figure 1.16	At various power levels/8 ohms
Efficiency of Power Stage (excluding signal processing circuits)	93%	100W/8 ohms (amplifier's +/−48VDC power only, excluding low-voltage rails)

a Resistive loads were used in place of speaker loads.

b An audio precision (AP) Class-D filter (AUX-0025) was used between the output of the reference amplifier and the input to the audio precision analyzer. This filter attenuates voltage ripple at the Class-D switching frequency that could potentially interfere with the analyzer's measurements of parameters within the desired audio frequency bandwidth.

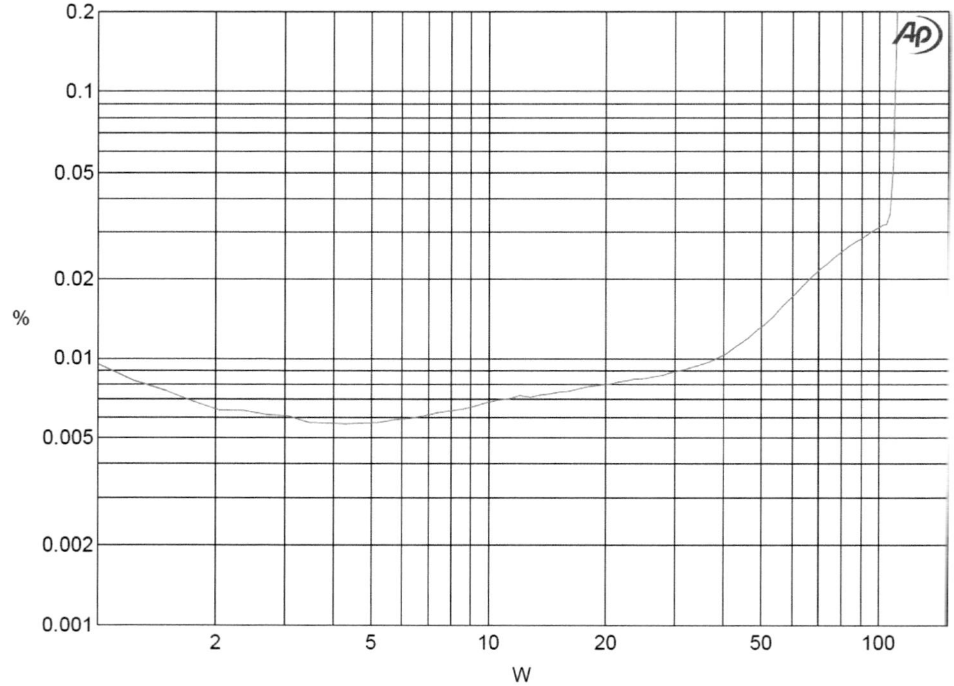

Figure 1.15 THD vs Output Power (8 Ohms)

are added as appropriate for creating a two-channel (stereo) implementation, as that is by far the most common configuration of medium-power Class-D audio amplifier designs. Signal names and reference designations ending in "A" are for channel 1 (typically the "LEFT" channel in a stereo implementation). Signal names and reference designations ending in "B" are for channel 2 (typically the "RIGHT" channel in a stereo implementation). It is important to note that any signal name ending in "A" will have a "B" counterpart in a stereo implementation. Similarly, any component having a reference designation ending in "A" will need to be duplicated in stereo implementation, where that corresponding component will have the same reference designation number but ending in the letter "B". For example, on schematic page 1 (Input/Preamp), all the reference designators, for all the components, end in "A". This means that this entire page will need to be duplicated to create channel 2 for a stereo design. All the components on that new page (for channel 2) will have the same reference designators, except they will end in the letter "B".

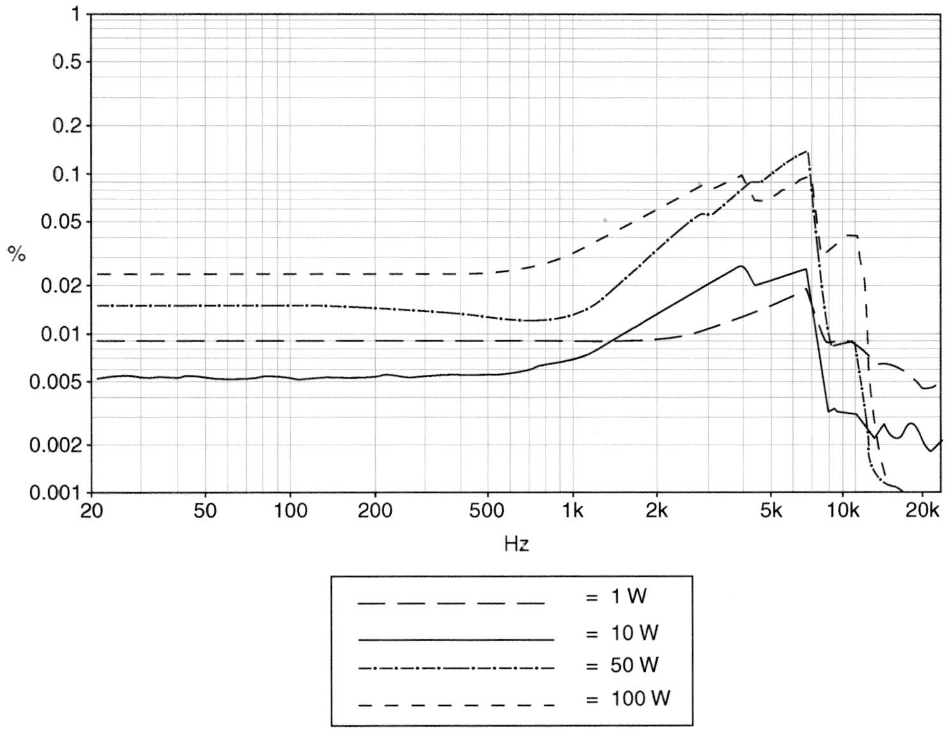

Figure 1.16 THD vs Frequency (8 Ohms, 1W, 10W, 50W, 100W)

Page-by-Page Summary of the Reference Design Schematic

Note: The reference design schematic is covered in detail in this book. Each chapter is devoted to a discussion of a functional circuit block of the design. The purpose of this summary is to give the reader an early overview of the schematic, its functional blocks, and its signal flow.

The page order of the reference design corresponds as closely as possible to signal flow through the amplifier, from input to output.

Page 1: Input/Preamp: The reference amplifier is designed to handle an input signal of up to 2VRMS. This input signal is conditioned and amplified by U1A. This op-amp is configured as a differential amplifier, in order to provide common-mode (CM) rejection, and amplification of the input signal, with a gain of 1.78×. This circuit also provides electrostatic discharge (ESD) protection and radio frequency interference (RFI) filtering at the input connector. The output of this circuit block feeds the error amplifier.

Page 2: Error Amplifier: The error amplifier accepts one input from the preamp circuit and compares it to the output signal from the amplifier (speaker output). It amplifies the *difference*, or *error signal*, so that the Class-D PWM stage which follows can respond in a manner to reduce this error (negative feedback). The error amplifier is an op-amp configured as a low-pass filter or integrator. The output of the error amplifier feeds the Class-D PWM comparator.

Page 3: PWM Comparator: A high-speed voltage comparator is used to compare the output voltage of the error amplifier to a triangular voltage RAMP signal. The output of this comparator is the Pulse Width Modulated (PWM) signal. This is a signal with varying duty cycle, which feeds the Class-D driver stage. This page also contains circuitry that provides minimum and maximum duty-cycle limit pulse widths for the PWM signal, so that this fixed-frequency Class-D amplifier always switches at the specified frequency and never misses a cycle. This circuitry also detects voltage clipping (i.e. minimum and maximum PWM duty cycle limits).

Page 4: Triangle Ramp Generator: This circuit block generates the fixed-frequency, fixed-amplitude, triangle ramp signal. This voltage signal is one of the signals fed to the PWM comparator. It is derived from a fixed-frequency clock oscillator, divided down to the desired Class-D switching frequency (500KHz). It contains a DC servo circuit that ensures that the triangle ramp voltage has zero DC offset.

Page 5: Gate Driver (Half-Bridge Power Stage): This circuit block accepts the PWM signal from the high-speed PWM comparator and level-shifts it to the drivers for the upper and lower mosfets in the half-bridge switching power stage. It also contains the bulk rail capacitance, for local energy storage. The half-bridge driver IC also has other features critical to the design. These are deadtime control and enable control. The output from the half-bridge driver is the switching node, which has the rail-to-rail switched voltage which is then fed to the output L-C filter.

Page 6: L-C Demodulation Filter: This circuit block is responsible for low-pass filtering the rail-to-rail switching output voltage from the half-bridge power stage. It also contains the Zobel network, 500KHz notch filter, output current sensing (used for overcurrent detection), and output CM EMI filter. Voltage feedback is taken from the output of the L-C filter and fed back to the error amplifier.

Page 7: Housekeeping Circuits: This circuit block contains protection circuitry that qualifies critical DC power supply voltages that must be present, and at the correct levels, for proper operation of the Class-D amplifier. If any of the sensed voltages are out of range, switching operation of the Class-D power stage will be prevented, by disabling the gate driver IC on page 6. This will happen if the low-voltage DC supply rails are in an undervoltage state, or if the high-voltage Class-D DC power rails are in an under- or overvoltage state. Voltage comparator ICs are used to sense these rail voltages and compare them to a precision voltage reference. A dedicated voltage detector IC is used to sense the +5VD rail. This functional block also contains digital logic ICs that process the voltage comparator and voltage detector outputs and enable the Class-D rail power supplies if all the low-voltage DC rails are qualified. These digital logic gates also enable the left and right (if present) amplifier channels if both the low-voltage DC rails are qualified (i.e. are above undervoltage threshold levels) *and* the Class-D power rails are qualified (i.e. are between undervoltage and overvoltage threshold levels).

Page 8: Protection and Fault-Sensing Circuits: This circuit block includes protection circuits for detecting overcurrent and overtemperature fault conditions, as well as DC voltage at the Class-D amplifier output(s). Dedicated voltage comparators and digital logic gates detect and process these faults. In addition, there is a discrete latch circuit. The response to these fault conditions is configurable by connection options. An overcurrent event in a Class-D output will result in a momentary interruption of PWM switching by the half-bridge driver for the effected channel. An overtemperature condition can result in either interruption of PWM switching, for the duration of the overtemperature event, with operation resuming upon cooldown, or latched shutdown, requiring the cycling of power to the amplifier to reset the latched fault condition. The output of the Class-D amplifier in this reference design should have a zero DC offset voltage, since this is a half-bridge design using bipolar (+ and −) power supply rails. Therefore, the presence of a DC voltage offset at the amplifier output is detected as a fault condition. This could result from a shorted (failed) mosfet, for example. In this case, the desired response is for the Class-D amplifier half-bridge driver(s) to be disabled (stop switching) and for the Class-D power rail voltage power supplies to be disabled. This circuit block accomplishes this latch off response. Another fault condition that will result in this latch off response is an overvoltage condition on the Class-D power rails, as might result from excessive rail pumping (a topic discussed in later chapters).

Note: The remaining schematic pages are the *low-* and *high*-voltage power supplies.

Page 9: AC-to-DC Power: This circuit block contains the AC-to-DC converters that generate the regulated +48V and −48V power rails for the Class-D switching power stage(s). These two converters have an enable control. In addition, there is a separate AC-to-DC converter dedicated to providing a 12V rail, that is used by the downstream POL (point-of-load) DCDC converters and LDO (Low DropOut) regulators. The +12V output will be present whenever AC is applied. This circuit block also contains fuse protection for the AC-to-DC converters as well as optional EMI filtering and input overvoltage protection.

Page 10: Low-Voltage Block Diagram: This is a block diagram illustrating the low-voltage power distribution scheme used in the reference design. Typical voltages and currents are included. The low-voltage distribution scheme is optimized for low noise and high efficiency. To that end, there are separate LDOs for analog (noise-sensitive) and digital (less noise-sensitive) circuitry. In addition, there is a separate dedicated DCDC that generates the 12V rail (referenced to the −48V Class-D power rail). This separate 12V rail ("−HV+12V") is the power supply rail for the gate driver IC(s).

Page 11: Low Voltage 1: This schematic page contains three DCDC converters. The first is a buck converter which generates the +7VA analog rail voltage from +12V. The second operates in inverting buck-boost mode, to generate the −7VA analog rail voltage from +12V. The third is referenced to the −48V Class-D voltage rail and operates as a buck converter, to generate a separate 12V rail (referenced to −48V) for the half-bridge driver IC(s).

Page 12: Low Voltage 2: This schematic page contains three LDO regulators. The first generates +5VA for analog (low-noise) circuitry. The second generates +5VD for digital circuitry. This separation into separate voltage regulators is necessary to prevent switching noise generated by digital circuitry from contaminating sensitive low-noise analog signal processing circuitry. The third generates −5VA for analog (low-noise) circuitry.

Note

1 Newer mosfets *do* have some SOA restrictions to be concerned about. This can be investigated further by reviewing device datasheets. It is evidenced by SOA curves that are distinctly non-linear.

Bibliography

Cordell, B. (2019). *Designing Audio Power Amplifiers* (2nd ed.). Abingdon: Routledge.

Exploding the efficiency myth of Class D amplifiers. (2010). *Audium Semiconductor*, AS-100204-WP, Issue 1.

Fletcher, T. (2020, November 23). History of Class-D amplifiers and smaller speakers. *Bass Musician Magazine*. https://bassmusicianmagazine.com/2020/11/history-of-the-rise-of-smaller-more-lightweight-bass-gear/

Gaalaas, E. (2006). Class D audio amplifiers: What, why, and how. *Analog Dialogue*, *40*(6), 1–7.

Jiang, X. (2017). Fundamentals of audio Class D amplifier design: A review of schemes and architectures. *IEEE Solid-State Circuits Magazine*, *9*(3), 14–25.

Texas Instruments. (2019, April). Benefits of Class-G and Class-H boost in audio amplifiers. Application Report SLAA888. Dallas, TX: Texas Instruments.

Zorpette, G. (2010, December 30). Class-D audio: The power and the glory: A quiet revolution is transforming audio electronics. *IEEE Spectrum*. https://spectrum.ieee.org/classd-audio-the-power-and-the-glory

Appendix 1A: Reference Class-D (Fixed-Frequency) Schematic

Input/Preamp

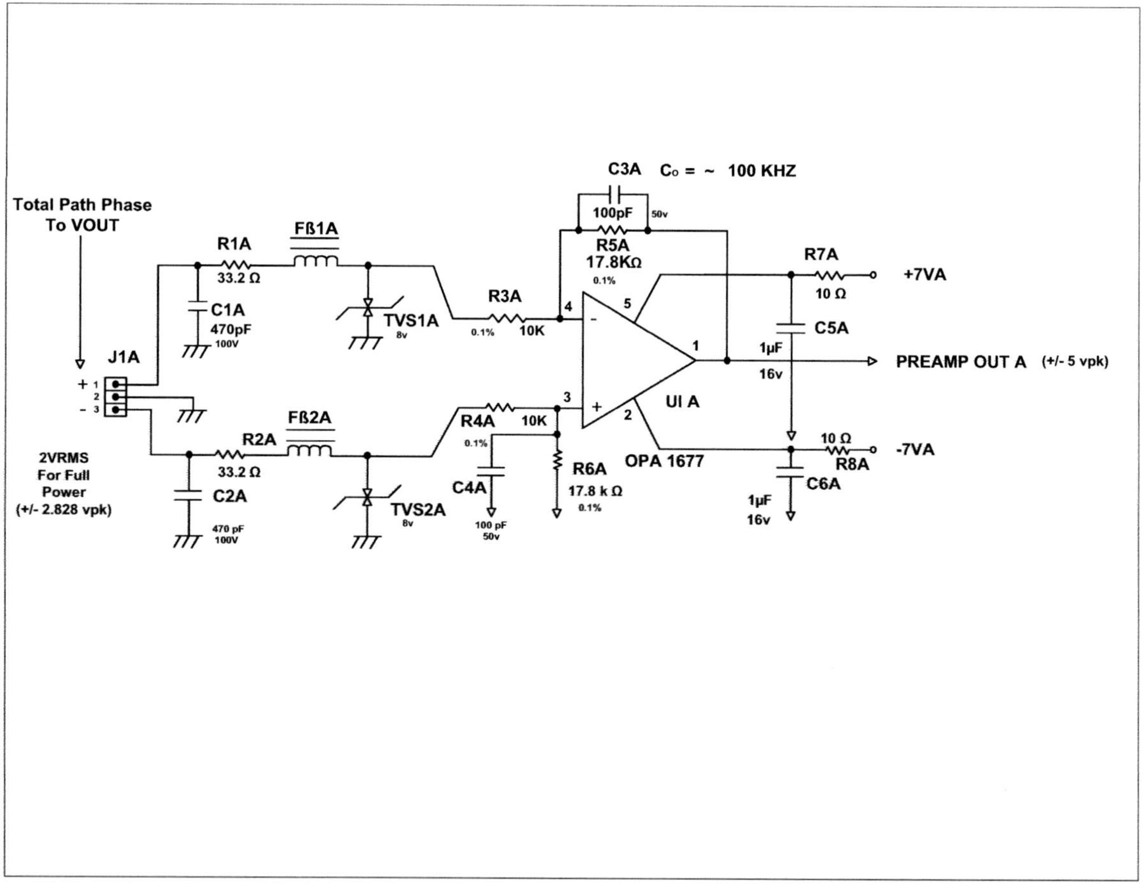

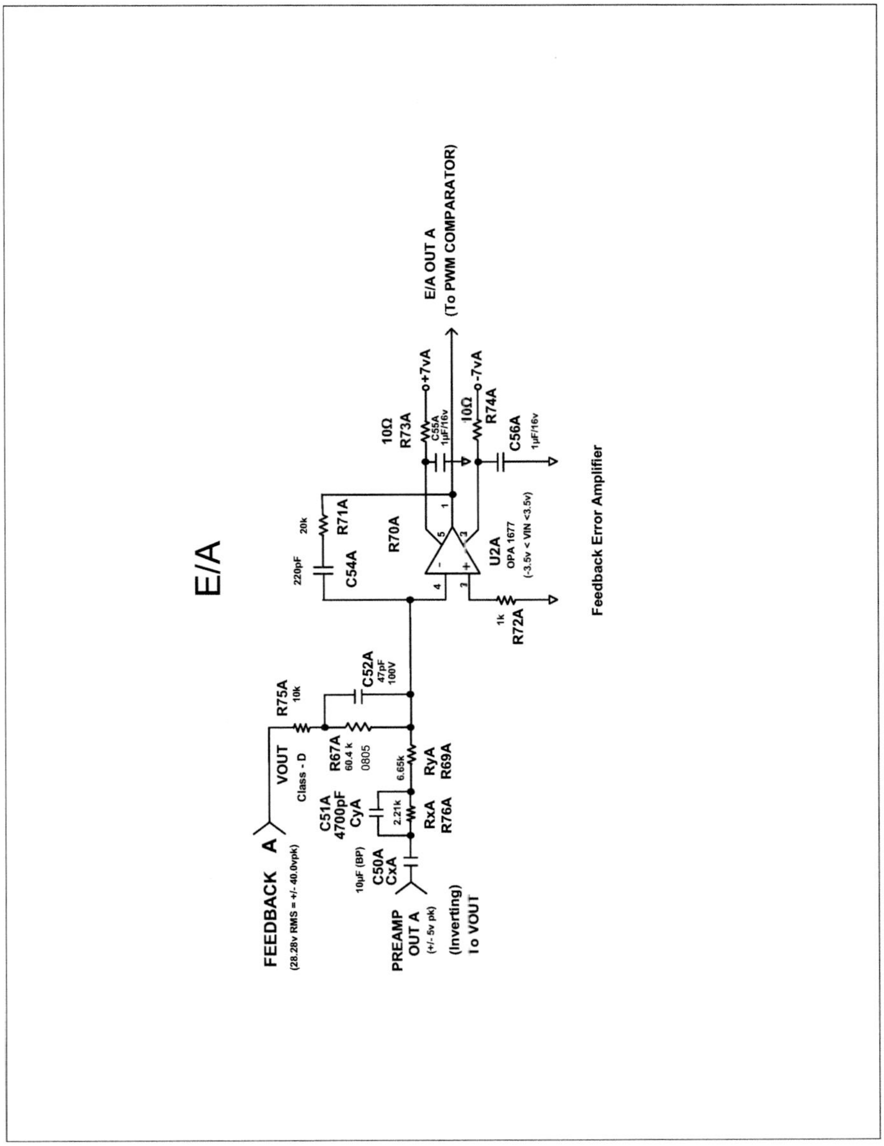

Feedback Error Amplifier

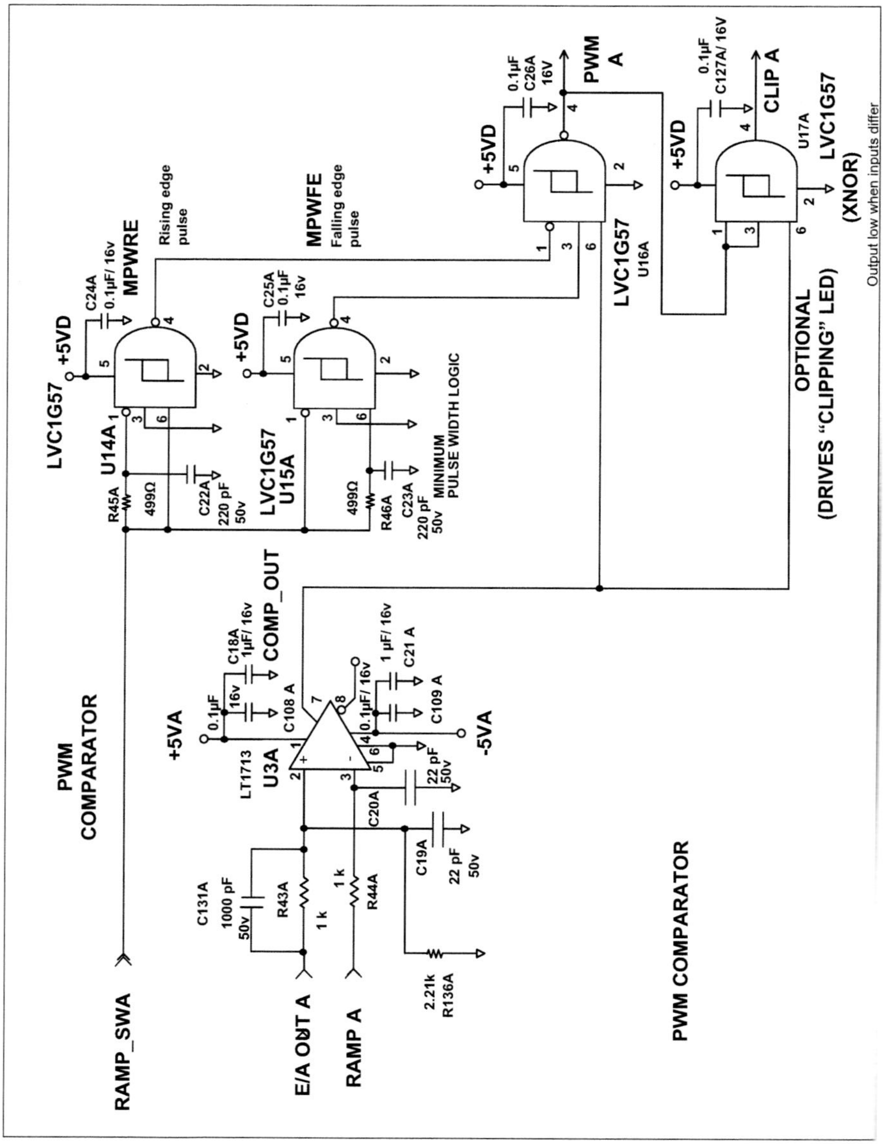

PWM Comparator

Triangle Ramp Generator

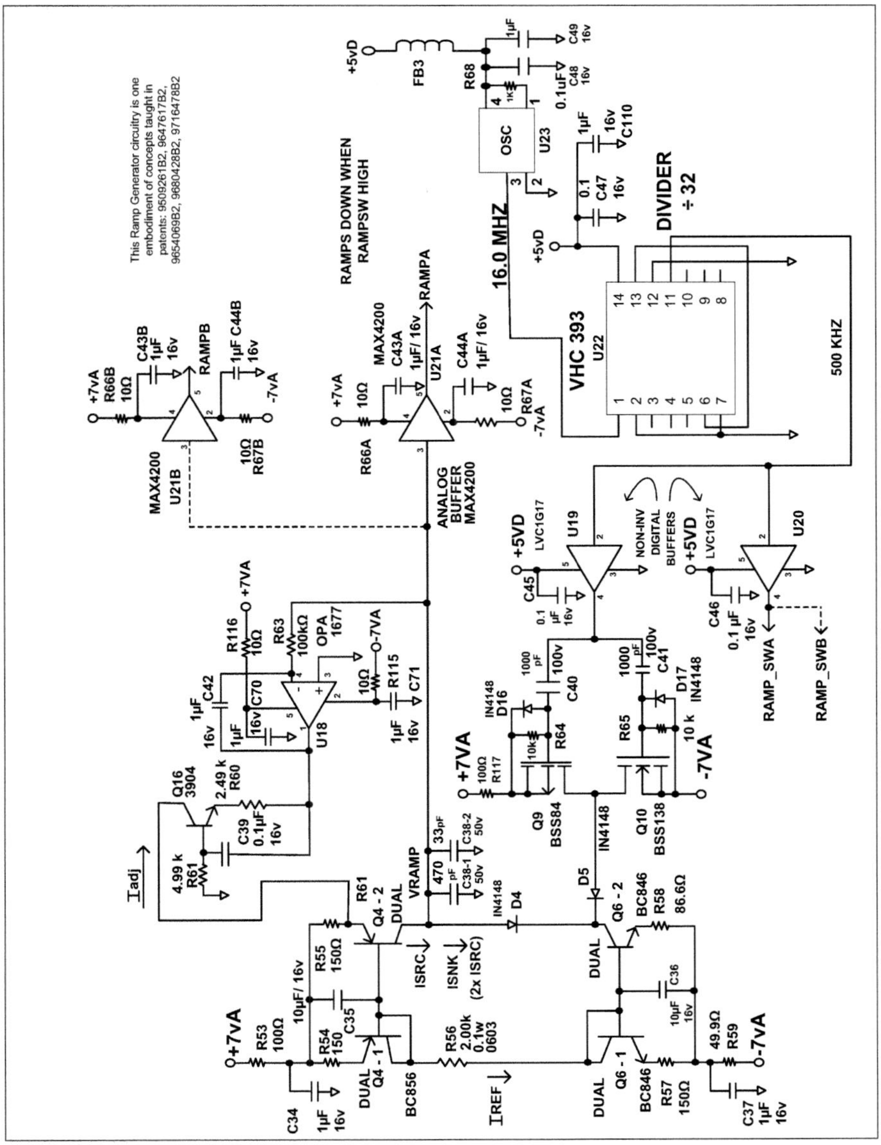

Gate Driver (Half-Bridge Power Stage)

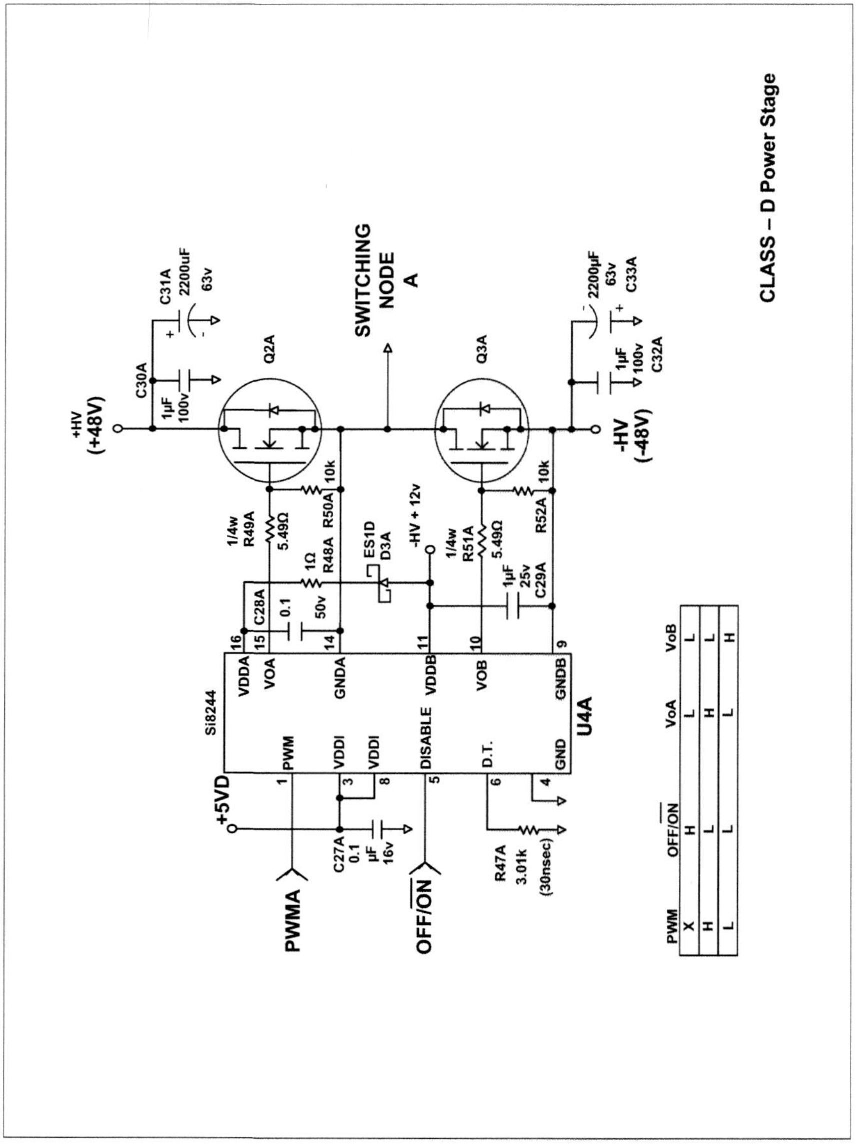

PWM	OFF/ON	VoA	VoB
X	H	L	L
H	L	H	L
L	L	L	H

L-C Demodulation Filter

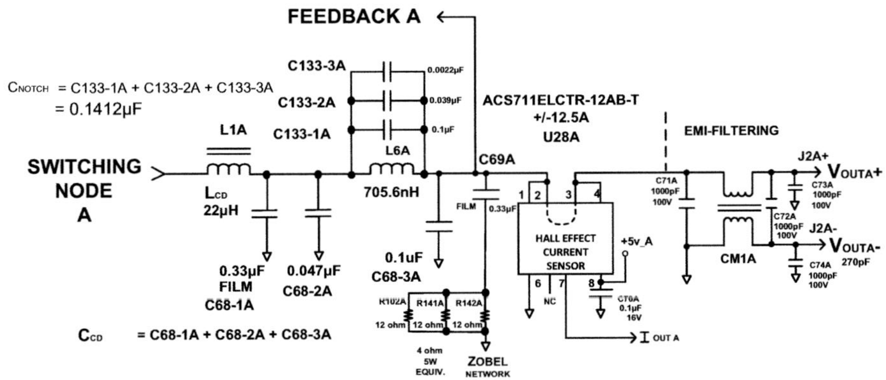

"C68" as used in the L-C Demodulation Filter Chapter (9), is the SUM of C68-1A + C68-2A + C68-3A, and is nominally equal to 0.47µF

L6A is 12 TURNS of 18 AWG Magnet Wire spaced evenly on a MICROMETALS T50-2 TOROIDAL CORE

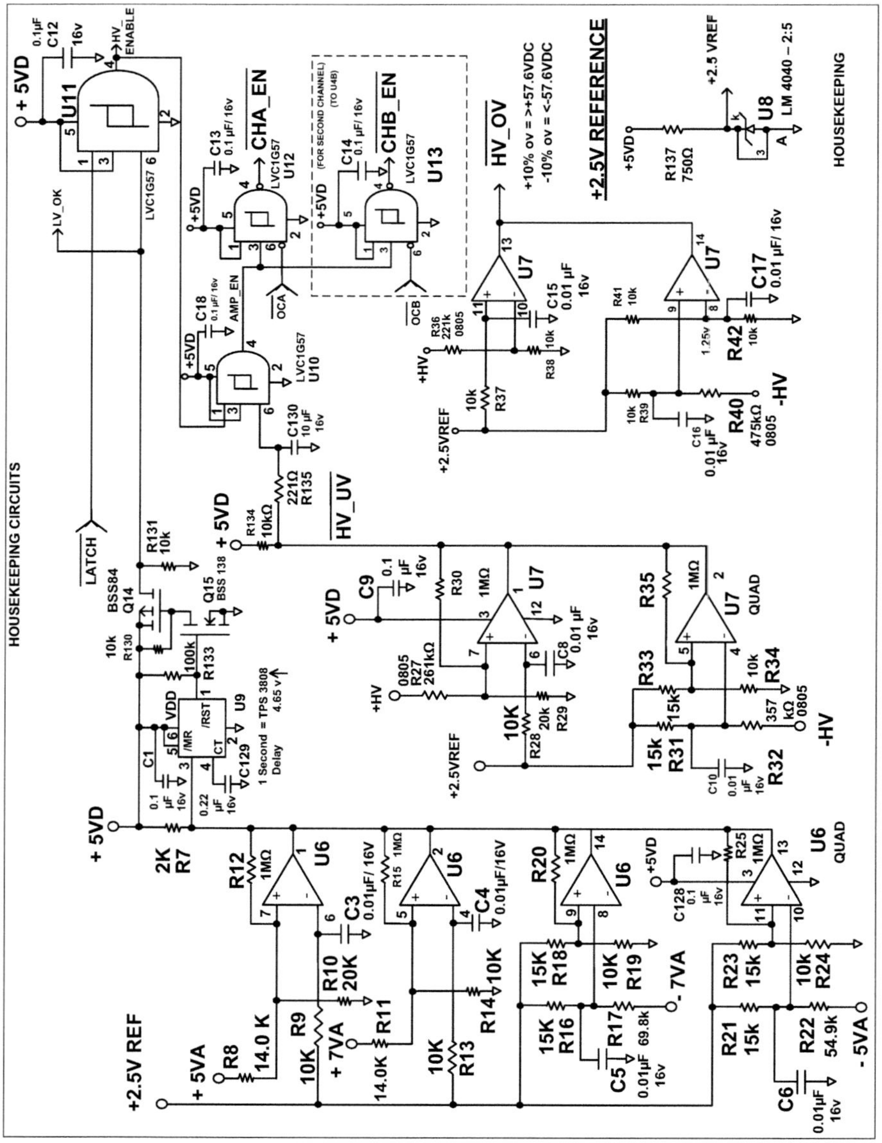

Housekeeping Circuits

Protection and Fault-Sensing Circuits

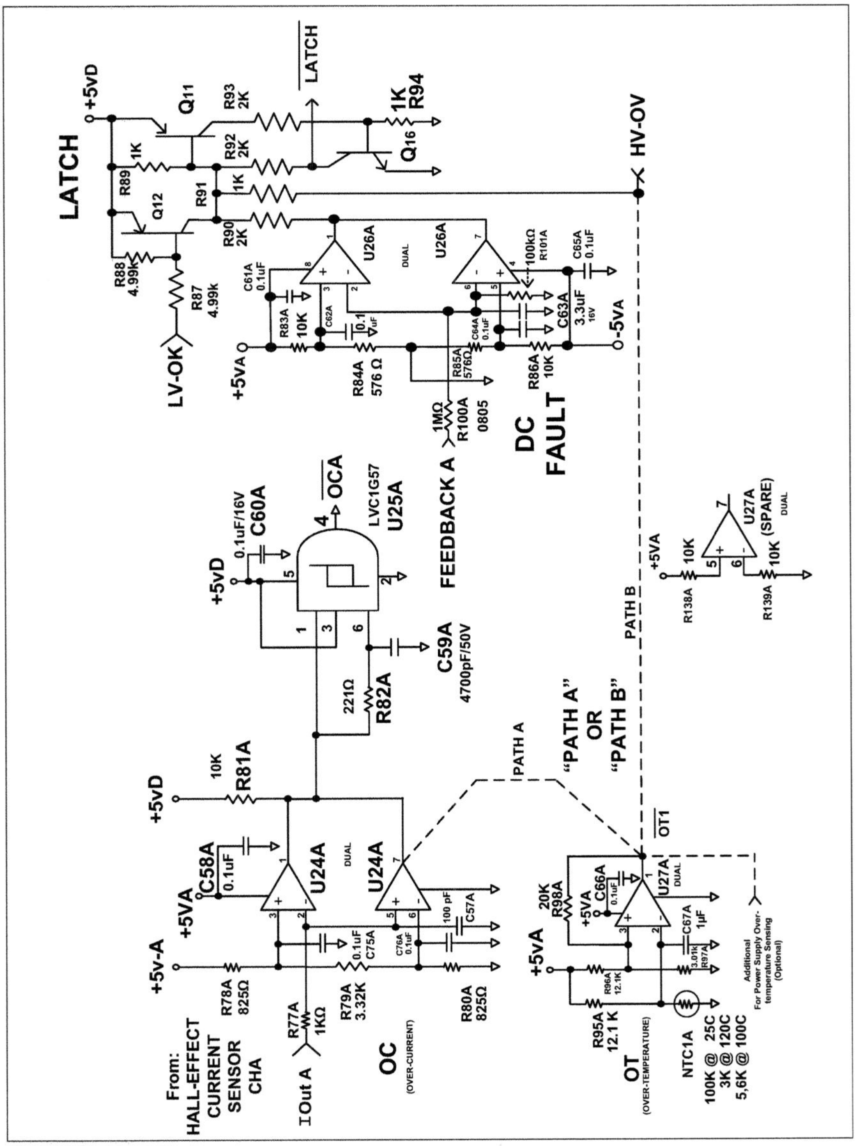

AC-to-DC Power

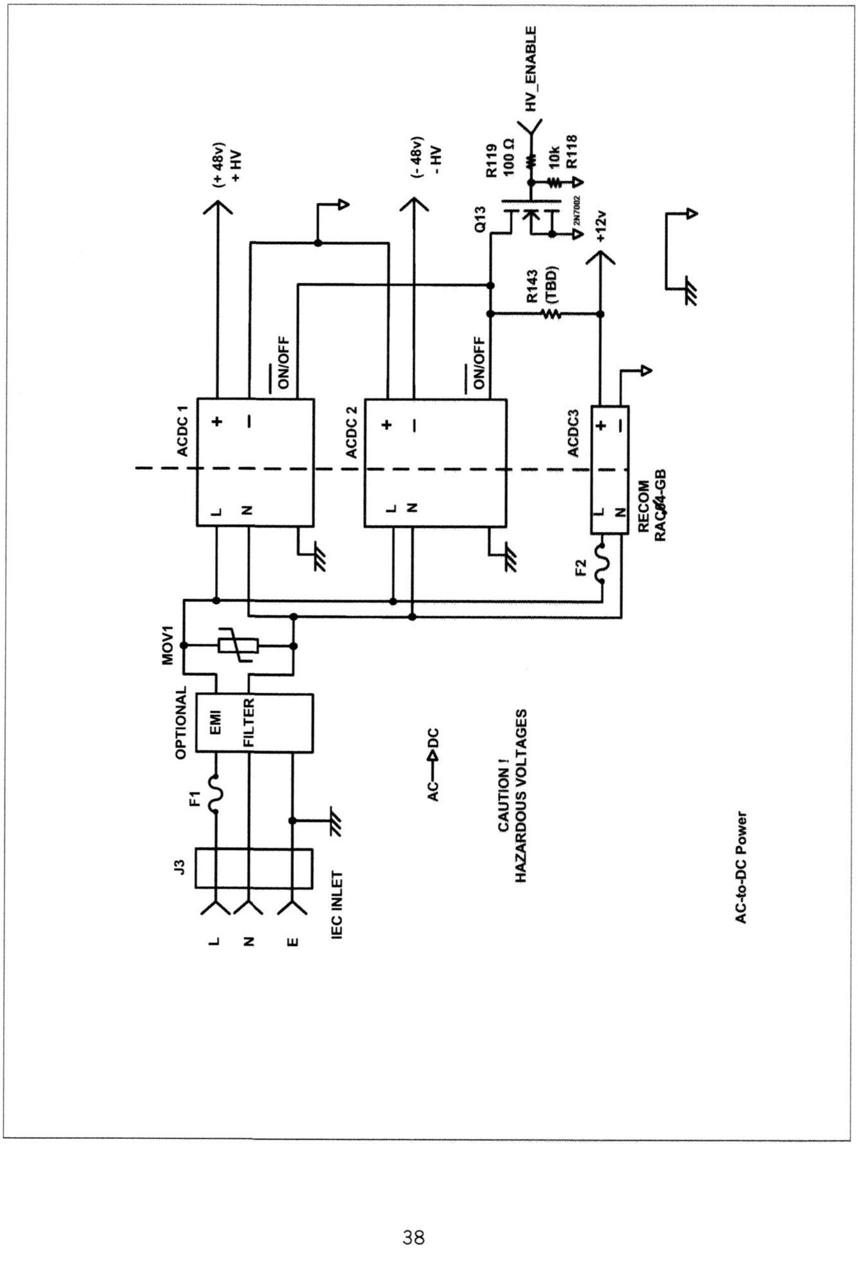

Low-Voltage Block Diagram

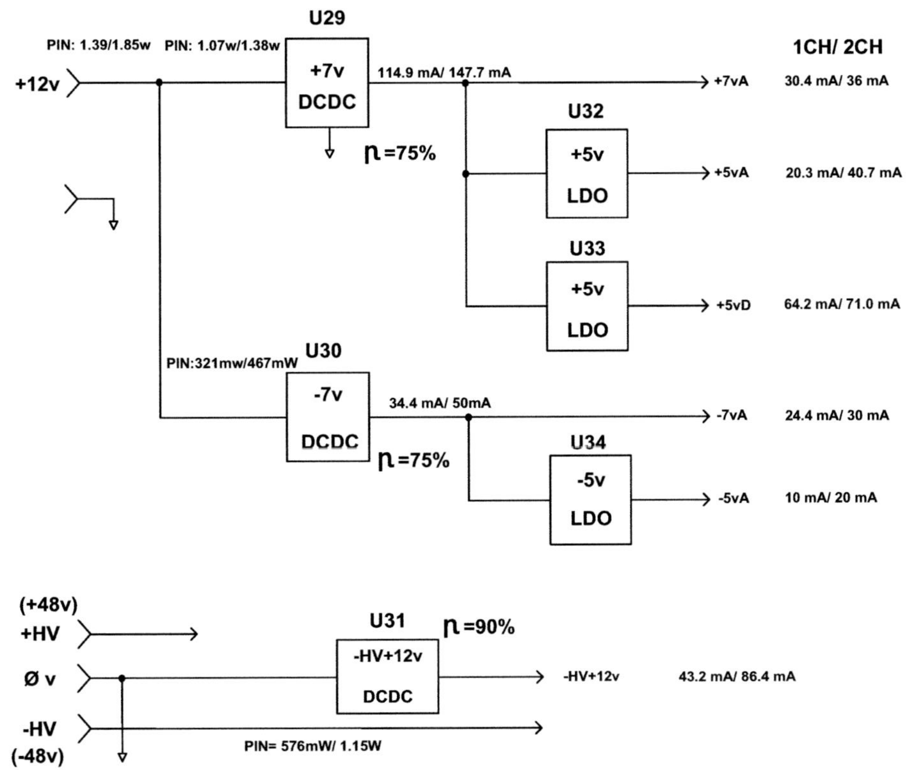

Low Voltage Block Diagram

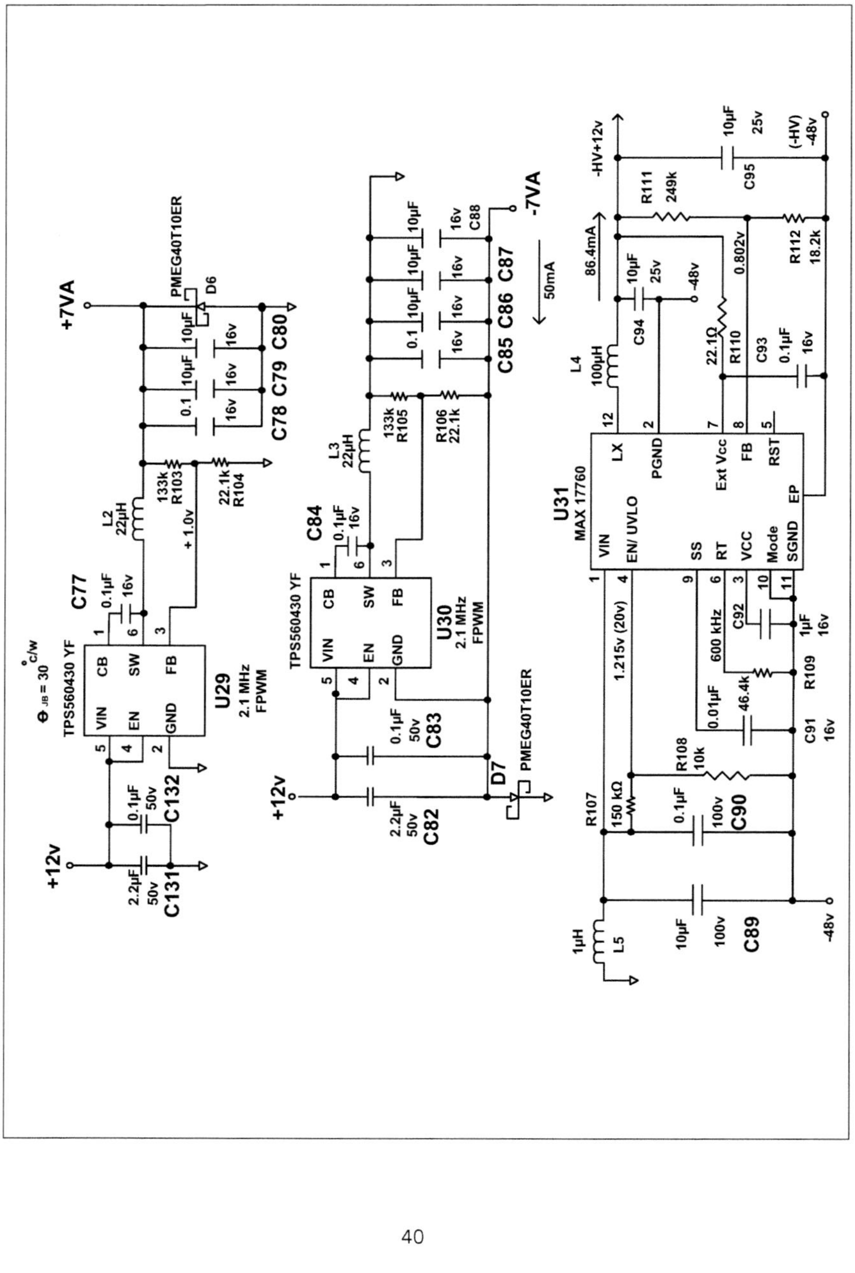

Low Voltage 1

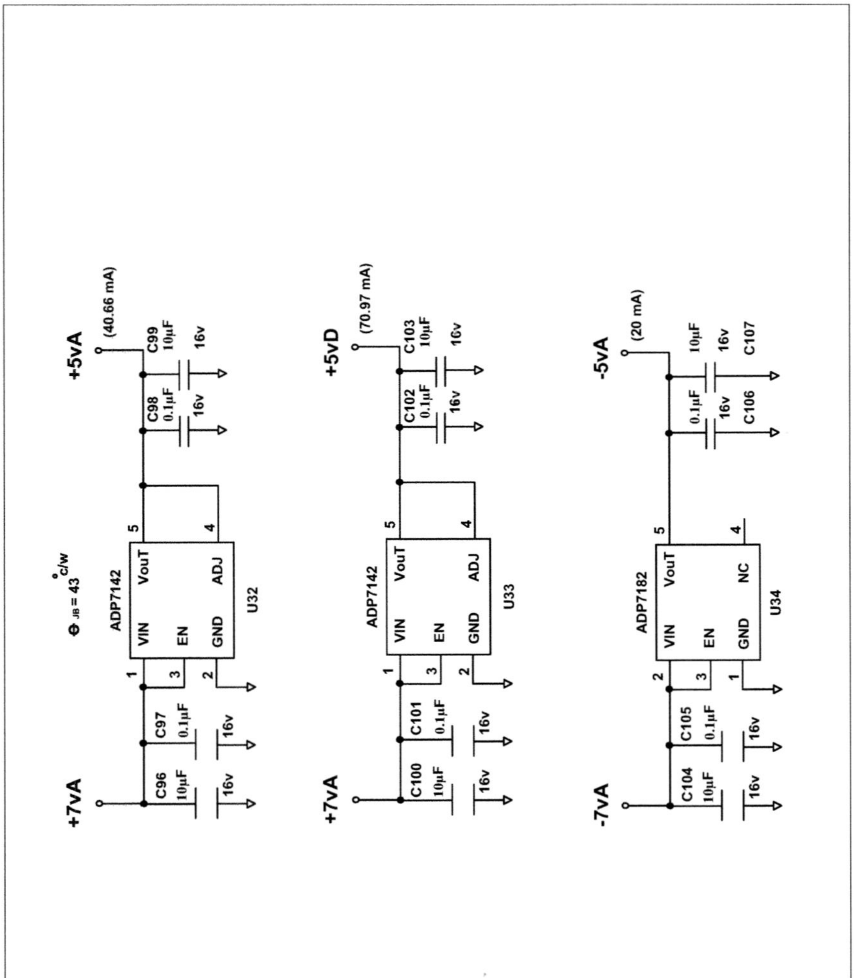

Low Voltage 2

Appendix 1B: Bill of Materials (Class-D Reference Design)

Reference Designator	*Description*	*Manufacturer*	*Manufacturer's Part Number*
C1A, C2A, C38-1	*470pF, Ceramic, C0G, 100V, 5%, 0603*[a]	*Multiple*	
C52A	*47pF, Ceramic, C0G, 100V, 5%, 0603*	*Multiple*	
C38-2	*33pF, Ceramic, C0G, 50V, 5%, 0603*	*Multiple*	
C3A, C4A, C57A	*100pF, Ceramic, C0G, 50V, 5%, 0603*	*Multiple*	
C31A, C33A	*2200μF, Aluminum Electrolytic, 63V, 105C, 18×40mm*	*Chemi-Con*	*EKYB630ELL222MM40S*
C82, C131	*2.2μF, Ceramic, X7R, 50V, 10%, 0805*	*Multiple*	
C28A, C83, C132	*0.1μF, Ceramic, X7R, 50V, 10%, 0603*	*Multiple*	
C90	*0.1μF, Ceramic, X7R, 100V, 10%, 0603*	*Multiple*	
C68-3A, C133-1A	*0.1μF, Ceramic, C0G, 100V, 5%, 1206*	*Multiple*	
C68-2A	*0.047μF, Ceramic, C0G, 100V, 5%, 1206*	*Multiple*	
C133-2A	*0.039μF, Ceramic, C0G, 100V, 5%, 0805*	*Multiple*	
C133-3A	*2200pF, Ceramic, C0G, 100V, 5%, 0603*	*Multiple*	
C89	*10μF, Ceramic, X7S, 100V, 10%, 1210*	*Samsung*	*CL32Y106KCVZNWE*
C35, C36, C79, C80, C86, C87, C88, C93, C96, C99, C100, C103, C104, C107, C130	*10μF, Ceramic, X7R, 16V, 10%, 0805*	*Multiple*	
C94, C95	*10μF, Ceramic, X7R, 25V, 10%, 1206*	*Multiple*	

Reference Designator	Description	Manufacturer	Manufacturer's Part Number
C3, C4, C5, C6, C8, C10, C15, C16, C17, C91	0.01µF, Ceramic, X7R, 16V, 5%, 0603	Multiple	
C22A, C23A	220pF, Ceramic, COG, 50V, 5%, 0603	Multiple	
C19A, C20A	22pF, Ceramic, COG, 50V, 5%, 0603	Multiple	
C1, C9, C12, C13, C14, C18, C24A, C25A, C26A, C27A, C39, C45, C46, C47, C48, C58A, C60A, C61A, C62A, C64A, C65A, C66A, C70A, C75A, C76A, C77, C78, C84, C85, C93, C97, C98, C101, C102, C105, C106, C108A, C109A, C127A, C128	0.1µF, Ceramic, X7R, 16V, 10%, 0603	Multiple	
C5A, C6A, C18A, C21A, C34, C37, C42, C43A, C43B, C44A, C44B, C49, C55A, C56A, C67A, C70, C71, C92, C110	1µF, Ceramic, X7R, 16V, 10%, 0603	Kemet	C0603C105K4RAC7867
C29A	1µF, Ceramic, X7R, 25V, 10%, 0603	Multiple	
C30A, C32A	1µF, Ceramic, X7R, 100V, 10%, 1206	Multiple	
C40, C41, C71A, C72A, C73A, C74A, C131A	1000pF, Ceramic, COG, 100V, 5%, 0603	Multiple	

(Continued)

Reference Designator	*Description*	*Manufacturer*	*Manufacturer's Part Number*
C50A	*10µF, Aluminum Electrolytic, Bipolar, 16V, 105C, 5.3×5.3mm*	*Nichicon*	*UUP1C100MCL1GS*
C51A	*4700pF, Ceramic, C0G, 50V, 5%, 0603*	*Multiple*	
C54A	*220pF, Ceramic, C0G, 100V, 5%, 0603*		
C68A	*0.47µF, Film, Metalized Polypropylene, 100V, 5%*	*Wima*	*MKP2D034701M00JSSD*
C68-1A, C69A	*0.33µF, Film, Metalized Polypropylene, 100V, 5%*	*Wima*	*MKP2D033301M00JSSD*
C59A	*4700pF, Ceramic, 50V, C0G, 5%, 0603*	*Multiple*	
C129	*0.22µF, Ceramic, X7R, 16V, 10%, 0603*	*Multiple*	
C63A	*3.3µF, Ceramic, X7R, 16V, 10%, 0805*	*Murata*	*GCM21BR71C335KA73L*
R1A, R2A	*33.2 ohm, 1%, 0603*	*Multiple*	
R3A, R4A	*10.00 Kohm, 0.1%, 0603*	*Multiple*	
R5A, R6A	*17.80 Kohm, 0.1%, 0603*	*Multiple*	
R8, R11	*14.0 Kohm, 1%, 0603*	*Multiple*	
R16, R18, R21, R23, R31, R33	*15.0 Kohm, 1%, 0603*	*Multiple*	
R17	*69.8 Kohm, 1%, 0603*	*Multiple*	
R22	*54.9 Kohm, 1%, 0603*	*Multiple*	
R12, R15, R20, R25, R30, R35, R100A	*1.00 Mohm, 1%, 0603*	*Multiple*	
R102A, R141A, R142A (Zobel Resistor(s))	*12 ohm, 2W, 1%, 2512*	*TE Connectivity*	*CRGP2512F12R*
R43A, R44A, R68, R72A, R77A, R89, R91, R94	*1.00 Kohm, 1%, 0603*	*Multiple*	
R82A, R135	*221 ohm, 1%, 0603*	*Multiple*	

Reference Designator	Description	Manufacturer	Manufacturer's Part Number
R45A, R46A	499 ohm, 1%, 0603	Multiple	
R59	49.9 ohm, 1%, 0603	Multiple	
R84A, R85A	576 ohm, 1%, 0603	Multiple	
R7, R56, R90, R92, R93, R140	2.00 Kohm, 1%, 0603	Multiple	
R47A, R97A	3.01 Kohm, 1%, 0603	Multiple	
R48A	1.0 ohm, 1%, 0603	Multiple	
R87, R88	4.99 Kohm, 1%, 0603	Multiple	
R49A, R51A	5.49 ohm, 1%, 0603	Multiple	
R53, R117, R119	100 ohm, 1%, 0603	Multiple	
R54, R55, R57	150 ohm, 1%, 0603	Multiple	
R58	86.6 ohm, 1%, 0603	Multiple	
R60, R61	2.49 Kohm, 1%, 0603	Multiple	
R63, R101A, R133	100 Kohm, 1%, 0603	Multiple	
R67A	60.4 Kohm, 1%, 0805	Multiple	
R7A, R8A, R66A, R66B, R67A, R67B, R73A, R74A, R115, R116	10 ohm, 1%, 0603	Multiple	
R69A	6.65 Kohm, 1%, 0603	Multiple	
R103, R105	133 Kohm, 1%, 0603	Multiple	
R71A	20.0 Kohm, 1%, 0603	Multiple	
R104, R106	22.1 Kohm, 1%, 0603	Multiple	
R107	150 Kohm, 1%, 0603	Multiple	
R9, R13, R14, R19, R24, R28, R34, R37, R38, R39, R41, R42, R50A, R52A, R64, R65, R75A, R81A, R83A, R86A, R108, R118, R130, R131, R134, R138A, R139A	10.0 Kohm, 1%, 0603	Multiple	
R78A, R80A	825 ohm, 1%, 0603	Multiple	
R79A	3.32 Kohm, 1%, 0603	Multiple	
R76A, R136A	2.21 Kohm, 1%, 0603	Multiple	
R137	750 ohm, 1%, 0603	Multiple	
R10, R29, R98A	20.0 Kohm, 1%, 0603	Multiple	

(Continued)

Reference Designator	Description	Manufacturer	Manufacturer's Part Number
R109	46.4 Kohm, 1%, 0603	Multiple	
R110	22.1 ohm, 1%, 0603	Multiple	
R95A, R96A	12.1 Kohm, 1%, 0603	Multiple	
R36	221 Kohm, 1%, 0805	Multiple	
R111	249 Kohm, 1%, 0603	Multiple	
R27	261 Kohm, 1%, 0805	Multiple	
R32	357 Kohm, 1%, 0805	Multiple	
R40	475 Kohm, 1%, 0805	Multiple	
R112	18.2 Kohm, 1%, 0603	Multiple	
R143	TBD, 0805, depending on ACDC Remote-On/Off requirements	Multiple	
NTC1A	100K @ 25C, 1%, 4200K, 0603	Panasonic	ERT-J1VR104FM
D3A	Rectifier, 1A, 200V, ultrafast, 25nsec, SMA	Diodes Inc.	ES1D-13-F
D4, D5, D16, D17	Rectifier, 75V, 150mA, 4nsec	Diodes Inc.	1N4148WS-7-F
D6, D7	Rectifier, 1A, 40V, SOD-123W	Nexperia	PMEG40T10ERX
CM1A	Common Mode Bead, 170 ohms @ 100MHz, 10A	Laird	CM2545X171B-10
L1A	22μH, Toroidal	CUSTOM	
L2, L3	22μH, 3.6A	Coilcraft	XAL5050-223ME_
L4	100μH, 1A	Coilcraft	LPS8045B-104MR_
L5	1μH, 1A	Coilcraft	LPO4815-102ML_
L6A	705.6nH	CUSTOM	12 Turns of 18 AWG Magnet Wire on Micrometals T50-2 Core
FB1A, FB2A, FB3	Ferrite Bead, 330 ohms @ 100MHz, 0603	Murata	BLM18KG331SN1D
TVS1A, TVS2A	TVS, 8V, SOD-323, Bidirectional, Low-Capacitance	Bourns	CDSOD323-T08C
Q11, Q12	PNP, 40V, SOT-23	Diodes Inc.	MMBT3906-7-F
Q16	NPN, 40V, SOT-23	Diodes Inc.	MMBT3904-7-F
Q2A, Q3A	Mosfet, 150V, DirectFET	Infineon	IRF6775MTRPBF
Q4	PNP, Dual, 65V, SOT-363	Diodes Inc.	BC856AS-7

Reference Designator	Description	Manufacturer	Manufacturer's Part Number
Q6	NPN, Dual, 65V, SOT-363	Diodes Inc.	BC846AS-7
Q9, Q14	Mosfet, P-channel, 50V, SOT-323	Diodes Inc.	BSS84W-7-F
Q10, Q15	Mosfet, N-channel, 50V, SOT-323	Diodes Inc.	BSS138W-7-F
Q13	Mosfet, N-channel, 60V, SOT-323	Diodes Inc.	2N7002W-7-F
U1A, U2A, U18	Op-Amp, Audio, SOT-23-5	Texas Instruments	OPA1677DBVR
U3A	Comparator, High-Speed, 11nsec, 8MSOP	Analog Devices	LT1713CMS8#TRPBF
U4A	Dual, Isolated, Audio Gate Driver	Skyworks Solutions Inc.	SI8244BB-D-IS1
U6, U7	Comparator, Quad, Open Collector, SOIC-14	Texas Instruments	LM339BIDR
U8	Voltage Reference, Shunt, 2.5V, SOT23-3	Diodes Inc.	LM4040D25FTA
U9	Voltage Detector, Adjustable Delay, Manual Reset, 4.65VTH	Texas Instruments	TPS3808G50DBVR
U10, U11, U12, U13, U14A, U15A, U16A, U17A, U25A	Multi-Function Logic Gate, LVC, SOT-23-6	Texas Instruments	SN74LVC1G57DBVR
U19, U20	Buffer, Non-inverting, Push-Pull, LVC, SOT-23-5	Texas Instruments	SN74LVC1G17DBVR
U21A, U21B	Buffer, 660MHz, High-Speed, Analog, Split Rails, SOT-23-5	Analog Devices	MAX4200EUK+T
U22	Dual 4-Bit Counter, 14TSSOP	Toshiba	74VHC393FT
U23	Oscillator, 16.0MHz, 5×7mm, Low-Jitter	IQD Frequency Products	LFSPXO019275

(Continued)

Reference Designator	Description	Manufacturer	Manufacturer's Part Number
U24A, U26A, U27A	Comparator, Dual, Open Collector, SOIC-8	Texas Instruments	LM393BIDR
U28A	Hall Effect Current Sensor, Bidirectional, +/−12.5A, 100KHz	Allegro Microsystems	ACS711ELCTR-12AB-T
U29, U30	DCDC, Buck, 600mA, SOT-23-6	Texas Instruments	TPS560430YFQDBVRQ1
U31	DCDC Buck, 300mA, HVin, 12TDFN	Analog Devices	MAX17760ATC+
U32, U33	Positive LDO, 5Vout, 200mA, TSOT-23-5	Analog Devices	ADP7142AUJZ-5.0-R7
U34	Negative LDO, −5Vout, 200mA, TSOT-23-5	Analog Devices	ADP7182AUJZ-5.0-R7
F1	Fuse, 5A, 277VAC, High-Breaking Capacity, Time-Lag	Schurter	3403.0281.11
F2	Fuse, 1A, 277VAC, High-Breaking Capacity, Time-Lag	Schurter	3403.0274.11
MOV1	MOV, 275VAC, 20mm	Littelfuse	V275LA40CP
ACDC1, ACDC2	ACDC, 85-264VACin, 48Vout, Baseplate, Screw Terminals, Remote ON/OFF	EOS	LFWLP225-1004-B-0
ACDC3	ACDC, 85-264VACin, 12Vout, 4W	RECOM	RAC04-12SGB
J1A	Connector, Input, 3-pos.	Application Specific	
J2A	Connector, Output, 2-pos.	Application Specific	
J3	AC Input Connector (IEC)	Adam-Tech	IEC-A-4

a Unless otherwise noted, all 0603 resistors are 1/10W.

Chapter Two

Input Circuits

Input Circuitry Differences between Class-D and Class-AB

Input circuitry for Class-D amplifiers is mostly differentiated from input circuitry for Class-AB amplifiers by its requirements for low-voltage DC power supply rails. In typical Class-AB designs, there is a motivation to power the input or "preamp" circuitry from the same high-voltage rails as are used for the power amplifier section. Typically, the rails used to supply the preamp section are filtered with an R-C (resistor-capacitor) filter to reduce noise on these rails. The other reason why the high-voltage power amp rails are used is that the design can be realized at low cost with a completely discrete design, where the preamp uses low-power, small signal transistors, albeit rated to handle a breakdown voltage exceeding the total supply rail voltage i.e. (|negative rail voltage|) + (positive rail voltage). *Note:* the absolute value signs bracketing "negative rail voltage". Since the preamp circuitry draws very low current (milliamps), there is only a small power dissipation penalty to be paid in this approach. Also, when Class-AB amplifier input circuits are designed using discrete components, instead of integrated ICs (such as operational amplifiers, "op-amps"), it is relatively easy to find small signal discrete components (transistors) with high voltage ratings.

However, in Class-D designs, there will typically be a requirement for a high-speed PWM comparator, low-noise op-amps, level shifter circuits, as well as gate driver circuits. All these components must be optimized for speed (low propagation delay), and therefore will likely be manufactured using a low-voltage process. For example, almost all high-speed analog comparators with propagation

DOI: 10.4324/9781003327943-2

delays of less than 40nsec have power supply rails rated for +/−7VDC or less. Additionally, there will be a requirement in the design for some type of low-voltage logic (digital circuitry), either for fault processing (discussed in Chapter 14) or for field-programmable gate array (FPGA) or central processing unit (CPU) control. These will typically require rails of 3.3VDC or 5VDC. With regard to op-amp circuits, there is such a wide selection now of low-noise, wide-bandwidth op-amps available, at low cost, that, given the requirement that has already been justified, for having low-voltage rails, it does not really make sense to pursue a discrete analog preamp design for a Class-D amplifier.

Requirements for Op-Amps Used in Analog Preamp Circuitry for Class-D

It is important to understand what specification parameters are critical for Class-D amplifier preamp design and which are less important. It is only through this process of differentiation that a suitable op-amp can be selected for the application. When starting a search, it is useful to begin by searching for "audio" op-amps, as a first selection filter. These types will have lower values of noise and total harmonic distortion (THD) than non-audio, or general purpose types.

Critical Specifications

1) Noise
2) THD
3) Bandwidth
4) Common-Mode Input Voltage Range (Rail-to-Rail input types or not)
5) Common-Mode Rejection Ratio (CMRR)
6) DC supply rail magnitudes (single or dual supply rail capable)
7) Output voltage range (Rail-to-Rail output types or not)
8) Output current drive capability
9) Unity-gain stability
10) Capacitive load drive capability
11) Input protections and input current handling capability
12) Overdrive response (output phase reversal protection; see Putzeys, 2014), and recovery characteristics
13) Crosstalk (for duals and quads).

Specifications of Secondary Importance

1) Input offset voltage (except as noted in text)
2) Input bias current and offset current (except as noted in text)
3) Voltage feedback or Current Feedback Amplifier (CFA) type
4) Package type (single, dual, quad packages).

Review of Critical Op-Amp Specifications

1) **Noise:** Noise performance is critical in all aspects of Class-D amplifier design. In most cases, the preamp noise should be low enough so that the overall amplifier noise is dominated by the Class-D power stage. This is true for a number of reasons, not least of which is that it is easier to find appropriate linear (i.e. non-switching) devices with very low noise performance, than it is to design a switching power stage that achieves low overall noise. In other words, there is a large selection of readily available, low-cost op-amps suitable for Class-D design, that have excellent low-noise specifications, such that the overall noise performance of the preamp design will make its noise contribution negligible when compared to that of the switching Class-D power amplifier stage. An op-amp input referred voltage noise specification of less than $5nV/\sqrt{Hz}$ should be selected.
2) **THD:** Typically, THD is not a limiting specification for op-amps selected for preamp duty. Most suitable "audio" op-amps will have a THD spec listed in the datasheet. An important determining factor of THD in IC op-amps is their output crossover distortion characteristic. Many op-amps have Class-AB output stages and therefore exhibit some crossover distortion inherent in that type of output stage. Interestingly, there is an easy method of pushing the output stage operation from Class-AB into Class-A, which inherently has no crossover distortion. This is usually done by adding either a pullup or a pull-down resistor, such that the output stage is always sourcing or sinking current (Texas Instruments, 2023). However, more modern "audio" op-amps do not require this addition to achieve low THD.
3) **Bandwidth (BW):** For most Class-D signal processing applications, a BW of 15MHz is adequate; however, some designs will benefit from using op-amps with wider BW and higher output slew rate specifications. Higher BW and slew rate in the signal processing circuitry (especially when negative

feedback is employed) can be helpful in decreasing Transient Intermodulation Distortion (TIM), which is discussed in later chapters.

4) **Common-Mode Input Voltage Range (Rail-to-Rail input types or not):** This specification determines over what range the input voltages to the op-amp can span, while the op-amp still delivers its rated specifications. The designer uses this specification to be sure that the op-amp inputs are not overdriven. Typical abbreviations used are RRI for Rail-to-Rail Input, and RRIO for Rail-to-Rail Input and Output. Op-amps that boast Rail-to-Rail input usually have input ranges that include the supply voltage rails, and sometimes even allow the input voltages to exceed the supply rails.
5) **CMRR:** There is little consistency in the way this specification is defined across the semiconductor industry. It is sometimes referred to as the ratio, expressed in dB, of the change in output voltage to the change in input common-mode-voltage. It is also sometimes defined as the ratio of the differential mode voltage gain to the common-mode voltage gain (Analog Devices, Inc., 2009b). Also, depending on whether numerator and denominator are swapped, the dB result will be either positive or negative. Either way, it is an indication of how well a device can reject signals that are common to both inputs. This is a critical specification for a device intended to be a differential amplifier, solely amplifying the *difference* voltage between the two inputs, and rejecting voltages in common to both. For example, if the input common-mode voltage were varied over an input voltage range of −5VDC to +5VDC (10V total), and the output was found to vary by 100μV, then the CMRR would be:

$$\text{CMRR} = 20\log(100\mu\text{V}/10\text{V}) = 20\log(100\mu\text{V}/10\text{V}) = -100\text{dB}$$

A similar specification is power supply rejection ratio (PSRR), which is the ratio of the change in output voltage to the change in supply rail voltage. It is a good indication of how well an op-amp will reject supply rail voltage noise. It also has a characteristic decreasing rejection versus frequency. This is another reason for the utility of supply rail bypassing. Typically, the effectiveness of supply rail bypassing needs to increase with frequency, to counteract the decreasing PSRR of the op-amp with frequency. This is why multi-layer ceramic capacitors (MLCCs) are used for supply rail bypassing. Their low internal inductance and resistance allows very low impedance noise bypassing at high frequencies.

6) **DC supply rail magnitudes (single or dual supply rail capable):** This is one of the first specifications that is reviewed in the selection process. The supply voltage range for the op-amp must be compatible with the rail voltages available in the Class-D amplifier design. Typically, Class-D designs use either a single low-voltage rail or bipolar (positive and negative) rails. An advantage of the single rail is cost, as a negative supply is not needed, but the signal path voltages must have AC coupling and a DC offset added, in order to utilize the full (unipolar) DC input voltage range of the op-amps in a single rail design. Presently, there is a plethora of op-amps designed for rails of 5.5V and lower, and fewer audio op-amps designed for higher rail voltages.
7) **Output voltage range (Rail-to-Rail output types or not):** It is important to know the guaranteed output voltage range of any op-amps selected. This output voltage is load dependent, decreasing with decreased load resistance. Some op-amps are termed "Rail-to-Rail Output Voltage" (RRO) types, in which the output voltage can come close to (but never quite equal to) the DC voltage rails. Most "RRO" types have outputs that can swing to within 800mV (or less) of the rails.
8) **Output current drive capability:** It is important to know how much current the op-amp is required to deliver in any application. Most op-amps for audio applications can drive loads of 2 Kohm and higher, at least. Be sure to cross check the output load resistance against the guaranteed output voltage level for that load resistance.
9) **Unity-gain stability:** Most op-amps that would be used in Class-D designs will be unity-gain stable. Decompensated types are usually intended for higher-BW applications, and are meant to be used with higher closed loop gains.
10) **Capacitive load drive capability:** This specification indicates the maximum capacitance value that the op-amp's output can drive, while still maintaining adequate gain and phase margins. It is usually easy to accommodate any capacitive loading limitations by inserting a low-value resistor between the op-amp's output and the capacitive load.
11) **Input protections and input current handling capability:** This is explained in greater detail below, in the reference amplifier schematic discussion.
12) **Overdrive response (output phase reversal protection), and recovery characteristics:** This op-amp characteristic is very important in Class-D

amplifier design. Some op-amp types (mostly older types) could experience an output phase reversal if one of the inputs were overdriven (Texas Instruments, 2022). This is particularly undesirable in an amplifier design featuring negative feedback, because a phase reversal of this type would convert the negative feedback to positive, resulting in a "latched" or railed output condition. Most modern op-amps include internal circuits to prevent output phase reversal.

13) **Crosstalk (for duals and quads):** A crosstalk specification in an op-amp is only applicable when there is more than one op-amp included in a single package. Op-amps are available as singles, duals, and quads. Some duals and quads have excellent crosstalk specifications; however, using single op-amp packages provides multiple benefits in Class-D design. The packages can be very small, and provide great freedom in terms of printed circuit board (PCB) layout, allowing the op-amp to be placed optimally in the layout. In contrast, if a dual package is used, the placement and trace routing are usually a compromise between what is optimal for each of the two devices in the one package. An important characteristic about crosstalk is that it usually increases with frequency due to capacitive coupling effects in the die. However, there is also a thermal effect, whereby the power dissipation in one device heats up that section of the die, which modulates the performance of the other op-amp in the same die. This is a low-frequency phenomenon, and is more prevalent in power devices, and less prevalent in small signal devices such as op-amps.

Review of Input Circuitry in the Reference Schematic

General Schematic Notes

Note: The reference schematic is presented as a single-channel design, in order to reduce the size and complexity of the schematic. In order to ease the conversion from a single-channel implementation to stereo implementation, any components with an "A" suffix in their reference designation should be duplicated to create the second channel in a stereo design.

In the reference schematic, the page designated Input/Preamp is the input circuitry. In general, the role of the input circuitry is to receive an audio input signal with a specified voltage range and amplify it to an acceptable voltage range required by downstream circuits. The amplifier designer must be aware of both the differential voltage range as well as the common-mode (CM) voltage range of the audio input signal for which the input circuitry should be designed.

U1A is a low-noise audio op-amp configured as a differential amplifier. In this topology, the following values should be equal (and tightly matched) for best CM rejection:

R1A = R2A

R3A* = R4A*

R5A* = R6A*

C3A = C4A

*It is recommended to use 0.1% values for the resistors in this circuit, for improved matching, and improved CM rejection. The use of 5% tolerance capacitors for C3A and C4A is also recommended. Capacitors C1A and C2A provide electrostatic discharge (ESD) protection as well as EMI benefits. It is important that C1A and C2A are connected to chassis ground (GND) and not to circuit ground. This provides ESD pulse currents with the shortest possible path, so that they cause the least damage. R1A, R2A and FB1A, FB2A attenuate radio-frequency (RF) noise pickup. The tranzorbs (TVS1A, TVS2A) provide overvoltage protection. Note that the resistors R1A, R2A, as well as the Ferrite Beads FB1A, FB2A, work in conjunction with tranzorbs TVS1A, TVS2A to help absorb fast overvoltage transients. Resistors R3A, R4A set the input resistance seen by the input source. Typically, an input resistance of at least 10 Kohm is expected by line-level audio sources. In addition to setting the input resistance, resistors R3A and R4A also protect the op-amp inputs from overvoltage damage.

Input Overvoltage Tolerance (Electrical Overstress)

This is a detail not normally discussed in op-amp designs (Texas Instruments, 2022). Typically, op-amps have maximum input voltage specifications, which state that the input voltage (at the op-amp input pins) must never exceed any supply rail by more than 0.3V. However, op-amp manufacturers have become more generous with their description and specification of input circuitry in recent years, and now qualify that input voltage specification. This is usually done with a footnote, stating something to the effect of: input voltages may exceed the supply rails, if the input current is limited to XmA. Typically, X is in the 5–10mA range for newer devices. If, for example, the input current must be limited to 10mA, then if R3A and R4A are sized at 10 Kohm, and the power supply rails for the op-amp are 0V (in the unpowered state), the input signal could be as high as +/−100Vpeak, before the +/−10mA input current spec would be exceeded. Please also note that it is very likely that an input signal will be applied to the input circuitry with the amplifier in both the powered and the unpowered state, because input signals are rarely disconnected when amplifiers are switched off. Therefore, this operating mode must be considered by the designer for a reliable product. In addition, the loading effect on the audio source must be analyzed when the amplifier is in the unpowered state. This is sometimes an overlooked detail. For example, if the amplifier's input protection circuits utilize diode clamps to the DC supply rails (see Figure 2.1), then these diode clamps will clamp any input voltages to 0V when the rails are unpowered, and any input current will flow through these diodes into the rails. It is prudent to check that the load impedance on these rails is low enough such that this extra current will not cause the rails to rise to an unsafe level. If that is a concern, then transient voltage suppressor (TVS) voltage clamping devices should be added to these rails.

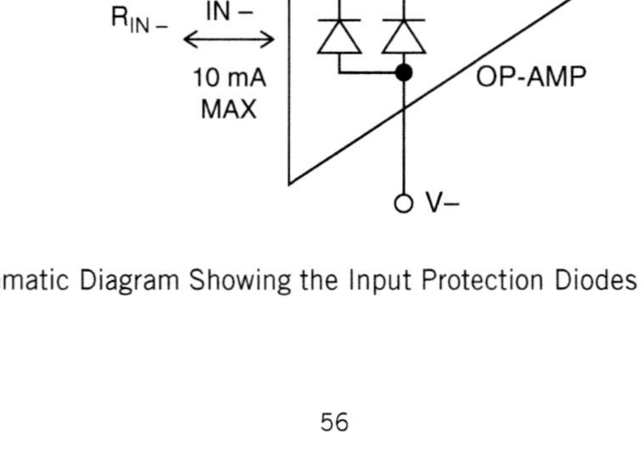

Figure 2.1 Simple Schematic Diagram Showing the Input Protection Diodes of a Representative Op-Amp Input Circuit

Be aware, though, that if this op-amp input protection scheme is used, any powered device driving these unpowered amplifier inputs will "see" Rin as a load. Another good alternative or additional overvoltage protection scheme is the use of ESD or TVS clamping diodes as input overvoltage protection devices (TVS1A and TVS2A in the reference schematic). This has the added advantage, in the case of ESD protection, of shunting ESD pulse currents to chassis close to the input connectors, instead of through the op-amp itself.

The gain of the op-amp is determined, knowing the input signal range, and the signal level required by the following stage, to drive the amplifier to full power. However, a good practice for best noise performance is to utilize the maximum output signal range of the op-amp. For the reference design, the power rails for the analog circuitry were chosen to be +/−7VDC. The maximum recommended output swing will be determined by the output specifications of the op-amp. Some op-amps (RRO types) have outputs that are specified to swing "rail-to-rail", whereas others may only be able to swing to within 2 or 3V of the rails, and this will also vary with the load resistance that the op-amp is driving. A typical op-amp running off +/−7V rails might be able to reliably drive a 10-Kohm load to within 2V of the rails, or +/−5Vpeak. Now, for example if a typical consumer audio input device is used, which is specified to be able to produce 2VRMS maximum (+/−2.822Vpeak), the gain of the difference amplifier circuit should be designed such that this maximum input signal level produces +/−5V at the op-amp outputs. This will result in the lowest overall noise performance for the input circuitry. Be aware, however, that the THD of the op-amp may increase at higher output levels, so the datasheet should be checked for that as well. In the reference circuit, this means that +/−2.822Vpeak should drive the op-amp output to +/−5V. This is a gain of 5V/2.822V = 1.77V/V = +4.97dB. The gain of this difference amplifier is R5A/(R3A + R1A) ignoring the minimal DC resistance (DCR) of ferrite bead FB1A. So 17.8K/(10K + 33.2) yields the closest result, using 1% resistor values (with 0.1% tolerance). C3A and C4A are selected to rolloff the high-frequency gain of the op-amp above the audio frequency range (i.e. > 20KHz). A subtlety in the value selection for C3A, C4A is that the value should allow for noise attenuation just above 20KHz, without significantly affecting 20KHz gain or phase. Typically, a corner frequency of 100KHz is selected, so that the gain rolloff at 20KHz is less than 0.5dB.

Supply Rail Bypassing

Even though IC op-amps intended for audio applications have excellent CMRR and PSRR specifications, IC manufacturers always advise adding bypass capacitors as close as possible to the power pins of the op-amp, to provide low-impedance noise bypassing to circuit ground, as well as a local charge "reservoir". Transient current demands of the op-amp can be delivered through the low-impedance connection to these "reservoir" capacitors, without having to travel through the longer, higher-impedance connection to the rail voltage source. In the reference schematic, the noise filtering of the op-amp rails has been enhanced further by using an R-C circuit instead of just C alone. This provides a single-pole noise filter for voltage noise on the rails and also provides some measure of isolation from other devices powered from the same rails. In the reference schematic, R7A/C5A and R8A/C6A provide a low-pass pole frequency of 15.9KHz. Remember that the op-amp's PSRR is very high at low frequencies and decreases with increasing frequency. So, the increasing attenuation with frequency, provided by these R-C filters, complements the op-amp's decreasing PSRR with frequency characteristic.

Common-Mode Input Voltage Range

There are two general types of input connections used for audio amplifiers. These are *balanced* and *unbalanced*, and are covered extensively in the literature (Macatee, 2002; Whitlock, 1995, 1996a). Balanced interconnects have a number of advantages over unbalanced, particularly regarding noise pickup and elimination of ground loops, and rejection of CM signals (high CMRR). The connection type used in the reference schematic is balanced. This differential (DIFF-AMP) op-amp configuration, in conjunction with the values used, allows CM signals in the range of +/−7.818Vpeak to be rejected. See Figure 2.2 for analysis.

Figure 2.2 illustrates the operation of U1A in the presence of a large CM voltage of +7.818V. In addition, there is the maximum rated level of differential signal (desired audio input signal of 2VRMS = +/−2.818Vpeak) riding on top of this CM voltage. Under these input conditions, U1A rejects the large CM voltage and produces the same output voltage as it would produce if the CM voltage were 0V. Note that Figure 2.2 depicts a positive CM voltage and a positive and negative differential signal swing. However, all voltages will remain within operational

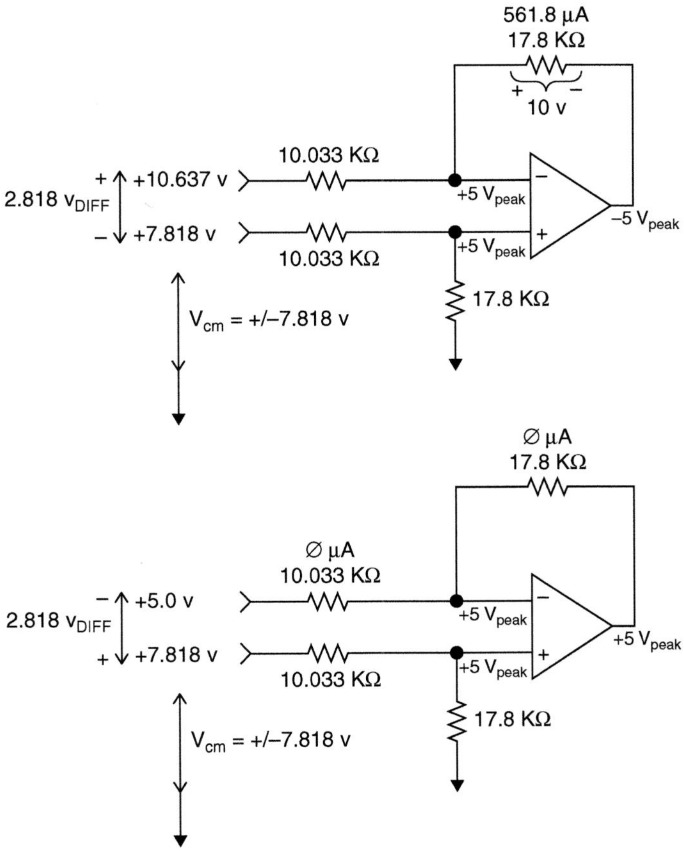

Figure 2.2 Common-Mode (CM) Voltage Calculations for Specific Input Voltage Conditions, Using a Differential Op-Amp (DIFF-AMP) Circuit Configuration

range for CM voltage of +/−7.818Vpeak and differential signal level of 2VRMS = +/−2.818Vpeak.

Input Coupling

The balanced input configuration of U1A in the reference schematic allows for the input signals to be DC-coupled at the amplifier input. AC-coupling is necessary at other points in the reference schematic and will be discussed in more detail when those sections are covered.

Phasing

Most audio amplifiers employ a non-inverting signal path from input to speaker output. Referring to the block diagram for Class-D amplifiers in Figure 1.14 of Chapter 1 (Introduction), the V_{IN} signal source feeding the error amplifier in the block diagram represents the output of the preamp circuitry. The phasing from that V_{IN} to the amplifier output is inverting. Therefore, in order for the entire signal path to be non-inverting, the phasing of the preamp must be inverting. This means that the *positive* input signal connection is at J1A-1, and the negative input signal connection is at J1A-3. Note that J1A-2 is connected to chassis. This follows the convention suggested in Whitlock (1995, 1996a), where the shield of balanced interconnect cables is connected directly to chassis ground (not circuit ground) at amplifier inputs.

Bibliography

Analog Devices, Inc. (2009a). Op amp input and output common-mode and differential voltage range. Analog Devices Tutorial, MT-041. Wilmington, MA: Analog Devices, Inc.

Analog Devices, Inc. (2009b). Op amp common-mode rejection ratio (CMRR). Analog Devices Tutorial, MT-042. Wilmington, MA: Analog Devices, Inc.

Brown, J., & Whitlock, B. (2003). Common-mode to differential-mode conversion in shielded twisted-pair cables (shield-current-induced-noise). AES preprint 5747. New York: Audio Engineering Society.

Macatee, S. (2002). Grounding and shielding audio devices. RaneNote 151. Mukilteo, WA: Rane Corporation. Originally published 1995; revised 2002. A version of this RaneNote was published in the *Journal of the Audio Engineering Society*, *43*(6), June 1995.

Putzeys, B. (2014, April 22). The G word, or how to get your audio off the ground. *EDN*.

Rane Technical Staff. (1985, 1995, 2006, 2007, 2011, 2015). Sound system interconnection. RaneNote 110. Mukilteo, WA: Rane Corporation.

Texas Instruments. (2022). Datasheet, OPA167x low-distortion audio operational amplifiers, SBOS855E. Originally published January 2017; revised December 2022. Dallas, TX: Texas Instruments.

Texas Instruments. (2023). Application design guidelines for LM324 and LM358 devices, SLOA277B. Originally published January 2019; revised July 2023. Dallas, TX: Texas Instruments.
Whitlock, B. (1995). Interconnection of balanced and unbalanced equipment. Jensen Application Note, AN-003. Van Nuys, CA: Jensen Transformers, Inc.
Whitlock, B. (1996a). Hum & buzz in unbalanced interconnect systems. Jensen Application Note, AN-004. Van Nuys, CA: Jensen Transformers, Inc.
Whitlock, B. (1996b). A new balanced audio input circuit for maximum common-mode rejection in real-world environments. AES preprint 4372. Van Nuys, CA: Jensen Transformers, Inc.

Chapter Three

Comparator and PWM Circuits

Comparator Characteristics for Class-D Usage

Note: The Error Amplifier page of the reference schematic is discussed in Chapter 12 (Feedback and Compensation)

Referring to the reference amplifier schematic, all circuitry up until the PWM comparator is continuous time, linear, analog circuitry. At the comparator, the circuitry becomes a sampled (quantized) system. In the reference amplifier, the comparator makes a PWM timing decision twice per period of the switching frequency. For example, if the switching frequency is 500KHz, the PWM comparator will make two PWM timing decisions within a period of 2μsec. The specification parameters that are critical for the selection of the PWM comparator are the following:

1) Supply voltage range
2) Propagation delay
3) Input voltage range
4) Input overdrive tolerance
5) Power dissipation.

Supply Voltage Range

The reference amplifier design powers the input circuitry components using positive and negative (split) low-voltage power supply rails. An alternate approach would

 DOI: 10.4324/9781003327943-3

be to use a single positive supply rail, but that would require DC input biasing of the signal to all op-amps, at one-half of the supply rail voltage, and AC coupling between all stages. By using split rails, all audio signals are referenced to 0V ground, eliminating the need for DC biasing of op-amp inputs.[1] An additional benefit is that this scheme eliminates one potential cause of pops, clicks, and thumps during turn-on/off of the amplifier. These pops, clicks, and thumps can occur when the biasing networks are charging and discharging. The split rail scheme avoids this potential problem, at the expense of requiring a negative voltage rail.

The supply voltage range should be maximized, to allow for larger signal voltage swing. This improves the signal-to-noise ratio (SNR) because, assuming noise voltage to be constant, increasing the available signal amplitude range increases the SNR. Presently available high-speed comparators are almost all limited to either a single +5V supply rail, or +/−5V supply rails. There are high-speed comparators available that run at +/−15V rails, but these are presently sole-sourced and expensive, or have longer than optimal propagation delay specifications. The comparator selected for the reference design (LT1713) (Analog Devices (Linear Technology), 2000) can handle up to +/−5V supply rails. Its input CM range is equal to the rails (+/−5V), and its output stage switches between the positive rail and GND.

Propagation Delay

In the context of this amplifier design, "high-speed" refers to propagation delays of less than 40nsec. Propagation delay is a time delay and appears as a phase-shift in the open-loop response of the Class-D amplifier, and must therefore be minimized. Propagation delay is measured as the time delay between the time when the input signal crosses the switching threshold, and the output voltage switches (measured typically at 50% of the rising edge). The comparator selected for the reference design (LT1713) has a propagation delay of 7nsec (Analog Devices (Linear Technology), 2000).

Input Voltage Range

Like the supply voltage range, it is important to maximize the available input voltage range in order to allow the largest input signal voltage swing, and SNR. The LT1713 allows rail-to-rail input voltage swing. The relevant specification here is input CM range. Other comparators such as TL3016 (Texas Instruments,

2000) have a CM range that is limited to within 2V of the positive rail. Therefore, if the positive supply rail is +5V, the comparator's input signals (RAMP and E/A output signals) cannot exceed +3V. Now, since *both* these signals are symmetrical about GND (0V) this will limit the RAMP signal to +/−3V peak. The LT1713, biased from +/−5V rails, will allow input signals up to the full supply voltage range, therefore +/−5V peak.

Some comparators also specify a maximum differential input voltage range. This is called out separately from the maximum voltage range of each of the individual inputs. In some comparators, the inputs are clamped to each other either through anti-parallel signal diodes or through zener devices. The purpose of the differential input voltage specification for these devices is to limit the current through these clamping devices. Devices that use this type of input architecture present a problem, because if these clamping devices conduct, they present a non-linearity in the input impedance of the comparator inputs and cause a loading change on the circuits driving the comparator inputs (the RAMP and error-amplifier circuits). This non-linear loading combined with the non-zero output impedance of these driving circuits will cause THD in the amplifier. Therefore, it is preferable to select comparators that do not have a differential input voltage limitation, or to ensure that the operational input voltage range never exceeds the differential input specification for the comparator selected.

Input Overdrive Tolerance

As mentioned previously, when discussing op-amp inputs, it is important to understand how the comparator inputs react when the inputs are driven above the CM input range, or above the supply rails (if it is possible for the comparator input signals to reach those levels). Some modern devices are specified in terms of an input current limit which should not be exceeded during overdrive. This specification makes it very convenient to simply limit the input overdrive current to a safe level, by the selection of an appropriate input resistor value. In terms of comparator behavior during overdrive, it is important to read the datasheet to determine how the comparator outputs behave during input overdrive. In some cases, especially for older comparators, there can be a phase reversal at the output, which is highly undesirable. The datasheet for the LT1713 specifically states that the inputs can exceed the supply rails without phase reversal. In the reference schematic, the RAMP and error amplifier are biased from +/−7V rails, and the

LT1713 comparator is biased from +/−5V rails. This means the absolute worst-case condition would be where the comparator inputs could be driven up to 2V above the comparator rail voltages. The input specification for the LT1713 allows a maximum input current of 10mA during this overdrive condition. Therefore, a resistor value of 2V/10mA = 200 ohm minimum is sufficient to limit the input current during overdrive, to 10mA, protecting the LT1713 inputs against overdrive damage.

Power Dissipation

Efficiency is the hallmark of Class-D design, and to optimize efficiency the power dissipation of *all* components in the design must be carefully checked. The PWM comparator datasheet must be perused for maximum power dissipation. Usually higher-speed operation (lower propagation delay) comes at the expense of greater power dissipation (supply rail current), but there are still many devices that can provide adequately low propagation delay with very low power requirements. The LT1713 is specified to draw 8mA maximum from the positive rail and 4.5mA maximum from the negative rail over its full operating temperature range.

Referring to the *PWM Comparator* page of the reference schematic, there are three logic gates used (U14A, U15A, U16A). Here, the versatile multifunction gate (LVC1G57) is used. U14A is configured to produce a rising-edge triggered pulse (Chenier, 2006). U15A is configured to produce a falling-edge triggered pulse. As will be seen later in this chapter, these pulses serve multiple functions. They prevent the PWM signal (PWMA) from reaching the full modulation extremes of 100% ON or 100% OFF. They thus set a minimum and a maximum duty-cycle limit for the PWM signal. This ensures that the "fixed-frequency" design does in fact always operate at the said fixed frequency. If, for example, the PWM signal were allowed to reach 100% ON or 100% OFF, switching of the power stage would stop, and the switching frequency would therefore drop to 0Hz. Another critical function that these pulses serve is that they ensure that each of the power mosfets switches during each period of the switching waveform. This is particularly important for the *lower* fet(s) in the half- or full-bridge power stage because the upper gate driver's boot-strap supply only gets charged when the lower fet switches ON. This will be discussed in more detail when the power stages are discussed. For now, it will suffice to say that the design goal for

Table 3.1 Input Switching Thresholds for LVC1G57

For Rising Input Voltage	*Vcc = 4.5V*	*Vcc = 5.5V*	*Extrapolated Values for 5V = Vcc*	*Mean of Extrapolated Values*
Min	2.16V	2.75V	2.455V	**2.71V**
Max	2.61V	3.33V	2.97V	
For Falling Input Voltage	*Vcc = 4.5V*	*Vcc = 5.5V*	*Extrapolated Values for 5V = Vcc*	*Mean of Extrapolated Values*
Min	1.41V	1.90V	1.665V	**1.87V**
Max	1.87V	2.29V	2.98V	

the rising- and falling-edge triggered pulses is a width of 100nsec. The rising-edge pulse is triggered by the RAMP_SW signal, and its pulse width is programmed by the selection of the R-C values of R45A and C22A. These values are determined by first checking the LVC1G57 datasheet (Texas Instruments, 2016). The datasheet lists the rising and falling input thresholds for the device. One of the characteristics that makes this logic gate so flexible is that each of its three inputs has hysteresis. This means the rising input voltage threshold is different (and higher) than the falling input voltage threshold. This hysteresis provides for clean switching and avoids multiple output transitions, as would occur if a single input voltage threshold were active for both rising and falling input voltages. Table 3.1 lists the values for positive and negative going inputs, versus supply voltage for the LVC1G57. Unfortunately, values are not provided for a Vcc value of 5V, which is used in the reference design, so some interpolation is necessary between the 4.5V and 5.5V Vcc values provided in the datasheet.

Next, the interpolated values are used in an LTspice simulation, to arrive at suitable values for R and C, such that a rising- and falling-edge pulse width of approximately 100nsec nominal can be obtained. The rising-edge triggered pulse is MPWRE (triggered by the rising edge of the RAMP_SW signal). The falling-edge triggered pulse is MPWFE (triggered by the falling edge of the RAMP_SW signal). *Note:* Due to the cumulative effects of component value tolerances, it may be necessary to shorten the minimum pulse width setting in order to be able to achieve high enough maximum duty cycle to achieve 100W out into an 8-ohm load, with +/−48VDC power supply rails. The 100nsec pulse can be shortened to 50nsec by reducing C22A and C23A from 220pF to 120pF.

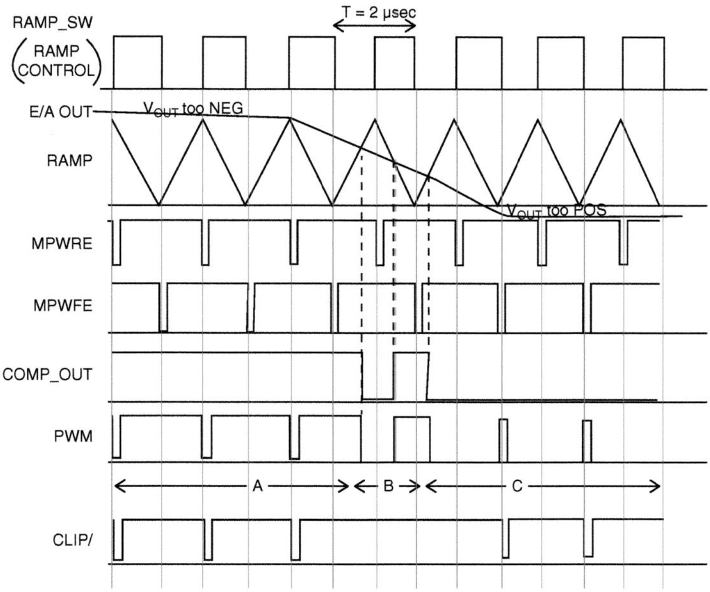

Figure 3.1 Timing Diagram of the Essential Signals Used in the Modulator Section

Figure 3.1 shows a timing diagram of the signals in the modulator section discussed so far. U16A receives the two pulses discussed above, as well as the COMP_OUT signal from the comparator, and produces the PWM signal which will propagate to the power switching stage. Please refer to Table 3.2 for the *logic table* for the LVC1G57, to understand how U16A produces the PWM signal from these signal inputs.

U17A is another LVC1G57 multifunction gate, but this time configured as an XNOR logic gate. It generates a CLIP/ signal output, using the PWM and COMP_OUT signals as inputs. Under normal (non-clipped) conditions, the COMP_OUT signal and the PWM signal will appear identical. However, during clipping, the PWM signal will differ from the COMP_OUT signal. The XNOR circuit (U17A) will produce a *low* output at CLIP/ whenever these signals differ. Referring to the timing diagram (Figure 3.1), regions A and C represent conditions where COMP_OUT and PWM differ, and region B shows when they are identical. U17A will produce negative going pulses during regions A and C. These pulses can be used to drive a Variable Gain Amplifier (VGA), muting

Table 3.2 LVC1G57 Logic Table

Inputs			*Output*
In2 (pin 6)	*In1 (pin 1)*	*In0 (pin 3)*	*Y (pin 4)*
L	L	L	H
L	L	H	L
L	H	L	H
L	H	H	L
H	L	L	L
H	L	H	L
H	H	L	H
H	H	H	H

(X) = Package pin number.

circuit, or limiter circuit, to reduce gain in response to clipping. This will be discussed in later chapters.

Clipping Detection

"Clipping", which is more accurately described as output voltage clipping, is a condition whereby the output voltage of an audio amplifier has exceeded its linear operating range. Typically, clipping is described as the point at which the output THD of an amplifier exceeds a certain threshold. For low-distortion, higher-quality amplifiers, this is typically when the output THD exceeds 1%. It should be noted that some lower-quality, higher-distortion amplifiers may exceed 1% THD even before reaching output voltage clipping. For these amplifiers, the clipping THD level will of necessity be a higher threshold. When clipping is reached, the output voltage has exceeded its operational range, and the output voltage no longer follows, nor is an accurate representation of, the amplified input voltage. Typically, the output peaks appear flattened or "clipped" on an oscilloscope (see Figure 3.2 for an example of a clipped waveform).

The most common causes of output voltage clipping in Class-D amplifiers are listed below:

1) Audio signal exceeding the supply rail voltages in other amplifier stages, such as the preamp stage or error amplifier stage.

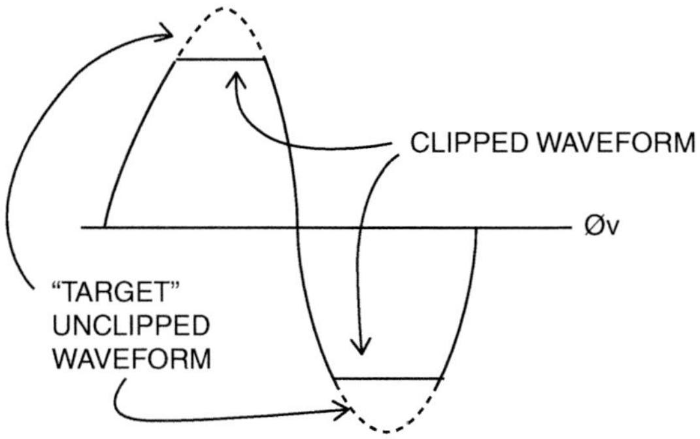

Figure 3.2 A Clipped Sinusoidal Waveform, as it Would Appear on an Oscilloscope

2) Maximum/Minimum PWM duty cycle limiting. This occurs when the minimum or maximum duty cycle is reached. This condition must be analyzed in conjunction with the DC rail voltages used in the design.

PWM duty cycle can be limited to other percentages besides the minimum and maximum mentioned above. Duty cycle limiting can be used as a method to limit output power in response to various fault conditions, for example.

Class-D Amplifier Output Voltage vs PWM Duty Cycle

There is a basic relationship between PWM duty cycle and amplifier output voltage. For a first-order approximation, it is assumed that the power switches are ideal, and the deadtime is zero. These assumptions allow for a simplified expression, depending on the type of power stage employed. (Power stage topologies will be discussed in Chapter 8.)

For a split supply system (positive and negative voltage rails), such as is used in the reference design, 0V quiescent output voltage occurs at 50% duty cycle.

The complete expression allows for independent positive and negative voltage rail values, and the expression of duty cycle (D) as a percentage. The percentage entered is the percentage of each switching period (entered as a number between 0 and 1, where 1 = 100%) during which the upper power switch is ON (closed).

$$\text{Duty Cycle} = (D) = t_{\text{ON UPPER MOSFET}} / T \qquad \text{EQ. 3.1}$$

$$D = 1900\text{nsec} / 2\mu\text{sec} = 0.95 = 95\% \qquad \text{EQ. 3.2}$$

$$\text{Vout} = (\text{Vp})(D) + (\text{Vn})(1 - D) \qquad \text{EQ. 3.3}$$

This expression can be simplified even further if the positive and negative voltage rails are *equal* = V (a single positive voltage value). For example, if Vp = +48V and Vn = −48V, then V in EQ 3.4 will be 48V:

$$Vout = V(2D - 1) \qquad \text{EQ 3.4}$$

Both of these above expressions show that the output voltage (Vout) can be limited by both the power supply rail voltages used *and* the duty cycle range (D).

A real-world example will be illustrative: Suppose the amplifier has a switching frequency of 500KHz (T = period = 2μsec) *and* the maximum duty cycle is 95% (D = 0.95). This means the upper power switch has a maximum ON time of 0.95 × 2μsec = 1.90μsec (1900nsec). Since we have assumed zero deadtime, this means the lower fet will be on for only 100nsec under this condition of 95% duty cycle, i.e. 1900nsec + 100nsec = 2μsec.

With power supply voltages of +48V and −48V, V is entered as 48.

$$Vout = 48(2(0.95) - 1) = 43.2V \qquad \text{EQ. 3.5}$$

The output voltage will be clipped at +43.2V.

If an output voltage of +45V were called for (based on the input voltage level and the gain of the amplifier), the output voltage will have been limited to +43.2V by the duty cycle restriction.

Similarly, if the other duty cycle extreme is examined, the upper mosfet would be ON for only 100nsec, and the lower mosfet would be ON for 2μsec − 100nsec = 1900nsec.

$$D = 100nsec / 2\mu sec = 0.05 = 5\% \qquad \text{EQ. 3.6}$$

$$Vout = 48(2(0.05) - 1) = -43.2V \qquad \text{EQ. 3.7}$$

Operation of the amplifier in clipping is best avoided for the following reasons:

1) High THD and unpleasant audible distortion, especially under heavily clipped conditions.
2) Excessive high-frequency signal levels due to fast dv/dt of high-amplitude output signal transitions.
3) Excessive output power (depending on loading), which can stress both the amplifier output stage and the attached loads.

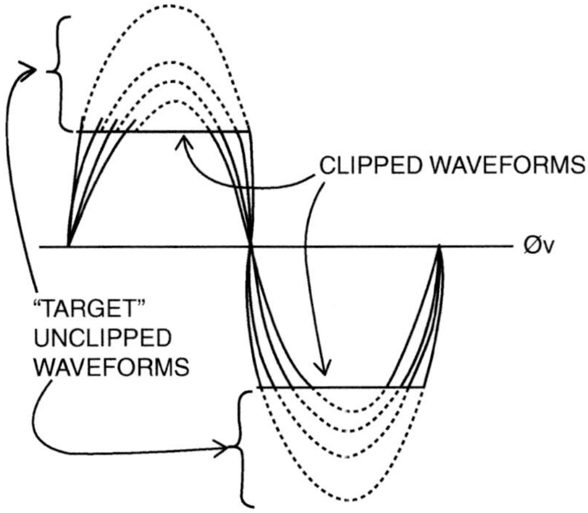

Figure 3.3 Clipped Sinusoidal Waveforms of Increasing Amplitude

Note: A sinusoidal signal with a peak voltage of 90V into 8 ohms represents an output voltage of Vpeak/SQRT(2) = 90/1.414 = 63.64VRMS. A square wave having the same 90V peak level represents an RMS voltage of 90V. Please note that as a sinusoidal signal is driven further and further into clipping, it becomes more "square" (see Figure 3.3 for sine wave to square wave clipping progression).

63.64V across an 8-ohm load represents **506W**
90V across an 8-ohm load represents **1012W**

So worst-case clipping can result in a doubling of output power! This high output power during clipping is one cause of damage to speaker drivers (Elliott Sound Products, n.d.).

4) Loss of DC servo (feedback) control of the output signal. Once the output voltage can no longer follow the (amplified) input signal, the error amplifier output will saturate, because the high-gain feedback loop can no longer control the output. When this happens, the output DC voltage level will no longer be controlled by the feedback loop. This can result in tripping of DC offset fault detection circuits (to be discussed in Chapter 14).

Now that clipping has been explained, it should be said that clipping detection circuitry is not an absolute operational necessity for an audio amplifier. An amplifier can operate safely into clipping (especially light clipping). However, for the reasons listed above, it is beneficial to include clipping detection and mitigation circuitry in any high-quality class-D audio amplifier design.

Alternative Design Approaches for the PWM Comparator Circuitry

An alternative and more sophisticated design approach to the generation of the RAMP_SW and minimum and maximum duty cycle signals, as well as detection and processing of clipping, would be to implement an FPGA to handle this PWM signal processing. This affords the design greater flexibility and features and will be discussed in other sections of this book.

Note

1 Although AC coupling is not needed to bias the audio signal at one-half of the positive supply rail, it *will* be used to decrease the DC offset voltage at the error amplifier, which will result in a reduction of the DC offset voltage at the output of the amplifier.

Bibliography

Analog Devices, Inc. (2009). Comparators. Analog Devices Tutorial, MT-083. Wilmington, MA: Analog Devices, Inc.

Analog Devices (Linear Technology). (2000). Datasheet, LT1713/LT1714. Single/dual, 7ns, low power, 3V/5V/±5V rail-to-rail comparators. Milpitas, CA: Linear Technology Corporation.

Chenier, G. (2006). Configurable logic gates' Schmitt inputs make versatile monostables. *EDN*, May 25, 94–95.

Elliott Sound Products. (n.d.). Speaker damage. https://sound-au.com/bi-amp-p1.htm

Texas Instruments. (2000). Datasheet, TL3016, TL3016Y, Ultra-fast low-power precision comparators, SLCS130D. Originally published March 1997; revised March 2000. Dallas, TX: Texas Instruments.

Texas Instruments. (2013). Comparing high speed comparators. Application Report: AN-87, SNOA658B. Originally published June 1973; revised April 2013. Dallas, TX: Texas Instruments.

Texas Instruments. (2016). Datasheet, SN74LVC1G57, Configurable multiple-function gate, SCES414P. Originally published November 2002; revised November 2016. Dallas, TX: Texas Instruments.

Chapter Four

Ramp Generator Circuits

Triangular Ramp versus Sawtooth Waveshape

One of the distinguishing features of a fixed-frequency, PWM-based, analog Class-D amplifier is the presence of a *ramp* or *sawtooth* waveform, which serves as one input signal to a PWM comparator. The other input to the PWM comparator is the output of an error amplifier, which is discussed in detail in Chapter 12. The triangle signal has a linear rising ramp and a linear falling ramp with the exact same slope (dv/dt) as the rising ramp, but of opposite polarity. The frequency of the triangle ramp is the switching frequency of the Class-D amplifier. The sawtooth waveform, in contrast, has only *one* linear ramp, either rising or falling, followed by an edge with ideally infinite slope (zero rise or fall time). See Figure 4.1 which shows triangle ramp and sawtooth ramp waveforms. An advantage of the triangle waveform over the sawtooth is that the triangle waveform allows the modulator (error amplifier and comparator) to make *two* PWM timing decisions (samples) per period, whereas the sawtooth waveform only allows *one* timing decision (sample) per period. This is illustrated in Figure 4.1. This effective doubling of the sampling rate using the triangle waveform increases the achievable bandwidth of designs using the triangle ramp waveform (Marco, 2009).

Figure 4.1 illustrates that the timing of *both* the rising and falling edges of the PWM waveform is variable within each *period* (T) of the switching frequency clock and depends on the instantaneous relationship between the ramp signal and the error amplifier output signal at the PWM comparator inputs. In contrast, however, when the *sawtooth* waveform is used, one of the PWM switching

DOI: 10.4324/9781003327943-4

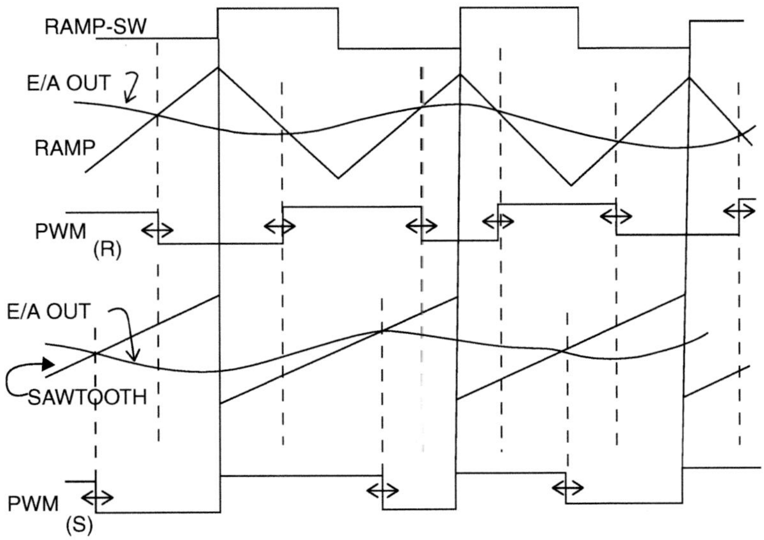

Figure 4.1 Timing Waveforms for a Triangular Ramp (R) and a Sawtooth Ramp (S)

transitions (the rising edge of the PWM waveform in the Figure 4.1 example) is *fixed* with respect to the switching frequency clock, and there is only one variable switch timing event per switching period (T) dependent on the instantaneous relationship between the *sawtooth* waveform and the error amplifier output waveform presented at the PWM comparator inputs. The type of sampling resulting from the triangular ramp and sawtooth waveform is termed "natural sampling".

The voltage gain of the modulator and power switching stage of the Class-D amplifier is defined by EQ. 4.1. The modulator stage generates the PWM pulse train that serves as input to the Class-D power switching stage. This will be discussed in more detail in other chapters, but is mentioned here because peak-to-peak ramp amplitude is discussed here and it is useful to know what other parameters in the Class-D design are affected by it.

$$\text{Modulator Gain} = V_{\text{Supply-Rails-Total}} / V_{\text{RAMP-peak-to-peak}} \qquad \text{EQ. 4.1}$$

EQ. 4.1 defines the gain of the circuitry that includes the PWM comparator, power switching stage, and L-C demodulation filter. This gain is roughly flat over the audio frequency range of 20Hz–20KHz. Therefore, the voltage gain is the total supply rail voltage divided by the peak-to-peak voltage of the triangle ramp waveform. An understanding of this relationship is important, and it will be explained in more detail in Chapter 12. But for now, it is important to realize that any noise, disturbance, or non-linearity in the triangle ramp waveform will be amplified by this "modulator gain." For this reason, when designing the ramp generator circuit, it is critical to achieve the following design goals:

- Minimize noise (for maximum Signal-to-Noise Ratio = SNR)
- Maximize linearity (for best Total Harmonic Distortion = THD)
- Maximize symmetry (falling slope = rising slope, for best THD)
- Maximize dv/dt at inflection points (sharp slope transitions, for maximum duty cycle range)
- Stabilize DC level (for maximum duty cycle range)
- Maximize frequency stability and minimize jitter (for lowest noise and best SNR)
- Maximize peak-to-peak voltage swing (for maximum SNR).

Detailed Discussion of the Reference Schematic Ramp Generator Circuitry

The operation of the ramp generator circuit in the reference schematic can best be understood by referring to Figure 4.2.

Figure 4.2 shows that the ramp generator circuitry is composed of sourcing and sinking current mirrors utilizing bipolar transistors.

Current Mirror Operation

To a first approximation, the circuit will be designed such that I_{REF} establishes a current in Q4-1 and R54, which is mirrored in R55 and Q4-2 and flows out of the Q4-2 collector into ramp capacitor C_{RAMP}. This current (X) is designed to be approximately 4mA. Likewise, I_{REF} establishes a current in Q6-1 and R57, which is mirrored in Q6-2 and multiplied by the approximately 2:1 ratio of R58 to R57, appearing as a sink current (2X) of approximately 8mA, from the collector of Q6-2.

Although Q4 and Q6 are not fabricated by the manufacturer to be "matched" transistors, they will nevertheless be matched well enough for this application, because they are identical (*dual*) transistors fabricated on the same die.

This four-transistor current source/sink circuit works as follows:

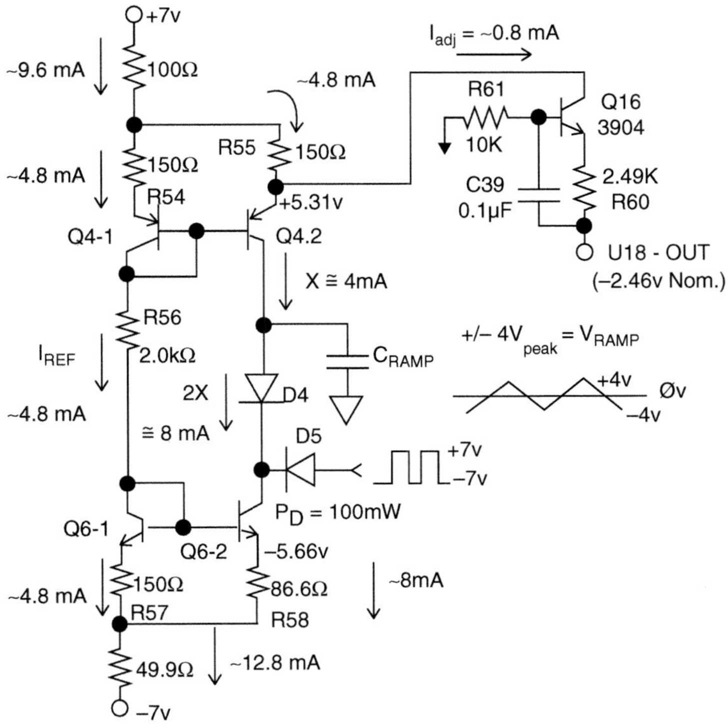

Figure 4.2 Annotated Schematic Diagram for the Current Mirror Circuit of the Triangular Ramp Generator

- Q4-2 will *source* a current X at all times (set to 4mA)
- Q6-2 will *sink* a current 2X at all times (set to 8mA). However, this 2X current will be steered by diodes D4 and D5, and the switching circuit comprised of Q9 and Q10.

Current Switching / Steering, in More Detail

The Q9 and Q10 circuit performs the "2X" current switching / steering function as follows: Crystal oscillator U23 generates a precise 16MHz clock signal. U22 divides this clock signal down to 500KHz. This 50% duty cycle, 500KHz clock is then buffered and level-shifted by capacitive coupling to the gates of P- and N-channel small-signal mosfets Q9 and Q10, which switch alternately ON/OFF in opposite phase. During the positive half-cycle of the RAMP_SW signal,

Q10 is ON, and pulls the voltage at the anode of D5 to −7V. This effectively disconnects Q9/Q10 from the collector circuit of Q6-2. This allows the 8mA (2X) current sunk by Q6-2 to flow through D4, discharging C_{RAMP} with a *net* discharge current of X (4mA). During the negative half-cycle of the RAMP_SW signal, Q9 is ON and pulls the anode of D5 up to the +7V rail. This causes the 8mA sink current of Q6-2 to flow through D5 and Q9, and back-biases D4. During this interval, there only exists a 4mA ("X") charging current into C_{RAMP} from Q4-2. By using diode current steering in this ramp generator circuit, all four bipolar transistors operate in a linear constant current mode, eliminating distortion that would be caused by bipolar transistor switching, with its associated delay time and storage time effects. The 1N4148 signal diodes have very low capacitance and low reverse recovery time, which provides for clean switching transitions at the peaks of the ramp waveform.

Therefore, during the charging interval of the ramp capacitors, they will be charged with X constant current, and during the discharge interval they will be discharged with X current. This will produce a linearly rising ramp voltage across the ramp capacitors during the charging interval and a similarly linear downsloping ramp (with the exact same dv/dt slope) during the ramp capacitor's discharging interval. The timing of the charging and discharging intervals is precisely controlled by the Q9, Q10 mosfet switching circuit, which is driven from the 500KHz clock. This then will be the switching frequency for the fixed-frequency Class-D amplifier and will be extremely precise. The peak-to-peak ramp voltage (V_{RAMP}) is then determined by the current X, the value of the ramp capacitors, and the charging interval (T/2), according to EQ. 4.2.

$$C_{RAMP} = (I_X)\,(dt)\;/\;dv \qquad \text{EQ. 4.2}$$

where:

- C_{RAMP} = the total value of the ramp capacitors (C38-1 plus C38-2)
- dv = peak-to-peak ramp voltage (8V) (*Note:* V_{RAMP} slews linearly from −4V to +4V during the C_{RAMP} charging interval which is equal to T/2 = 1μsec = dt.)
- dt = T/2 = 1μsec.

Note: In order to provide greater flexibility in adjusting the peak-to-peak ramp amplitude, the ramp capacitor is split into two separate capacitors, consisting of the main capacitor (C38-1) and a "shim" capacitor (C38-2) of a much smaller value.

In the reference schematic, charging current "X" is designed to be 4mA, and discharging current is therefore 8mA. It is important to understand that noise would decrease if larger current values could be used, but so would the power dissipation in Q6. Ultimately a value of X = 4mA is selected, with Q2 dissipating 100mW.

It will be evident that if the discharging and charging currents are not in an exact 2:1 ratio, then there will be a DC drift in the ramp voltage (V_{RAMP}), and either the positive or negative peaks of the ramp voltage waveform will be clipped. To prevent this, a DC servo circuit is added, which adjusts or "fine-tunes" the value of sourcing current "X" to be exactly the value necessary to maintain an average DC voltage of 0V for V_{RAMP}. This is accomplished by integrator U18 and associated circuitry. U18 is configured so that it will maintain the average DC voltage on the ramp capacitors (V_{RAMP}) at 0V. It does this as follows.

Referring to Figure 4.2, component values are selected such that the current through R55 is approximately 20% higher than the desired value X. U18 will siphon off this excess current through R60 and Q16, leaving precisely X = 4mA as the charging current sourced by collector Q4-2. In this way, U18 uses negative feedback to maintain V_{RAMP} at 0VDC. If V_{RAMP} drifts positive, the output of U18 will swing negative, siphoning off more current and reducing charging current X. Likewise, if V_{RAMP} drifts negative, the output of U18 will swing positive, siphoning off less current and therefore increasing charging current X. U18 is configured as an integrator, with extremely high DC gain, with gain rolling off with a single-pole response. The low-frequency pole is set by C42 and R63 at 1.59Hz. The AC gain in the audio frequency BW is therefore very low, and U18 does not disturb the fidelity of the ramp signal over the audio frequency BW. If it did, this would result in increased THD. The action of U18 maintains a 0V average voltage across the ramp capacitors. This effectively centers the triangle ramp at 0VDC, with +/−4Vpeak (8Vpeak-to-peak) linear excursions. The heavy filtering capacitance values in this circuit (C34–C37), along with the long time constant of the integrator U18, mean that this circuit takes up to 1 second to reach its DC operating point and stable operation at power-up. This is another reason for requiring turn-on delay blocks in the housekeeping circuitry, as is discussed in those sections of the schematic. With a low-noise, linear waveform established across ramp capacitors C38-1 and C38-2, all that remains is to buffer it so that it can be distributed to the PWM comparators in each Class-D channel.

Other design constraints in this ramp generator circuit are achieving the desired voltage swing across the ramp capacitors (C38-1 and C38-2) of +/−4Vpp, *and*, at the same time, an acceptable minimum Vce voltage across Q4-2 and Q6-2. This is called the compliance voltage of these current mirror transistors. It is desired to maintain a minimum Vce voltage across Q4-2 and Q6-2 of 600mV, so that these transistors do not saturate at the peaks of the V_{RAMP} waveform, and therefore maintain high current gain (β), which maintains ramp linearity. LTspice is used to simulate this circuit and define the operational voltage limits at the critical circuit nodes. Figure 4.2 illustrates the critical currents and node voltages of the ramp generator circuit.

Note: In multi-channel designs, the distribution of the ramp signal should be approached with care. Since this same ramp signal is used by multiple channels, it is a critical avenue for propagation of *crosstalk* interference. Therefore, distribution of this ramp signal to the individual channels is best accomplished with dedicated voltage-follower buffers for each channel. It is critical not to *filter* the ramp signal excessively, in an attempt to minimize noise and crosstalk, because it is also essential that the ramp signal should have very sharp inflection points at its transitions. Any "rounding" of the ramp near the inflection points, or any change in ramp slope, will cause a level-dependent gain change, which will increase THD.

Ramp Buffer Circuits

The ramp voltage signal generated at the timing capacitor (C_{RAMP}) has a high source impedance. The reason is that it is fed by an ideal current source, which has characteristically high impedance. This leaves the impedance of the timing capacitor itself as the source impedance of the ramp signal (V_{RAMP}). Due to this high source impedance, any loading of this signal is undesirable, and will cause a change in gain and/or an increase in THD. The solution to this potential problem is to buffer this ramp signal with a high input impedance voltage buffer. This buffer circuit block not only solves the problem of loading on C_{RAMP}, but also provides protection against potential crosstalk between channels (discussed further in Chapter 19, Multi-Channel Designs). It has already been mentioned that V_{RAMP} is a critical circuit node, and any noise there is amplified by the channel gain. Because of this sensitivity, switching-induced noise from other channels

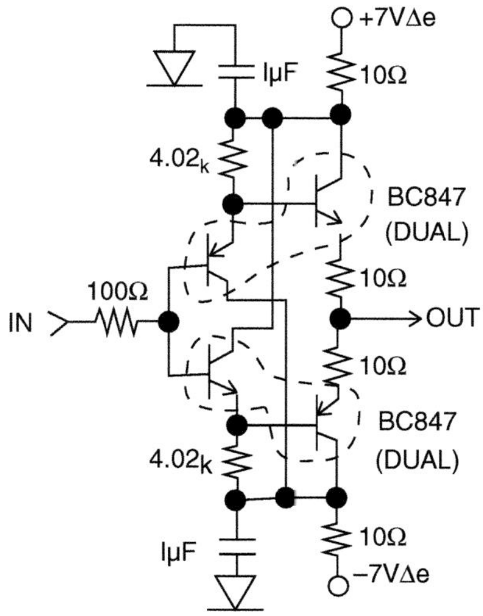

Figure 4.3 Discrete Voltage Buffer Circuit

can crosstalk due to the common usage of V_{RAMP} and its associated PWM ramp across multiple channels. One such mechanism for this type of crosstalk is noise induced in V_{RAMP} from the comparator input circuits. Some PWM comparators have input circuitry which exhibits small changes in impedance at the switching transition. This can create a noise on V_{RAMP}, which can then crosstalk to all other channels using the PWM ramp signal. Using a voltage buffer between V_{RAMP} and the PWM comparator input isolates the noise effects of the PWM comparator input switching circuits from V_{RAMP}, and this decreases crosstalk.

The voltage buffer function can be implemented by either an IC or discrete implementation. One IC implementation is the MAX4200 (Maxim Integrated (Analog Devices), 2017). A discrete implementation is shown in Figure 4.3. The specification requirements for the buffer are:

1) A supply rail voltage capability large enough to accommodate the peak-to-peak ramp voltage swing of V_{RAMP} with adequate headroom
2) Wide BW (ideally >100MHz)
3) High input impedance, with low input (bias) current specification
4) Low power consumption.

Note: Exceedingly low voltage offset is *not* a requirement (as long as negative feedback is used in the Class-D design, to correct for any output DC offset voltage).

Bibliography

Buono, R. (2017a). Reduced crosstalk and matched output power class D audio amplifier with oppositely polarized triangle waves. US Patent 9654069B2.

Buono, R. (2017b). Reduced crosstalk and matched output power audio amplifier using a triangle wave generator. US Patent 9680428B2.

Buono, R., & Fink, D. (2016). Reduced crosstalk and matched output power audio amplifier. US Patent 9509261B2.

Buono, R., & Fink, D. (2017a). Method for delivering a duty cycle limited signal to a speaker using a class D amplifier. US Patent 9647617B2.

Buono, R., & Fink, D. (2017b). Reduced crosstalk class D audio amplifier matching output power based on channel gain. US Patent 9716478B2.

Marco, T.P. (2009). A comprehensive study of Class-D amplifier technology. MS paper, The Pennsylvania State University, The Graduate School, College of Engineering, Department of Electrical Engineering.

Maxim Integrated (Analog Devices). (2017, December). Datasheet, MAX4200-MAX4205, Ultra-high-speed, low-noise, low-power, SOT23 open-loop buffers, Revision 4. San Jose, CA: Maxim Integrated Products, Inc.

Chapter Five

Level-Shifter Circuits

Desired Features and Characteristics of Level-Shifter Circuits for Class-D Amplifier Design

The reference Class-D amplifier design uses a gate driver that has three separate galvanically isolated sections. The input section is powered by a 5V rail and referenced to circuit ground. The second and third sections are independent gate drivers for the upper and lower power switches of the half-bridge power section. The lower gate driver circuitry is referenced to the negative high-voltage DC rail, and the upper gate driver circuitry is referenced to the switching node. Therefore, each section has a different "0V" reference.

Note: For the purposes of this discussion, the reference schematic implementation of the gate driver will be used to discuss its functionality.

The gate driver used in the reference design employs internal level-shifter circuitry which works through a gated RF coupling mechanism, which spans the isolation barriers. It has the essential features of excellent common-mode rejection, and transient (dv/dt) immunity. Its role is to couple the logic signal from its input section to both the lower and upper gate driver circuits. In this Class-D application, the coupling from the input section to the lower gate driver circuit is less challenging, because *both* the input section and lower gate driver section are referenced to DC voltages, albeit with a large difference in magnitude (in the tens of volts). But the relative magnitude of the voltage difference between these two reference points is fairly stable, with relatively small dynamic voltage change. However, the situation is drastically different for the upper gate driver circuitry.

DOI: 10.4324/9781003327943-5

It is referenced to the switching node of the half-bridge power circuitry. This node has the highest voltage swing and slew rate of the design. The gate-drive logic signal received by the upper gate driver circuitry must be undisturbed by the dynamics of the reference voltage differences between these sections. Making this goal even more challenging is the fact that the signal must propagate from input to both the upper and lower gate drive sections with ideally equal and very low propagation delay.

There are a large number of half-bridge gate drivers which have a different topology than the Si8244 driver used in the reference design. These other gate drivers, for example, have the input circuitry and the lower gate driver circuitry sharing a ground reference, or allowing just a small voltage difference (a few volts) between the 0V reference levels for these two sections. In these drivers, the input section must basically be considered to be referenced to the negative high-voltage rail. When these types of gate drivers are used in a half-bridge Class-D design utilizing split (positive and negative) rails, a level-shifter circuit must be employed to level-translate the ground-referenced PWM logic signal(s) to the negative-rail-referenced gate driver IC input circuitry. There are various circuits that are appropriate for implementing this level-shift function.

Figure 5.1 shows a discrete approach, whereas Figure 5.2 shows an approach using an isolated logic IC.

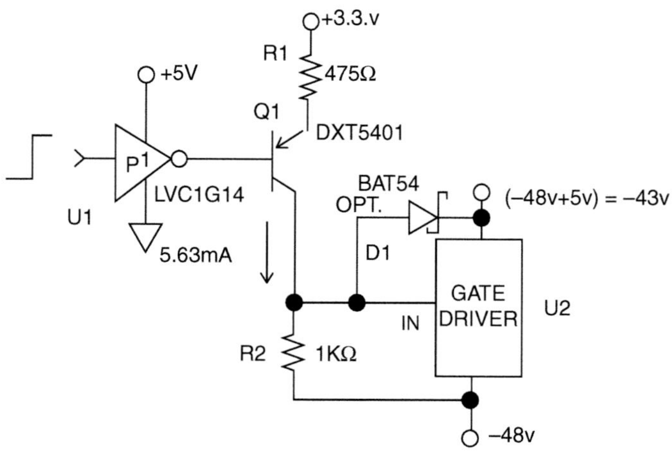

Figure 5.1 Level Shifter Composed of Discrete Components

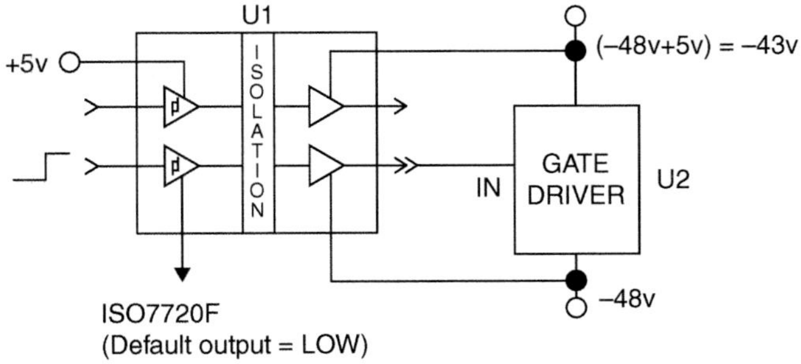

Figure 5.2 Level Shifter Constructed Using a High-Speed Isolated Logic IC

In Figure 5.1, the 5V logic signal driving the input to inverting logic gate U1 could be the PWM signal from the PWM comparator, for example. The output of U1 is applied to the base of PNP transistor Q1. Note that the emitter resistor of Q1 is connected to a +3.3V rail. This means that the base-emitter junction of Q1 is reverse biased when the output of U1 is positive at +5V. This reverse biasing provides for rapid turn-off of Q1. When the output of U1 is low, Q1 will be switched on, and will sink a constant current from the +3.3V supply. This current is calculated as follows: When Q1 is on, the voltage across R1 will be +3.3V − 0.6V = 2.7V. The current flowing through R1, and thus the emitter current of Q1, will be: 2.7V / R1 = 5.68mA. Assuming a current gain (beta) of 100 for Q1, this emitter current produces approximately (5.68mA × 0.99 = 5.63mA) collector current, which then flows through R2, which is connected to the negative rail (in this example = −48VDC). This 5.63mA collector current produces a voltage of 5.68V at the input of U2. Schottky diode D1 is optional, and is added to prevent overvoltage at the input of U2. This overvoltage could occur from tolerance buildup in the resistor values, deviation in the +3.3V supply rail, and Vbe ON voltage variation of Q1 over temperature. D1 is only necessary if the voltage developed across R2 can exceed the recommended operating range value of the input pin of U2. Some newer gate drivers allow a very wide range of input voltage. For example, some devices can have a Vdd range up to 18V and allow an input voltage up to Vdd +0.3V, yet an input will be received as a valid logic high at only 1.8V (for Transistor-Transistor Logic (TTL) input threshold types). For this type of gate driver, D1 can be omitted, and the capacitance at this node reduced to a minimum. Minimum capacitance at this node is critical for fast switching, since it is driven high by a constant (limited) current, and only passively pulled low with a pulldown resistor (R2). It is for this reason that the turn-off of Q1 must be fast, and the capacitance at the input pin of U2 must be minimized. The capacitance of the BAT54 Schottky diode is only 10pF, but even that level of capacitance is significant, as the calculations which follow will show. The output capacitance of Q1 (Cobo) is 6pF max. The input capacitance of the gate driver input pin is rarely specified by manufacturers, but can be approximated as 11pF. This yields a total capacitance at this node of 13pF. In addition, the value of R2 (and therefore R1) must be as small as possible, consistent with power dissipation limits of Q1. The voltage across Q1 (when ON) will be the full value of the negative voltage rail, minus the voltage across R2 (5.63V) plus Vbe (0.6V). So 48V − 5.63V + 0.6V = 42.97V. Power dissipation in Q1 (when ON) = Ic × Vce = 5.63mA × 42.97V = 241.9mW. In the interest of conservative design, it will be assumed that the ON time (duty cycle) of Q1 can be very close to 100%, so this level of power dissipation will be taken as is, and not reduced by a calculation of the expected average duty cycle of less than 100%. Q1 is selected in a SOT89 package, which can handle this level of power dissipation. If U2 has TTL input thresholds and is driven to +5.63V as a high, the voltage will need to decay to 0.8V to guarantee a logic low level during Toff. The capacitive discharge equation will be used to calculate the time delay for this voltage decay to 0.8V, starting with an initial value of 5.63V:

$$V_{initial}\, e^{-t/\tau} = V_t$$

$$e^{-t/\tau} = V_t/V_{initial}$$

$$-t/\tau = \ln\,(V_t/V_{initial})$$

$$-t/\tau = \ln\,(0.8V/5.63V)$$

$$-t/\tau = \ln\,(0.142)$$

$$-t/\tau = -1.95$$

$$-t = (\tau)(-1.95)$$

$$t = 1.95\tau$$

where $\tau = R \times C$. R = 1 Kohm and C is the total capacitance at the input node of U1 = 11pF. So t = 1.95(1K × 11pF) = 21.45nsec. This is a sufficiently small value of delay time for this application.

Note the design tradeoffs involved here. In order to reduce this delay time further, either a reduction in total capacitance at this circuit node is required, or a reduction in the value of R2, which would mean an increase in the power dissipation of Q1, R1, and R2.

Figure 5.2 shows another approach for implementing the level-shifter function. It utilizes a digital isolator IC (U1 in Figure 5.2). This family of devices transmits high-speed digital signals between circuits which share different ground references. In this example, a dual-channel device is used: ISO7720F. The important specifications for this device are extremely high-speed operation (100Mbps (mega bits per second) operating rate), and a common-mode transient immunity (CMTI) of >100KV/μsec. The "F" suffix indicates that the default state for the outputs (such as during undervoltage on either the input supply voltage or the output supply voltage) is *low*. This device has a propagation delay of only 16nsec maximum, using a 5V supply. This solution is slightly more expensive than the discrete solution, but operates with lower power dissipation: 4.2mA × 5V = 21mW, compared to 241.9mW for the discrete solution.

Bibliography

Diodes Inc. (2018). Datasheet, DXT5401 150V PNP small signal transistor in SOT-89. Document number: DS31226. Rev. 4 – 2. Plano, TX: Diodes Incorporated.

Fairchild Semiconductor Corporation (Onsemi). (1993). Datasheet, BAT54 / BAT54A / BAT54C / BAT54S, Rev. 2.3. Aurora, CO: ON Semiconductor.

Texas Instruments. (2018). Datasheet, SN74LVC1G14 Single Schmitt-trigger inverter, SCES218Y. Originally published April 1999; revised November 2018. Dallas, TX: Texas Instruments.

Texas Instruments. (2020). Datasheet, ISO772x High-speed, robust EMC, reinforced and basic dual-channel digital isolators, SLLSEP3E. Originally published November 2016; revised March 2020. Dallas, TX: Texas Instruments.

Chapter Six

Gate Driver Circuits

Gate Charge

Gate driver circuits are responsible for driving the power switching devices. The characteristics of these drivers are tailored to the characteristics of the specific power switching devices selected. For example, in the reference design, power mosfets are the power devices selected. More specifically, these are N-channel enhancement mode power mosfets. Mosfets of this type are offered with various gate drive voltage limits. For mosfets used in medium- to high-power amplifiers, the gate drive voltage levels are nominally 0 and 10V, or they may be "low-threshold" types, requiring a voltage of approximately 5V for full gate enhancement. It should be noted that N-channel mosfets requiring less than 5V gate drive are available, but these are usually lower-voltage and lower-power devices. These types may find usage in very low power / low voltage Class-D amplifiers.

Referring to the reference design, the IRF6775 (Infineon (International Rectifier), 2014) directfet is used. This is nominally a 10V drive mosfet, which has a unique package, allowing it to be conveniently top-side cooled if necessary. The mosfet is a voltage-driven device, having a capacitive input characteristic. The role of the gate driver is to rapidly charge the gate capacitance of the mosfet to the full enhancement voltage level, to turn ON the device, and to rapidly discharge this capacitance to 0V in order to turn OFF the device. Although this sounds simple, there are many nuances that must be understood and dealt with in order to design an effective gate driver circuit.

DOI: 10.4324/9781003327943-6

The first characteristic to be understood and dealt with is *gate charge.* Each mosfet has a gate charge specification which indicates the amount of charge required to fully enhance the gate. This is necessary, because driving the gate is more complicated than simply driving a fixed capacitance. There are, in fact, several different device capacitances that must be charged and discharged. The gate charge specification is broken down into component parts, which indicate how much charge each of these capacitive elements requires. The total gate charge specification, as well as the separate gate charges of which it is comprised, are used to calculate gate drive timing and power dissipation. For timing, it is critical to understand how fast the mosfets will be switched in the design. Since the mosfet has a capacitive gate, its switching speed is determined by how fast the gate capacitance can be charged and discharged. This means the drive current level of the gate driver determines the switching speed of the mosfets. To understand this better, it is necessary to analyze the particular mosfet's gate drive curve. Figure 6.1 is a generic gate drive curve, used to explain the concepts.

The gate drive curve has three principal regions. These can be defined as the pre-threshold region (0nC–6nC on the curve in Figure 6.1), the post-threshold or Miller region (6nC–15nC on the curve in Figure 6.1), and the excess charge region (15nC–30nC on the curve in Figure 6.1). These regions can be described as follows.

The pre-threshold region is the region starting from 0V = V_{GS}, to the point where the gate voltage threshold has been reached. Consulting the gate drive curve in the datasheet for the IRF6775 (Infineon (International Rectifier), 2014), the gate threshold voltage is 3V min. to 5V max. and the gate charge for this pre-threshold region is 5.8nC typically.

The next region is the post-threshold region, or Miller region. During this region, the mosfet is beginning to turn-on and conduct, because the gate threshold voltage has been reached. The drain voltage (V_{DS}) of the mosfet begins to fall, and as it falls, the voltage across the mosfet's drain to gate capacitance is decreasing. This drain-to-gate capacitance is being discharged, and the discharge current must flow through the gate circuit. This current flows during this interval, and the gate voltage remains constant at approximately the gate threshold voltage, while the drain voltage falls to its final ON-state voltage (V_{DS}ON). When the drain voltage has reached its final on-state voltage, the current through the drain-to-gate capacitance will fall to zero, and the mosfet is essentially ON.

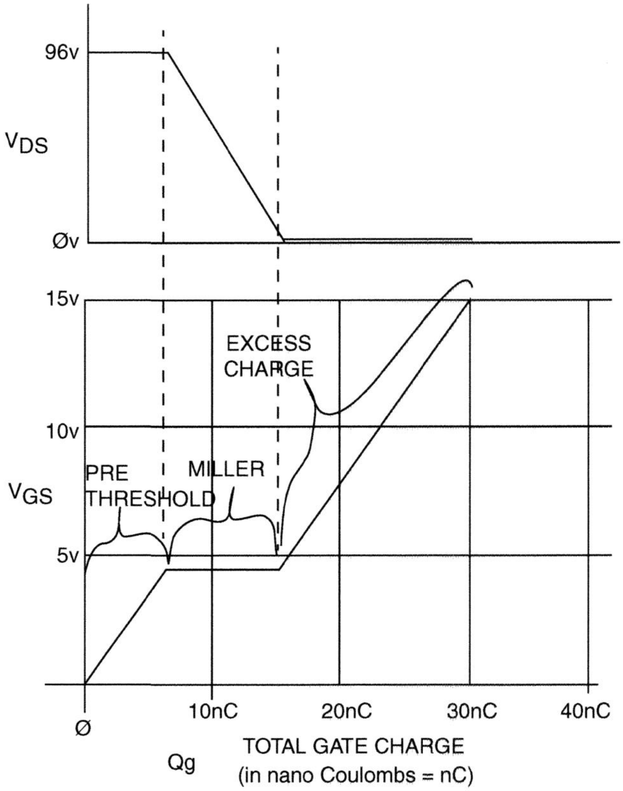

Figure 6.1 Generic Gate Charge Curve for a Power Mosfet, during Turn-ON

The next interval is the excess charge interval. The reason for this interval is that it is advisable to charge the mosfet gate beyond the minimum charge necessary to traverse the Miller interval. It is prudent to provide excess charge to ensure the mosfet is fully enhanced. This will provide the lowest values of ON resistance for the mosfet, and also makes it possible to ensure that the mosfet is fully enhanced, taking into account all specification tolerances in the mosfet and gate driver components.

Now, returning to the switching speed calculation, it can be seen that the switch turn-on time is the total time to traverse the pre-threshold and post-threshold (Miller) intervals of the gate charge curve. For this calculation (and referring to the reference schematic), let us assume a value of 10.0 ohms for the external gate drive resistor. The total resistance in the gate charging current "loop" is 2.7 ohms (Si8244 (Skyworks Solutions Inc., 2021) sourcing resistance)

+ 10.0 ohms (external gate resistor) plus 3.0 ohms (internal IRF6775 gate resistance) = 15.7 ohms. The voltage across this total "charging resistance" is the Si8244 Vdd voltage (12V) minus the average gate voltage (2V) of the mosfet during the pre-threshold interval (estimating 2V from the gate charge graph in the IRF6775 datasheet) = 10V.

The general equation:

$$q = i \times t \qquad \text{EQ. 6.1}$$

is used, where q = the pre-threshold charge of IRF6775 = 5.8nC; i = charging current (10V/15.7 ohms = 637mA); and t is the charge time for this interval.

Rearranging, we get:

$$t_{\text{pre-threshold}} = q/i = 5.8\text{nC}/637\text{mA} = 9.1\text{nsec} \qquad \text{EQ. 6.2}$$

For the Miller interval, the gate voltage will remain relatively constant at the threshold voltage (3V min., 5V max. = 4V average), so the voltage available across the total charging loop resistance is 12V − 4V = 8V, and the charging current is therefore 8V/15.7 ohms = 510mA. Gate charge during this post-threshold (Miller) interval increases from 5.8nC to approximately 15nC. So 9.2nC of charge must be provided during this interval.

$$t_{\text{post-threshold}} = q/i = 9.2\text{nC}/510\text{mA} = 18.0\text{nsec} \qquad \text{EQ. 6.3}$$

Turn-on time will therefore be 9.1nsec + 18.0nsec = 27.1nsec.

Now, it is important to realize that even though the drain-to-source voltage of the power mosfet will not change much during the excess charge interval, this is an important interval to consider, because it impacts gate drive power dissipation (to be discussed in the next section). In the reference design, the IRF6775 will be charged to a gate voltage of 12V. Therefore, the gate voltage during this interval increases from 4V to 12V, averaging 8V. So the average available charging current during this interval is (12V − 8V)/15.7 ohms = 255mA and the gate charge of the IRF6775 increases from 15nC to 30nC (at V_{GS} = 12V). Therefore, the gate charge required to be delivered during this interval is 15nC.

$$t_{\text{excess-charge}} = q/i = 15\text{nC}/255\text{mA} = 58.8\text{nsec} \qquad \text{EQ. 6.4}$$

Table 6.1 Mosfet Gate Drive Discharging Currents and Discharging Interval Durations

Interval	*Discharging Current (i)*	*Interval Duration (t)*
$t_{excess\text{-}charge}$	8V/14 ohms = 571mA	t = q/i = 15nC/571mA = 26.3nsec
$t_{post\text{-}threshold}$	4V/14 ohms = 286mA	t = q/i = 9.2nC/286mA = 32.2nsec
$t_{pre\text{-}threshold}$	2V/14 ohms = 143mA	t = q/i = 5.8nC/143mA = 40.6nsec

So even though the mosfet has effectively switched from OFF to ON in 27.1nsec, the gate driver (Si8244) is delivering current into the mosfet's gate for 27.1nsec + 58.8nsec = 85.9nsec.

The same calculations are followed during turn-off, except that the sinking resistance of the Si8244 is used (1 ohm instead of 2.7 ohms), and the driver is pulling down towards ground (0V) instead of towards 12V. This results in a total turn-off loop resistance of 14.0 ohms instead of 15.7 ohms, and gate discharging currents and interval times as in Table 6.1.

It is critically important to realize that during turn-off, the excess charge interval is extremely important, because the mosfet does not even begin to turn-off, and therefore the drain voltage does not even begin to rise, until the end of the excess charge interval. This must be taken into account in order to ensure that the proper value of deadtime is used (to be discussed later), so cross conduction through the power switches can be prevented.

Gate Drive Power Calculations

Compared to bipolar transistors which require base current to flow continuously during any ON-time, to keep the device in the ON state, power mosfets only require drive current to charge and discharge the gate. Once the gate is fully charged, or fully discharged, gate drive current and power are zero. This is one of the great benefits of using enhancement mode power mosfets. However, although gate drive power can be described as "low", it must be calculated, and becomes more significant as switching frequency increases.

EQ. 6.5 is used to calculate the gate drive *power*:

$$\text{Gate Drive Power} = QgVgF \qquad \text{EQ. 6.5}$$

where Qg is total gate charge in Coulombs; Vg is *final* gate voltage (i.e. maximum on-state gate voltage reached); and F is frequency in Hz.

For the reference design, using IRF6775, Qg = 36nC *max* at 10V, but, since the gate is driven to 12V in the reference design, the gate charge graph must be consulted, and gate charge up to the 12V level must be checked. In this case, the graph indicates a gate charge of 30nC at a gate voltage of 12V. This is less than the datasheet *max* listed for a 10V gate drive, so we will use the 36nC value in order to be conservative.

Switching Frequency = 500KHz

Gate Drive Power = (36nC)(12V)(500KHz) = 216mW

This power (216mW) is provided by the power rail biasing the gate driver and is dissipated primarily by the resistive elements in the gate drive circuit loop. Referring to Figure 6.2, it can be seen that the resistive elements in the gate drive loop are comprised of the internal source and sink resistance of the gate driver itself, the internal gate resistance of the IRF6775 mosfet, and any external gate drive resistor.

In the reference design, the Si8244 gate driver has an internal sourcing resistance of 2.7 ohms typical and an internal sinking resistance of 1 ohm typical. It is important to note that most gate drivers display this asymmetrical source and sink resistance characteristic. This is the case for two reasons. The first is that the pulldown (sinking) device internal to the gate-driver IC is an N-channel device, and the pull-up device is commonly a P-channel device. N-channel devices have lower on-resistance per unit die area than P-channel devices. The second reason is that it is most desirable to turn the mosfet OFF in less time than it takes to turn the mosfet ON. This asymmetry in ON and OFF times helps to prevent cross conduction in half- and full-bridge power switching stages. The internal gate resistance of the IRF6775 DirectFet mosfet itself is specified as 3.0 ohms maximum.

This section will demonstrate how to calculate an accurate value of power dissipation in the various elements of the gate drive circuitry, but the next section will discuss a real-world conservative design approach for dealing with this gate driver power dissipation.

The gate drive power is dissipated in the gate drive loop resistances. From above, we have calculated the total gate drive loop resistances to be:

- 15.7 ohms charging
- 14.0 ohms discharging.

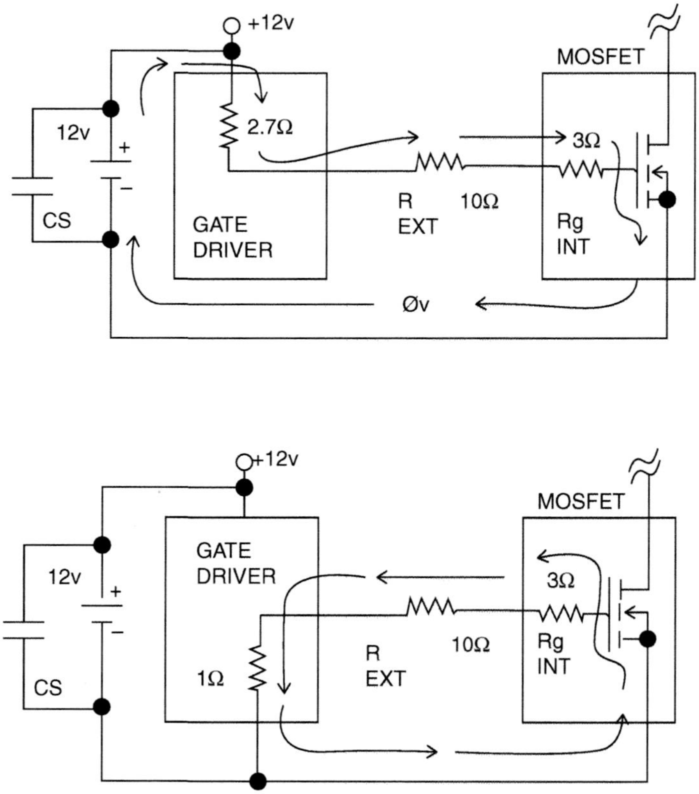

Figure 6.2 Gate Current Flow through Internal and External Resistances, during Turn-on (Sourcing Current into Mosfet Gate) and Turn-off (Sinking Current from Mosfet Gate) Events

The 216mW total gate drive power is split evenly between turn-on gate drive losses and turn-off gate drive losses of 216mW/2 = 108mW for turn-on gate drive losses in the charging (sourcing) loop, and 108mW for turn-off gate drive losses in the discharging (sinking) loop.

The power dissipated by the individual resistive components comprising the charging and discharging circuit loops will be proportional to the resistance values of each of these individual resistances.

Apportioning the power dissipation among the component resistances of the charging (sourcing) and discharging (sinking) circuit loops:

- For the Si8244 (sourcing resistance = 2.7 ohms, sinking resistance = 1.0 ohm):

  (2.7 ohms / (2.7 ohms + 10 ohms + 3 ohms)) × 108mW = 18.6mW (1.0 ohms / (1.0 ohms + 10 ohms + 3 ohms)) × 108mW = 7.7mW Total power dissipation in (*one*) Si8244 gate driver = **26.5mW**

- For the external gate resistor (first-pass assuming 10-ohm value):

  (10 ohms / (2.7 ohms + 10 ohms + 3.0 ohms)) × 108mW = 68.8mW (10 ohms / (1.0 ohms + 10 ohms + 3.0 ohms)) × 108mW = 77.1mW Total power dissipation in external 10-ohm gate resistor = **145.9mW**

- For the internal gate resistance of the IRF6775 DirectFet:

  (3.0 ohms / (2.7 ohms + 10 ohms + 3.0 ohms)) × 108mW = 20.6mW (3.0 ohms / (1.0 ohm + 10 ohms + 3.0 ohms)) × 108mW = 23.1mW
  Total power dissipation in internal gate resistance of IRF6775 DirectFet = **43.7mW**

Effects of Reducing or Eliminating the External Gate Resistor

From the above example, it is obvious that the power dissipation of each resistive component in the charging and discharging gate drive loops is proportional to the resistance value of that element. However, it is important to also realize that the total gate drive power of 216mW (for the specific example above) will be dissipated independently of the absolute value of loop resistance. That is because the power dissipation value derives from the fact that a "capacitor" is being charged and discharged, and the capacitive element itself is lossless. Therefore, the power dissipation associated with the charging and discharging of this capacitance must occur in the resistive elements in those charging and discharging loops, and this power is provided by the gate drive bias supply.

For the example above, the 10-ohm external gate drive resistor is much greater than both the internal resistances (sourcing and sinking) of the driver IC (Si8244) (Skyworks Solutions Inc., 2021) and the internal gate resistance of the IRF6775 mosfet. Therefore, it dissipates the majority of the 216mW of gate drive power (145.9mW). If a power derating factor of 50% is used for resistors, in order to achieve a highly reliable, conservative design, then a 10-ohm resistor with a maximum power rating of at least 2 × 145.9mW = 291.8mW should be specified. The next standard power rating for surface-mounted chip resistors is 500mW, but this would require a 1206 size resistor, which is relatively large. As will be discussed in Chapter 16 on layout, it is beneficial to keep the gate drive current loops as small and tightly routed on the PCB as possible, to minimize loop inductances that can cause ringing and EMI. The value of the external resistance will be recalculated now, with the goal of achieving a power dissipation of 125mW, allowing the selection of an 0805 size chip resistor, with a power rating of 250mW, for 50% derating. This iterative exercise illustrates a common theme in Class-D audio amplifier design, which is that there are many interrelationships between design parameters that need to be met, and meeting them all simultaneously is sometimes an iterative process. Perhaps the most crucial role to be considered in value selection for the external gate resistor is that it affects mosfet switching speed. A lower value allows for increased gate charging current and faster rise and fall times for the mosfet V_{DS} voltage. This switching speed affects efficiency and EMI, as well as Total Harmonic Distortion (THD). The THD effect is due to the fact that faster rise and fall times approach the "ideal" zero switching time, which means that the switching output waveform of the power stage will more accurately follow the PWM timing from the modulator. This translates into less open-loop error and better THD performance. For now, a recalculated value for the external gate resistor will be used, based on the 125mW power dissipation limit, with the caveat that this resistor value may have to be adjusted up or down in the actual design, to achieve the desired THD, efficiency, and EMI performance.

Recalculation proceeds by using the turn-off power dissipation equation above, since that represents the higher value of power dissipation in the external resistor. In order to achieve a *total* power dissipation of less than 125mW, the power dissipation of the turn-off event alone will need to be less than 125mW/2 = 62.5mW.

$$(X \text{ ohms} / (1.0 \text{ ohm} + X \text{ ohms} + 3.0 \text{ ohms})) \times 108\text{mW} = 62.5\text{mW}$$

$$(X/(X + 4)) = 62.5\text{mW} / 108\text{mW}$$

$$X = (X + 4)(0.579)$$

$$X = 0.579X + 2.316$$

$$X - 0.579X = 2.316$$

$$0.421X = 2.316$$

$$X = \mathbf{5.5\ ohms}$$

The closest 1% value of 5.49 ohms will be used as the "initial value" for the gate resistors.

UVLO

A detailed discussion about UnderVoltageLockOut (UVLO) can be found in Chapter 13 (Housekeeping and UVLO Circuits). In that chapter, it is noted that the UVLO feature of the gate driver is a critical protection. This feature allows for flexibility in the undervoltage behavior of other circuits. As long as the gate driver is disabling the switching of the power fets, until a power rail or rails is/are qualified, then other circuits are allowed to have indeterminant behavior while those monitored rails are in an undervoltage condition. This is a basic tenet of the UVLO scheme presented in that chapter. The details of the UVLO protection will vary depending on the type of gate driver IC used. For example, in the reference design, an Si8244 gate driver is used. This type of driver has three separate galvanically isolated sections. These are the input section (referenced to circuit ground) and two independent output sections (A and B). In the reference design, the "B" output section is referenced to the negative rail, and drives the lower mosfet in the half-bridge power circuit. The "A" output section is referenced to the switching node and drives the upper mosfet in the half-bridge power circuit. Each one of these sections of the IC has its own UVLO protection circuit. The critical aspects of UVLO which must be reviewed during the design process are:

- Rising voltage thresholds and tolerances
- Falling voltage thresholds and tolerances
- Behavior of the gate driver upon entering and exiting UVLO.

This last specification should be given special consideration.

Gate driver ICs usually specify a *power-up* delay time. This is the time delay from when the Vdd rail exceeds the rising UVLO threshold, to the time the output state is valid. This delay can be significant. For the Si8244 this delay can be 7μsec maximum. This means the output state of the driver will remain low, and not respond to a high-level input, until 7μsec after the driver's Vdd voltage exceeds the rising UVLO threshold. For example, considering the half-bridge topology in the reference design, suppose the Vdd supply (−HV+12V) is rising and exceeds the rising UVLO voltage threshold for section "B". It will be prudent to maintain that section's output level as a *low*, until at least 7μsec after the rising UVLO threshold is exceeded. This is usually not a problem as the *disable* pin can be used to keep *both* outputs disabled, and this can easily be maintained in the disabled state much longer than the necessary 7μsec start-up delay time. The more complicated issue is the power-up delay of the upper gate driver circuitry. The upper gate driver is powered by the BOOT capacitor C_{BOOT}. C_{BOOT} will not be charged until the lower fet is switched ON. Normally then, it would be expected that as soon as the lower fet is switched off (allowing for deadtime), the upper fet will be switched on, and PWM operation will commence. However, due to the power-up delay spec, the upper fet driver output will not switch high until 7μsec after the voltage across C_{BOOT} has risen above the rising UVLO threshold of section "A". This delay may cause an audible "pop" on Class-D start-up and, if so, may need to be addressed. This is one potential benefit that can be provided by adding digital control (such as an FPGA) to the design (see Chapter 22). This power-up delay can be managed in the context of a popless start-up PWM sequence.

The driver datasheet also details what the behavior will be for the output sections when the input section experiences UVLO. All these behaviors are critical to consider for a robust Class-D design. Another detail to consider is whether the output circuits of the upper and lower driver sections can sink any current when in UVLO. Ideally, it is desired that these outputs be able to sink current and pull down when their bias supplies are between 0V and the rising UVLO threshold voltage. Of course, any *active* pulldown circuitry will require some minimum bias voltage to function, but some circuits can accomplish this active pulldown even with bias voltages as low as 0.6V–1V. Typically, as long as the output voltage is held below 2V maximum (the lowest value of gate threshold voltage for most N-channel power mosfets used in Class-D designs), it will be low enough to keep the mosfets turned OFF. The circuitry in Figure 6.3 is an example of the type of circuitry that can accomplish this function.

The circuitry depicted in Figure 6.3 does not replace the internal switching circuits in a gate driver IC. Instead, it is an additional circuit block connected to the output pin, that ensures that Vout remains *below* 1.2V when the IC is unpowered, or when the IC's Vdd is at any voltage below the UVLO threshold. The idealized circuit operates as follows. In the absence of drive to Q4, the base of Q3 will be pulled high by any voltage present on Vout (SELF-BIAS). If Vout rises above 0.6V, Q3 will begin to turn on. It will then pull low on the base of PNP

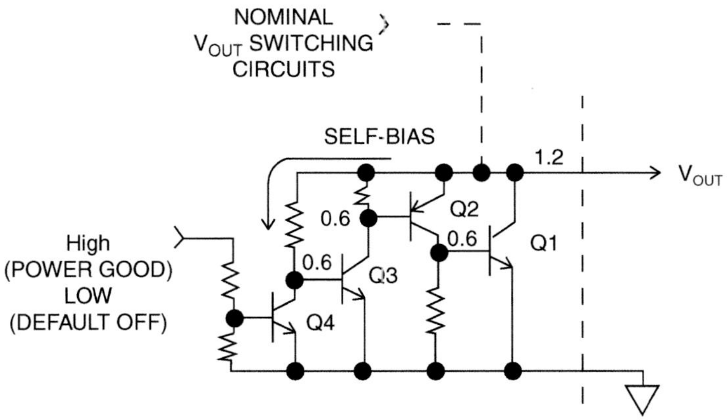

Figure 6.3 Schematic for a Discrete Output Voltage Pulldown Circuit

transistor Q2. When Vout rises as high as +1.2V, there will be enough voltage available for Q2 to turn on, and apply 0.6V to the base of Q1, turning it on. In this manner, the voltage at Vout servos to approximately +1.2V, which is sufficiently low to ensure any mosfet connected to Vout will be OFF. When Vdd exceeds the UVLO threshold, internal circuitry in the IC (not shown) will drive the base of Q4 low. This in turn switches Q3, Q2, and Q1 OFF, and allows Vout to be controlled by the normal output switching circuitry in the IC (not shown).

Absent an active pulldown capability, it is usually prudent to add a pulldown resistor to the mosfet gate, to drain off any leakage charge which could accumulate on the gate and turn the mosfet on. These resistors are usually in the 10K range (R50A, R52A in the reference design) and, even with a 12V gate drive level, dissipate only 14.4mW, so an 0402 size (1/16W) resistor can be used and still easily maintain a 50% power derating.

Different Types of Drivers (Dual Half-Bridge, Independent Dual, PWM, Isolated, or Referenced to Negative Rail)

Dual gate drivers and half-bridge drivers come in a number of different varieties. As mentioned previously, they can have up to three isolated or independently referenced sections, allowing the lower driver to be referenced to the negative rail, the upper driver to be referenced to the switching node, and the input section to be ground referenced. This is the arrangement in the Si8244 used in the reference design. The second common arrangement is for the input section and the lower driver section to *both* be referenced to the negative rail. If this type of driver is used in a half-bridge design, with positive and negative supplies, the input PWM signal will need to be level-shifted from the (usually) ground-based modulator circuits down to the negative-rail-referenced input section of this gate driver.

Gate drivers also offer a variety of input logic schemes. These have names such as "half-bridge driver", "high side / low side", "dual driver", and PWM. Half-bridge or high-side / low-side drivers usually incorporate a non-overlap or cross-conduction prevention logic which prevents both output drivers from having high output voltages at the same time. This prevents both mosfets from turning on at the same time and experiencing catastrophic cross conduction. Sometimes they will incorporate a deadtime feature (fixed or adjustable) which adds adjustable deadtime, enhancing the non-overlap feature. In contrast, "dual drivers" have no overlap protection and do not prevent both output drivers from having high outputs at the same time. If this type of driver is used in a half-bridge Class-D design, extreme care must be taken to ensure that there will be no circumstance where both mosfets could be switched on at the same time. Sometimes, external cross-conduction (overlap) prevention logic is added to designs using this type of driver. The selection of this type of driver is probably due to it having some other overriding feature that made it desirable, in spite of the fact that it lacked internal cross-conduction prevention. The Si8244 half-bridge driver in the reference schematic uses a single PWM input. The *logic table* shown on the schematic shows how the output logic states are controlled by the PWM and the DISABLE inputs. It is important to note that for this particular gate driver IC, *both* outputs are off only during UVLO or the DISABLE high input state.

dv/dt CM Ratings

A critical specification to be checked as part of the selection criteria for the gate driver is its dv/dt or common-mode (CM) transient specification. This parameter specifies the maximum dv/dt that can be sustained without inducing a logic state error. This is important, because the upper gate driver circuitry is referenced to the switching node of the Class-D amplifier. This node swings from rail to rail (between positive and negative rails) at a very high dv/dt. The high dv/dt is necessary to maintain high efficiency. The mosfet switches need to spend as little time as possible in the switching transition state, and ideally should only be in the fully on or fully off states. This CM transient immunity is usually specified in units of KV/μsec or V/nsec. The desired response is no logic state errors, and no latchup due to the high voltage rate of change. A typical good value for this CM transient immunity is 50V/nsec, with newer devices rated to 100V/nsec or more. This spec is necessarily improving, as the adoption of wide bandgap (WBG) devices such as GaN and SiC progresses, and dv/dt rates continue increasing, due to the faster switching speeds of these newer power semiconductor device types.

VS Transient Voltage Undershoot

Another important feature of gate drivers used in half-bridge and full-bridge Class-D applications is their ability to handle voltage undershoot on the V_S pin. Here we define the "V_S pin" as the pin that is connected to the switching node. This pin is the "ground" reference for the high-side driver circuitry and also connects to the source of the high-side mosfet. This pin will necessarily experience a transient voltage undershoot relative to the COM pin, which is defined as the "ground" reference of the low-side driver circuitry and low-side mosfet source.

Built-in Protection Features

Most gate driver ICs incorporate protection features. The most common of these is cross-conduction prevention, also called overlap protection. This is internal logic that prevents the low-side and high-side mosfets from being switched on at the same time.

Some gate drivers offer a selectable deadtime adjustment. In some ICs (such as the Si8244 used in the reference design), the deadtime is continuously adjustable within a defined timing range. Other devices may offer deadtime adjustment in discrete steps. Often, the deadtime adjustment is set via a resistor value selection. In the reference design, a deadtime of 30nsec is selected by the value of R47A.

Another useful protection feature incorporated in some gate driver ICs is overcurrent protection for the power mosfets of the half- or full-bridge circuit. This can take various forms, but the most common implementation uses V_{DS}ON sensing to determine that an overcurrent condition is occurring. This method uses the voltage developed across the mosfet (drain-to-source) during its ON state, to determine the current level, and works as follows: When the mosfet is OFF, it will have a high voltage from drain to source. This OFF-state drain–source voltage is either blocked by a reverse-biased diode internal to the IC, or ignored in some other way. It is only when the mosfet is driven ON that the drain–source voltage is sensed. In the ON state, the drain–source voltage of the mosfet will be Id × R_{DS}ON. Assuming that R_{DS}ON is known, the drain–source voltage can be calculated for different current levels of Id through the mosfet. Typically, gate driver ICs that incorporate this type of overcurrent protection allow the voltage trip level of Id × R_{DS}ON to be set by the designer. This is usually accomplished by a voltage divider at selected pins of the IC. One complication of this type of overcurrent protection is that the R_{DS}ON of the mosfet varies from device to device over production lots, and varies greatly with junction temperature. Typically, power mosfets of the type used in Class-D amplifiers can exhibit a 2.5:1 variation in R_{DS}ON for a junction temperature swing from 25 degrees C to 150 degrees C. Luckily, the temperature coefficient for R_{DS}ON is positive. This means that R_{DS}ON and the sensed drain–source voltage will *increase* as temperature increases. In other words, the mosfet will reach the overcurrent trip threshold at lower values of drain current as it heats up. This is an advantageous direction for this protection.

Important Specifications for Gate Drivers

Propagation and Turn-on Delays

Propagation delay is one of the key selection parameters for a gate-driver IC. Usually the lowest possible propagation delay is desired. Propagation delay appears as a phase shift in the feedback loop in closed-loop systems. It is desirable to minimize this source of phase shift (as there are many others in the forward and feedback loops). In some self-oscillating amplifier topologies (to be discussed in other chapters), the phase shift resulting from the propagation delay is a critical design parameter used in determining the unity gain frequency and switching frequency.

Another critical parameter is *delay matching*. It is one thing to have low propagation delay, but it is equally important to have good delay matching between upper and lower gate driver circuits. If there is poor or variable delay matching, this will degrade the accuracy of deadtime adjustments, and in the extreme could result in cross-conduction between high- and low-side mosfets.

Quiescent Power

Quiescent power may not be thought of as a primary selection parameter, but can be critical in extremely low power designs, or designs where there are tight constraints on the boot supply for the upper gate driver. For example, some designs employ a design technique where the bootstrap supply for the upper fet is precharged. This means it is charged *before* switching commences. It should be noted that in Chapter 7 on bootstrap circuits, it will be explained that the upper gate driver's bootstrap supply does not normally get charged until the first time that the switch node is pulled low by the lower mosfet switching on. Under some circumstances this turn-on sequence, coupled with start-up delays for the upper gate driver (also discussed in Chapter 7), can sometimes result in an audible, and undesired, start-up "pop". In these cases, a precharge circuit is sometimes used so that the bootstrap supply can be fully charged *before* the PWM activity starts. This makes the start-up delay for the upper gate driver irrelevant, because the bootstrap supply will have been fully precharged, and the start-up delay expired, before the upper mosfet is driven on for the first time. This allows for tighter timing control over PWM start-up sequences, allowing for audibly "popless" PWM start-up. Figure 6.4 shows an example precharge circuit.

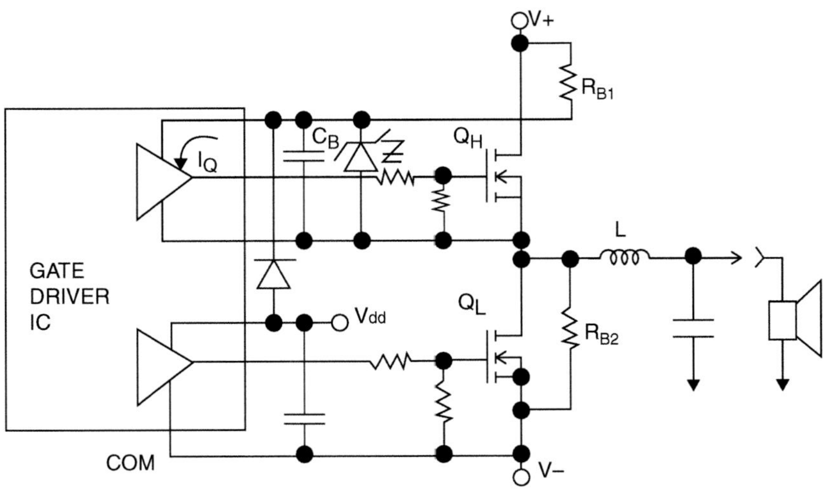

Figure 6.4 Schematic Illustrating a Method to Precharge the Upper Mosfet's Boot Supply in a Half-Bridge Circuit Topology

For this circuit to work, the quiescent current of the upper gate driver must be specified. Assume for this example that the quiescent current of the upper gate driver is 300uA maximum (in the active state, output low condition). Further, assume that a precharge voltage of 10V is desired for C_{BOOT} (C_B). (*Note:* C_B must be charged up to a voltage greater than the rising UVLO threshold voltage in order for the upper gate driver to start-up.) Assume the power supply rails for this half-bridge example are +/−48VDC (such as in the reference design). Another requirement is that it is desirable for the switch node voltage in the quiescent state (static, non-switching state) to be as close to 0V as possible, even when no external speaker load is connected. This ensures minimum "pop" from residual output voltage, if a speaker is connected to an output when the amp is in the off state. This means there will be 48V across RB2 and 48V − 10V = 38V across RB1. Both of these resistors should be sized to provide 300uA. RB2 = 48V/300uA = 160 Kohm = 158 Kohm, 1% value, and RB1 = 38V/300uA = 126.7 Kohm = 124 Kohm, 1% value. These resistors must be sized correctly to handle the expected power dissipation. In a conservatively designed circuit, it would be prudent to size these resistors for the maximum power dissipation which could occur if the output were "stuck" at either supply rail. Granted... this is an unrealistic condition. DC fault protection circuits and AC input signal coupling should prevent that from ever occurring, but in this case, it does not impact component size drastically to use this worst-case calculation. RB2 can "see" 96V across it, so Pd = 96^2/158 Kohm = 58.3mW (use a 0.1W rated resistor). Pd for RB1 will be $(96 - 10)^2$/124 Kohm = 59.9mW (use a 0.1W rated resistor). Be sure to always check the *voltage rating* of the resistors selected. Most 0603 size resistors are rated for a maximum "rated continuous working voltage" (RCWV) of 50V, whereas 0805 size resistors are rated for 150V. For this circuit function, therefore, 0805 size resistors are recommended.

Note: It is important to use minimum sizes for all components in the area of the switch node, to keep trace loop areas small, to reduce inductances and *radiating antenna loops*. Smaller surface areas also reduce capacitances, which in turn reduce power dissipation caused by switch node capacitance (to be explained later).

The reference design does not include this precharge circuitry. It utilizes a conventional bootstrap circuit arrangement, comprised of D3A, R48A, and C_{BOOT} (C28A). This circuit charges C_{BOOT} each time the lower mosfet switches ON, pulling the switching node (A) below the −HV+12V rail. Each time this happens, D3A is forward-biased, and charging current flows from −HV+12V through D3A and R48A into C28A. Thus, C28A is kept charged by repetitive charging pulses at the switching frequency (500KHz). R48A limits the charging current to a safe value that is less than the surge current rating of D3A.

Input Voltage Thresholds and Voltage Range

Another critical datasheet specification to check when selecting a gate driver IC is the input voltage thresholds and input range. Generally speaking, gate driver ICs are designed to accept either TTL voltage logic levels or complementary metal oxide semiconductor (CMOS) voltage logic levels. However, in addition to accepting these typical signal levels, they sometimes can accept input voltage "HIGH" levels much greater than TTL and CMOS voltage levels. However, this depends on the type of gate driver IC selected. "Half-bridge" driver ICs that have input sections that share their low-side, "ground" reference with the low-side driver circuitry, also typically share the same Vdd rail that biases the low-side driver circuitry. This low-side driver rail is typically in the 10–12V range, because the driver output will be rail-to-rail, and a gate drive voltage of at least 8V will be needed to drive "standard" gate voltage mosfets. In these types of drivers, the input voltage level is sometimes allowed to be as high as this Vcc. The (TI) UCC27210 (Texas Instruments, 2014) driver is a good example of this type of driver. The low-side gate driver circuitry has a maximum rail voltage of 20V, and this is the maximum allowed input voltage as well. If the half-bridge driver is the type of driver that has a separate 0V referenced input section (having an independent ground reference from the low-side driver circuitry), this type of driver usually has a lower bias voltage rail for the input section. This applies to the Si8244 driver used in the reference design. Skyworks Solutions Inc. (2021) specifies TTL input signal thresholds of 2V for a logic high, and 0.8V for logic low, with 0.45V typical hysteresis. The maximum signal input level is 6.0V.

Input Voltage Threshold Hysteresis

Another input specification of importance is whether the inputs feature hysteresis in their switching voltage thresholds. These are sometimes referred to as "Schmitt trigger" inputs, because the rising input voltage threshold is higher than the falling input voltage threshold. The voltage difference between these

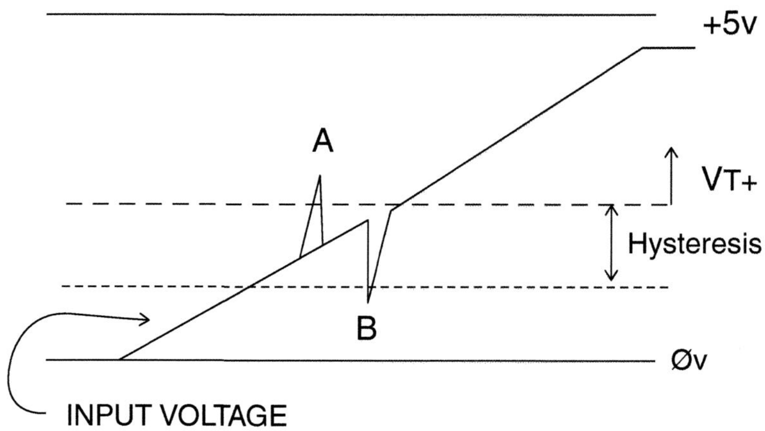

Figure 6.5 Input Voltage Hysteresis and its Effect on Switching Thresholds

two different thresholds is termed the hysteresis voltage. The greater the hysteresis voltage, the greater immunity the inputs have to spurious switching when the input voltage is near the switching threshold. Figure 6.5 illustrates this concept.

In Figure 6.5, a slowly rising input voltage will cause the output to switch logic states when it crosses the rising input voltage threshold (VT+). Once this threshold is crossed, the input voltage can drop, but as long as the drop is *less* than the specified hysteresis voltage, the output logic state will not change. This feature prevents "chattering", or multiple output switching events, as a noisy input voltage crosses the input voltage threshold. In the Figure 6.5 example, noise spike "A" exceeds the VT+ threshold voltage and will cause the output to switch logic states. If noise spike "B" does not occur, the output would remain in the state it switched to at "A". However, if noise spike "B" occurs, the output will switch back to its original logic state, then switch again as the rising input voltage again crosses the VT+ threshold. This happens because noise spike "B" exceeded the hysteresis voltage window. The multiple output switching events that occur in this example are undesirable, because they increase power dissipation in the output stage, and will increase THD due to non-ideal switch timing.

Frequency Range and Pd

Gate driver ICs intended for Class-D usage are designed for high-frequency operation, and their datasheets usually include a graph of either power dissipation versus frequency, or supply current versus frequency, for the various circuit blocks making up the driver. These graphical data must be consulted during the design, to make sure the driver will not experience thermal stress beyond its design limits.

Drive Current

Drive current capability is a core selection criterion for the gate driver. Since mosfets require high drive currents to deliver gate charge rapidly for fast switching, gate driver ICs must have high pulse current capability. These will typically be in the 0.5A–4A range for devices used for Class-D amplifier design. As with most specifications applied to Class-D design, there are nuances to this current spec as well. For example, different gate driver ICs have different output structures. Some older devices have output structures that deliver maximum current at the beginning of the gate drive voltage switching event. Some newer devices have output structures that have been optimized to deliver highest output current when the output voltage is at the "Miller" plateau of the gate charge curve, allowing for faster mosfet switching.

Isolation and Level-Shift Technologies

Half-bridge gate driver ICs also differ in the technology used for level-shifting the drive signal from the control section of the IC to the upper or lower gate drive circuitry. If the IC is one where the input section and lower gate drive circuitry share a common "ground" reference, then the level-shift is necessary between this lower section circuitry and the upper gate driver circuitry. If the driver is of the type that has an independent input section, that has its own floating "ground" reference, such as the Si8244 used in the reference design, or the IRS20957 (Infineon (International Rectifier), 2016), then different level-shifting schemes are required. The Si8244 has a level-shift function from the input circuitry to the lower gate driver, and also from the input circuitry to the upper driver. In fact, in this device, the upper and lower driver designations are interchangeable. The two drivers are identical independent drivers, with flexible referencing. The signal transmission technology used in this level-shift, isolation function is an ON/OFF keyed RF carrier (see Figure 7, page 12 of Si8244 datasheet – Skyworks Solutions Inc., 2021). Other isolation / level-shifting technologies used are capacitive coupling (TI) and magnetic coupling (ADI). These technologies have been developed and optimized to maximize noise immunity, allowing them to work well in the presence of extremely fast CM slew rates (high CM transient immunity). They have also been optimized for EMI immunity as well as low EMI emissions. As mentioned above regarding V_S undershoot, some devices specify on their datasheets the maximum V_S undershoot that can be tolerated without introducing logic errors. Sometimes this specification is broken down further, into a V_S undershoot level during which logic functions normally, versus a greater V_S undershoot level during which a logic state is maintained, and a new logic state transition (programmed from the input section) is ignored. This illustrates how critical it is to design for and control V_S undershoot in the design.

SiC and GaN Driver Types

SiC and GaN power switch technologies will be discussed in other chapters. However, in this chapter it is important to say that gate driver ICs are available that are optimized to drive these newer, wide bandgap (WBG) power switch devices. In the case of SiC, which requires higher gate voltages than standard mosfet devices, gate drivers for these devices tolerate higher bias voltages for powering the output driver sections of the gate driver ICs (typically 20V at least). In the case of GaN drivers, the gate drive requirements are for lower voltages (typically less than 6V) and with much tighter overvoltage limits on gate drive, to protect the voltage-sensitive gates of GaN devices. Gate driver ICs optimized for GaN typically incorporate Low DropOut (LDO) regulators for controlling the rail voltages for both high- and low-side devices. They may also incorporate voltage

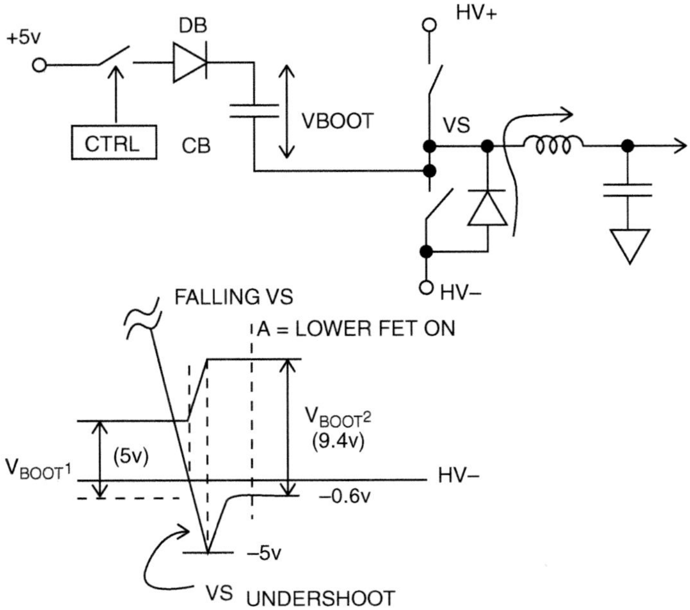

Figure 6.6 C_{BOOT} Overcharge Due to Excessive V_S Undershoot

control on the bootstrap voltage in the form of a clamp or controlled switch, in addition to the normal high-voltage bootstrap rectifier D_B (see Figure 6.6).

Figure 6.6 illustrates how the boot capacitor for the upper gate driver can be overcharged due to excessive V_S undershoot (in the absence of a controlled boot switch). The waveform depicted in the lower diagram is for the conventional case, where the bootstrap rectifier (D_B) has its anode connected to the +5V bias supply. In that case, for "normal" V_S undershoot of approximately 0.6V, clamped by the body diode of the lower mosfet, the voltage V_{BOOT} on C_B will be approximately 5V. However, in the case of severe V_S undershoot of −5V peak, for example (due to high current switching of inductor (output) current), the voltage on C_B could be charged as high as 9.4V, which would exceed the safe gate voltage range for GaN devices. To prevent this C_B overcharge situation, some gate drivers incorporate a controlled switch, in series with the bootstrap rectifier (D_B), depicted in the upper schematic in Figure 6.6. This switch is turned OFF when the lower mosfet is OFF. Its operation and effectiveness can be understood by again referring to the lower waveform diagram in Figure 6.6. Initially, the lower mosfet is off, and the upper mosfet has been on, and current is flowing in the direction shown, through the output inductor. When the upper mosfet is switched OFF, the switch node voltage falls ("Falling V_S"). Now, *both* mosfets are off during this deadtime interval. In this example, falling V_S then undershoots severely to −5V below HV−. However, since the lower mosfet is OFF, the "CTRL" block keeps the boot switch OFF, and V_{BOOT} will not be charged through D_B. This prevents overcharge of V_{BOOT}. Now, after the V_S undershoot event, the lower mosfet will be switched ON at time "A". The "CTRL" block will switch ON the boot switch, and V_{BOOT} will be charged to its nominal safe level through D_B.

Note: The V_{BOOT} supply for the upper mosfet is a critical part of the design. It will be discussed again, in greater detail, in the next chapter.

Higher-Frequency Operation

Gate drivers optimized for these Wide BandGap (WBG) devices (especially GaN) are also capable of higher-frequency operation as compared to gate drivers intended for mosfets or insulated-gate bipolar transistor (IGBT) devices, and have lower propagation delays, better propagation delay matching, and tighter deadtime control (if available as a feature) than drivers intended for mosfets.

Discrete Gate Driver Circuits

Even though there is now available a great variety of IC gate drivers that are suitable for Class-D amplifier applications, there will always be designs where a discrete driver may be suitable. This could be due to cost considerations, or where a gate driver IC has been used for the level shift, UVLO, and input signal processing, but a small amount of extra output current boost is required, or when an asymmetric drive is desired. Asymmetric drive is when the drive current differs substantially for turn-on (sourcing) compared to turn-off (sinking). When an asymmetric drive is required, it usually involves making the turn-off (sinking) drive current greater than the turn-on (sourcing) drive current. This produces mosfet turn-off speed enhancement compared to turn-on, and, if used in *both* the upper and lower mosfet drives in a half- or full-bridge configuration, will help to prevent overlap and cross-conduction. *Note:* Of course, there are half-bridge driver ICs that have the gate drive current capability desired, but perhaps the combination of the other features desired could not be found in a commercial

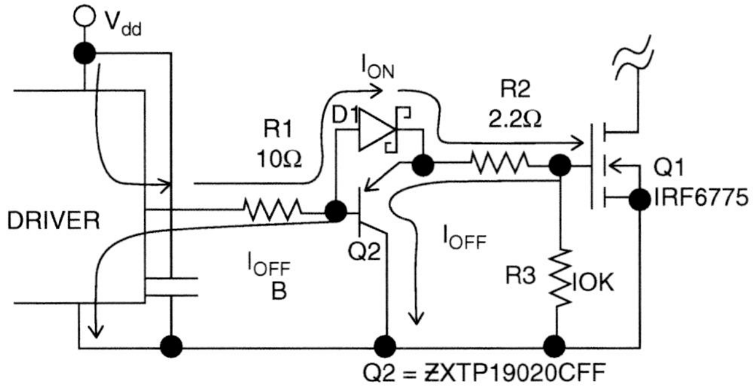

Figure 6.7 Schematic of a Discrete Gate Driver Circuit Using a PNP Transistor for Enhanced Pulldown Current Sinking, for Faster Mosfet Turn-off

gate driver IC, and so this one missing requirement (sink current enhancement) can be achieved by a simple discrete circuit addition to the driver IC circuitry. Figure 6.7 depicts a simple discrete PNP circuit that boosts turn-off current.

In this figure, Q2 is OFF during turn-on of mosfet Q1. The turn-on sourcing current path to the gate of Q1 consists of R1 (10 ohms) in series with R2 (2.2 ohms) and the forward-biased D1. Note how when D1 is ON, it reverse biases the Vbe junction of PNP transistor Q2, ensuring that Q2 is off. During turn-off, when the output of the driver IC (depicted on the left side of Figure 6.7) goes low, Q2 will switch ON, and D1 will switch off. Emitter-Follower Q2 will not saturate, so its current gain (β) will remain high throughout the turn-off event. As an example, if Vdd were 12V, then the Q1 gate would charge to a final voltage of 12V − 0.5V = 11.5V, assuming D1 is a Schottky rectifier with Vf = 0.5V typical. The maximum sink current demanded of Q2 would be (initially at the beginning of the turn-off event) = (11.5V − 0.6V) / 2.2 ohms = 4.95A. *Note:* This calculation is fairly accurate, but still approximate. A circuit simulator, such as LTspice, will quickly solve for the driving voltages and currents more accurately, but this calculation will illustrate the current-boosting capability of this simple discrete PNP follower circuit. Q2 maintains a very high β of 100 even with Vce as low as 1V, so we can assume base current will be a *maximum* of 4.95A/100 = 49.5mA. This base current flows through R1 = 10 ohms at a voltage drop of (49.5mA × 10 ohms) = 495mV. So if the driver output can pulldown all the way to 0V, while sinking 49.5mA (which should be no problem for a driver with CMOS output structure), then the base of Q2 will be at +495mV. Vbe-ON for

Q2 at a collector current of 5A is 0.8V (see datasheet for Q2 – Diodes Inc., 2016), so the emitter will be at (0.495V + 0.8V) = 1.295V. So, iterating, the actual peak gate pulldown current will be (11.5V − 1.295V) / 2.2 ohms = 4.64A instead of 4.95A. Further, if we look at the pulldown current provided during the critical "Miller" region of the gate charge curve, we can see that this simple PNP circuit still supplies a substantial current boost during this region as well, ensuring a quick transition through this region. Referring to the gate charge graph from the IRF6775 datasheet (Infineon (International Rectifier), 2014), the gate voltage during the Miller region (plateau) is 4.3V. Approximating the emitter voltage of Q2 to be 1V, this will mean that the pulldown gate current for Q1 will be (4.3V − 1V) / 2.2 ohms = 1.5A during the Miller interval. At this level of drain current, β for Q2 is still easily 100. Therefore, base current for Q2 will be 1.5A/100 = 15mA. Therefore, the voltage at the base of Q2 will be 150mV, and (iterating) Vbe-on for Q2 will be about 0.75V, so our original assumption of 1V for the emitter voltage was close to the more accurate (150mV + 0.75V) = 0.90V. Again, LTspice would yield more accurate results, but this simple hand calculation example yields accuracy adequate for design purposes.

There is also a gate driver IC that incorporates simple NPN and PNP emitter follower transistors for just this current-boosting reason. Although this is an IC (Diodes Inc., 2020), it illustrates how a single or dual discrete transistor can be used to implement a source or sink current boost into the gate drive circuit.

A Word about Gate Drive Transformers

Gate drive transformers are typically not used for driving the power switches (mosfets) in Class-D audio amplifiers. The reason is that the duty cycle of the power switches varies over a very wide range (at least 20%–80% and often greater than 5%–95%). Transformers have the requirement that the positive volt-second product and the negative volt-second product across every winding must be equal. This means that, as the duty cycle changes, the positive and negative winding voltages will change. This is not good for driving a device (like a mosfet) that requires +10V as an ON-state gate drive voltage, and 0V (or less) for an OFF-state gate drive voltage. An example will better illustrate this. Suppose a transformer secondary winding is switching at 50% duty cycle, and the positive voltage swing is +10V, and the negative voltage swing is −10V. The volt-second products will be balanced. Now suppose the duty cycle changes to 5%. This means the secondary winding voltage will be positive for 5% of the period and negative for 95% of the period. Since the volt-second products must be equal, the positive state voltage will be +19V (for 5% of the period) and the negative state voltage will be −1V (for 95% of the period). Obviously, it is not advisable to drive a mosfet gate over so wide and dynamic a voltage range as this. Capacitor coupling, and diode clamping circuits can be used with transformers to limit this drive voltage range, but this adds complexity, and is not suitable for circuits (like Class-D power switching circuits) where the duty cycle range is so wide and dynamic and where very tight timing constraints apply.

Bibliography

Begue, M. (2020). External gate resistor design guide for gate drivers. Texas Instruments Application Note, SLLA385A. Originally published May 2018; revised March 2020. Dallas, TX: Texas Instruments.

Diodes Inc. (2016, June). Datasheet, ZXTP19020CFF, 20V PNP low saturation voltage transistor in SOT23F. Document number: DS33730, Rev. 2 – 2. Plano, TX: Diodes Incorporated.

Diodes Inc. (2020, February). Datasheet, ZXGD3003E6, 40V 5A gate driver in SOT26. Document number: DS33458, Rev. 4 – 2. Plano, TX: Diodes Incorporated.

Infineon (International Rectifier). (2014, February 26). Datasheet, IRF6775MTRPbF Digital audio mosfet. El Segundo, CA: International Rectifier.

Infineon (International Rectifier). (2016, September 19). Datasheet, IRS20957 Protected digital audio driver, No. PD. El Segundo, CA: International Rectifier.

Infineon (International Rectifier). (n.d.). Use gate charge to design the gate drive circuit for power MOSFETs and IGBTs. Infineon (International Rectifier) Application Note, AN-944. El Segundo, CA: International Rectifier.

Skyworks Solutions Inc. (2021, July 31). Datasheet, Si824x Class D audio driver with precision dead-time generator, Rev. 1.0. Irvine, CA: Skyworks Solutions Inc.

Texas Instruments. (2014). Datasheet, UCC2721x 120-V boot, 4-A peak, high-frequency high-side and low-side driver, SLUSAT7F. Originally published November 2011; revised December 2014. Dallas, TX: Texas Instruments.

Chapter Seven

Bias Supply for Upper Gate Driver

Objectives and Challenges of Bias Supply Design

The circuit for generating the upper gate driver's bias supply has been given its own chapter because this circuit block is deceptively critical. At first it may seem like a trivial circuit function, but it is not. Looking at the half-bridge topology used in the reference design for example, the upper gate driver bias supply floats with the upper mosfet. The low side of this supply is referenced to the source pin of the upper mosfet, which is the switching node. The switching node has extremely high dv/dt and its voltage is switching from the negative supply rail (−48VDC) to the positive supply rail (+48VDC).

One challenge in the design of this "floating" bias supply is that it does not share a common reference with other supply rails in the design, so different techniques are used to generate this supply. The most common technique is a bootstrap technique where the bias supply for the upper mosfet Q_H (in a half-bridge topology) is generated (charged) from the bias supply powering the lower mosfet Q_L. In Figure 7.1 it can be seen that each time the switching node voltage is less than Vdd by more than one diode drop, the potential exists to charge the storage capacitor C_{BOOT}. This capacitor (called the boot capacitor) will provide charge storage and maintain a "constant" supply voltage for the upper gate driver circuitry. In a first-order approximation, assuming that the switch node exactly equals V_{NEG} when the lower mosfet is on, C_S will be able to charge to Vdd − V_D (where V_D is the forward voltage drop of diode D1). Once the charging mechanism for C_{BOOT} is understood, it becomes apparent that it is critical for the lower

DOI: 10.4324/9781003327943-7

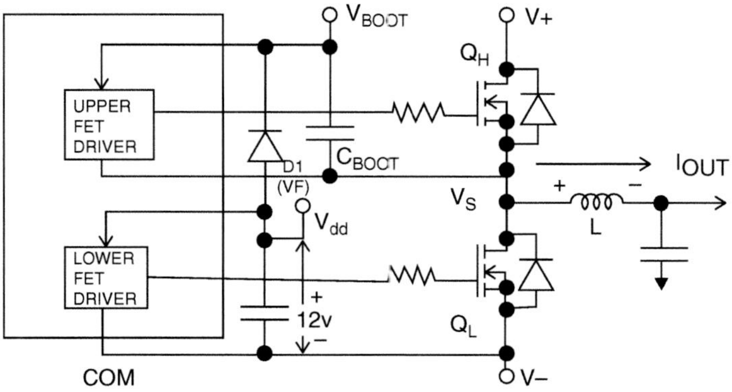

Figure 7.1 Block Diagram/Schematic of a Conventional Bootstrap Circuit as Used in a Half-Bridge Class-D Switching Power Stage

mosfet to be ON for some finite time or minimum duty cycle, in order to maintain an adequate voltage across C_{BOOT}. If the voltage across C_{BOOT} droops, it may eventually drop below the turn-off UVLO threshold for the upper gate driver, and the upper mosfet will switch OFF. This will result in output voltage collapse during the amplifier's positive output voltage peaks, and severe distortion. In a fixed-frequency design such as the reference design, the minimum on-time for the lower mosfet and the switching deadtimes both affect this.

Bootstrap Circuit Design

First, the power required by the upper gate driver must be calculated. This consists of the following:

- Quiescent (on-state) power of the upper gate driver
- Dynamic gate drive power required by the mosfet = 166mW
- Any other loads on the upper gate driver supply
- Power required due to the bootstrap diode's reverse recovery characteristic and leakage.

Bootstrap Circuits with Internal Switch

Some IC gate drivers (Texas Instruments, 2019) incorporate an active switch as part of the bootstrap circuit (see Figure 7.2).

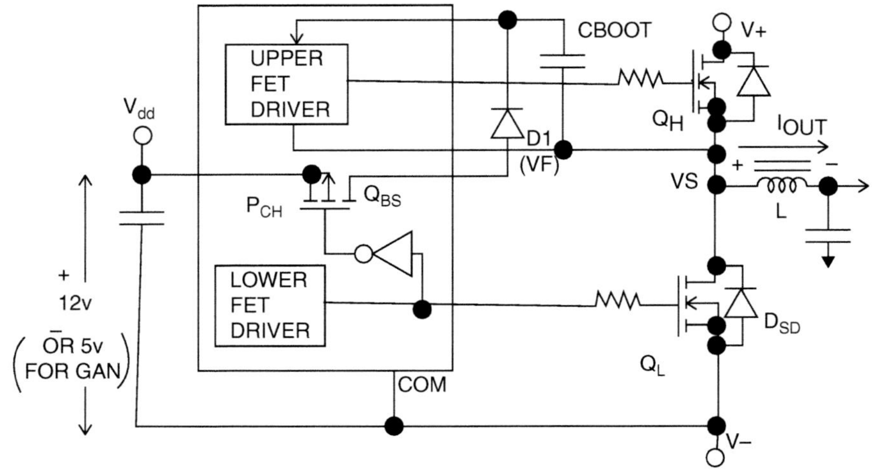

Figure 7.2 Block Diagram/Schematic of a Conventional Bootstrap Circuit as Used in a Half-Bridge Class-D Switching Power Stage, with the Addition of an Internal Switch (P-Channel Mosfet)

The bootstrap rectifier is still required, to block reverse voltage, but instead of the anode of this rectifier being connected directly to the bias supply that powers the lower gate driver (Vdd), it is connected to this bias supply through an active switch. This switch can prevent overcharging of the bootstrap capacitor. This is important, because the bootstrap capacitor voltage ***is*** the bias voltage for the upper gate driver. Therefore, an excessive voltage on the boot capacitor can cause an excessive gate voltage on the upper mosfet, when the upper gate driver drives the upper mosfet into the ON state. This overvoltage can possibly damage the gate of the upper mosfet.

The details of the operation of this boot switch will now be explained. Some simplifying assumptions will now be made to illustrate the circuit operation. A half-bridge topology is assumed, with positive and negative rails of V+ and V−. The lower gate driver circuitry is powered by a 12V DC source, referenced to V−. Whenever D1 is forward biased, it will have an assumed forward voltage of V_F = 1V. It is clear that whenever V_S (the voltage at the switch node) is lower than Vdd − V_F, C_{BOOT} will be charged. If the instantaneous voltage at V_S were equal to V− (which would occur if Q_L were on and had an ideal on-resistance (R_{DS}ON) of 0 ohms), then C_{BOOT} would charge to 11V. Now we will add some real-world considerations to this circuit discussion. Figure 7.3 depicts a voltage waveform

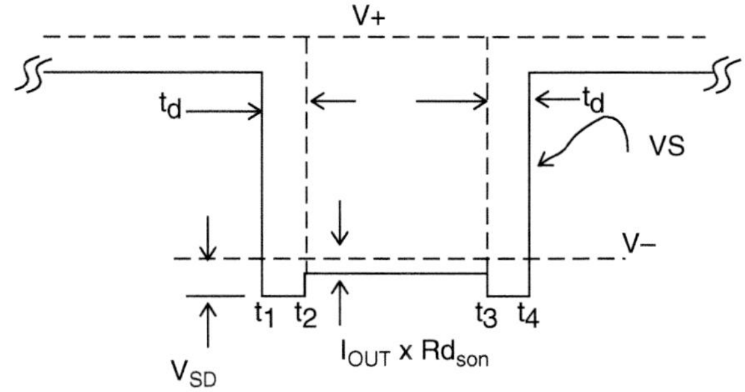

Figure 7.3 Idealized Switching Node Waveform (V_S)

at V_S (switch node) that would occur when the half-bridge circuitry is *sourcing* a large positive current, as depicted by the arrow labeled Iout.

In this circumstance, when Q_H is switched off at time t_1, the current direction through L will be maintained, since output inductor L will resist any current change, by virtue of its inductance. Now, since the interval t_1–t_2 is a deadtime (t_d), *neither* Q_H nor Q_L is on, therefore the only way the current through inductor L can be maintained is for the body diode D_{SD} to conduct. The typical forward voltage for a mosfet body diode (source-to-drain diode) is typically in the 1–2V range, and will vary with current level. This means that the switch node will be *more negative* than V_{NEG} by at least 1V. This extra volt will add to the charging voltage available for C_{BOOT}. Now this scenario depicted a rather "tame" example where there were no parasitic or stray inductances involved. But it should be appreciated that typical Class-D circuits do not always have ideal layouts, with low parasitic and stray inductances, and typically exhibit very high dv/dt and di/dt switching events. Furthermore, P-N junction rectifiers (both discrete and mosfet body diodes) do not have instantaneous turn-on and turn-off times, but rather experience reverse recovery time and forward recovery time. In this case, it is forward recovery time, forward voltage, and ringing that are critical. At the instant that Q_H switches off, there will be a very high falling dv/dt at the switch node. This will cause an undershoot of voltage, as V_S continues to drop below V−, until the body diode of Q_L starts to conduct in the forward direction. The time it takes for this body diode to turn on is the forward recovery time, and the *peak* forward voltage is called the forward recovery voltage. Unfortunately these are rarely specified parameters. The peak forward recovery voltage will also increase as falling dv/dt of the switch node increases. Also, any ringing at the switch node, caused by parasitic inductances internal to components or external in circuit traces, will add to the peak voltage undershoot at V_S. It is entirely possible, and common, for V_S undershoot to exceed 5V. In this case, the amount of voltage undershoot will *add* to the voltage across C_{BOOT} and could result in an overvoltage on C_{BOOT}. It is also for this reason that gate driver ICs normally have a specification for the amount of voltage undershoot that can be tolerated at the V_S pin of the driver. This is because there are typically protection diodes at the pins of gate drivers, and turn-on of these protection diodes should generally be avoided (Infineon (International Rectifier), 2013). An example of the arrangement of these protection diodes is shown in the Infineon (International Rectifier) (2013) datasheet. Another specification parameter related to this is the maximum dv/dt spec for the V_S pin that can be tolerated by the driver IC. Now that the mechanism for possible overvoltage generation on C_{BOOT} is understood, the utility of the added internal boot switch can be explained. Referring to Figure 7.2, an added P-channel mosfet switch can be seen, connected between the Vdd supply and the anode of rectifier D1. Now, instead of C_{BOOT} being charged whenever V_S is less than Vdd − 1V, there is an additional condition necessary; Q_{BS} must be ON. The presence of Q_{BS} allows better control of the charging of C_{BOOT}. For example, if Q_{BS} is kept OFF during the deadtimes (t_1–t_2 and t_3–t_4), and *only* allowed to be on when Q_L is driven on, then the overcharging of C_{BOOT} that could occur during the deadtimes is avoided. EPC (2016) shows an alternate approach that uses a small enhancement mode GaN switch as the boot switch. This allows for an extremely fast boot switch.

Completely Floating Gate Drive Supplies

In the sections above, the bias supply for the upper switch in the half-bridge topology was generated by "bootstrap" action, from a bias supply referenced to the negative supply rail (and the lower mosfet's source terminal). This arrangement gives rise to the complexities and subtleties of design addressed above. Care must be taken to ensure that the bias supply for the upper switch does not droop excessively due to insufficient charging time, and does not "pump-up" excessively due to V_S undershoot.

One way to avoid these complexities and non-idealities in the upper gate drive supply is to employ a completely floating bias supply for the upper switch. Figure 7.4 shows one possible implementation of a floating bias supply.

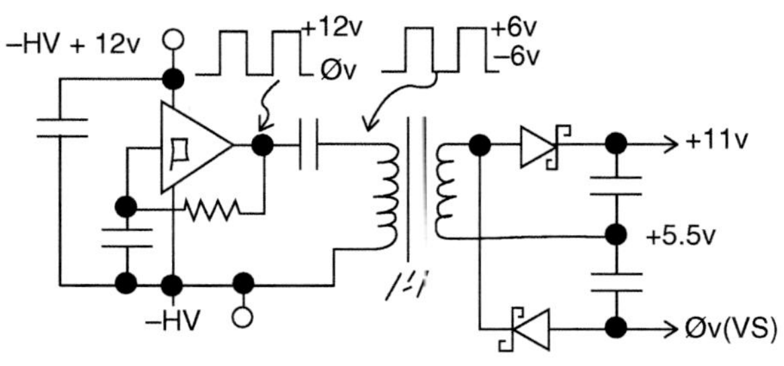

Figure 7.4 Floating Bias Supply Constructed Using an Inverting Gate Driver IC and a Gate Drive Transformer

This circuit uses a small gate drive transformer for isolation. This type of transformer is ideal for this application, because it is already designed to handle the voltages present in this design, both isolation voltage rating (1500V is common) and V-μsec rating. High-frequency gate drive transformers have V-μsec ratings, provided on the datasheet, which make it easy to select appropriate parts for this application. This is defined as the Voltage × Time product that the primary winding can handle without saturating.

In this example, it is desired to create a floating 12V supply from the lower mosfets 12V bias supply, which is referenced to the negative supply rail. So this is a 1:1 DCDC converter application. The circuit in Figure 7.4 uses a conventional *inverting* gate driver IC (Fairchild Semiconductor (Onsemi), 2007) with input voltage hysteresis, to implement a self-oscillating transformer driver. The gate driver IC is well suited for this application, because its output stage is already designed to drive capacitive loads. This means it can deliver high pulsed currents for short duration, and this is exactly what is necessary to charge the capacitance of the bias supply via a transformer-rectifier (peak-detecting) circuit. Since only a single gate driver IC is used in this example, a gate drive transformer with a single primary winding is required. The secondary side may consist of either a single secondary, or dual secondaries. If a dual secondary type is selected, these secondaries will be connected in a center-tapped arrangement, if not already pre-wired that way. Figure 7.5 illustrates how the dual secondary of a 1:1:1 transformer can be configured to produce the same output voltage as the transformer configuration of Figure 7.4. Some examples of transformers with single and dual secondaries are found in the Pulse Electronics (2020) datasheet.

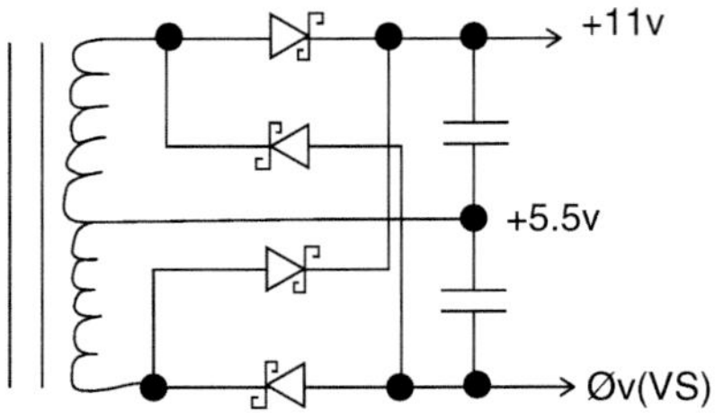

Figure 7.5 Center-Tapped Secondary Winding, Configured with Four Rectifiers, for Full-Wave Rectification

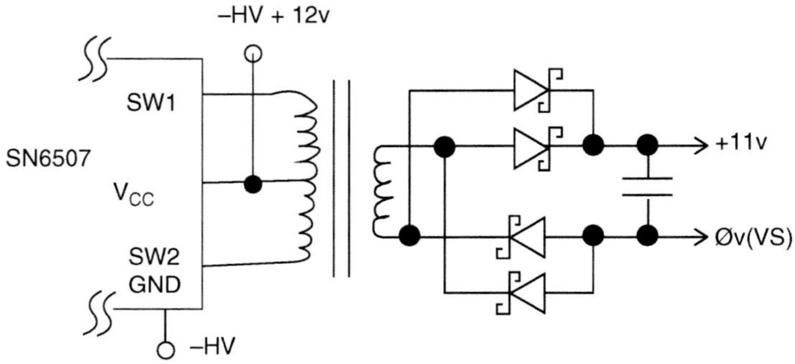

Figure 7.6 Floating Bias Supply Constructed Using a Transformer with a Center-Tapped Primary Winding and a Single Secondary Winding, Driven by the SN6507 Transformer Driver IC

Another approach to generating a floating gate drive supply is to use a push-pull transformer driver such as the TI: SN6507 (Texas Instruments, 2022). This IC is designed to drive a push-pull transformer (see Figure 7.6).

The datasheet references numerous possible transformers that use a center-tapped primary winding. While this primary winding arrangement is common in transformers intended for isolated power applications, it is *not* common among gate drive transformers, which almost always have a single primary winding. There are advantages to using a gate drive transformer in this application. These are:

1) The V-μsec rating and isolation voltage ratings for the transformer are appropriate for, and well matched to, the application.
2) The transformer construction has already been designed with the goals of small size, low inter-winding capacitance, and low leakage inductance.

Advantages Explained

Gate drive transformers have already been designed to drive the gates of mosfets requiring 10V gate drive levels, at frequencies appropriate for Class-D designs (200KHz–1MHz typically). The V-μsec rating of the primary winding indicates how much voltage can be impressed across the primary winding and for what time duration, without causing *saturation*. This specification is typically given, assuming a unipolar drive (worst-case topology). If instead, the winding is driven in both directions (bipolar drive), the unipolar V-μsec rating can be doubled. In addition, since gate drive transformers are already designed to operate in exactly this circuit location (upper mosfet drive), they have already been designed to handle the voltages involved. For example, the primary-to-secondary voltage rating will need to be greater than the voltage difference between the positive and negative supply rails. Taking the reference design as an example (with +48V and −48V rails), a "functional" voltage rating exceeding 100V (as an absolute minimum) is required. Gate drive transformers are available with many different voltage ratings, and are usually much higher than 100V. This means that there will be a wide range of gate drive transformers suitable for this DCDC application. Gate drive transformer manufacturers should be consulted concerning optimizing a center-tapped primary type of transformer, suitable for the SN6507 type driver (see Figure 7.6), specifically for this floating-gate-driver application, where low interwinding capacitance and low leakage inductance are critical.

Another goal of the design of the gate drive supply is small size. Of course small size is always desired on its own; however, it takes on even more importance in gate drive circuitry, because the upper gate drive components are referenced to the switching node, and experience high dv/dt switching. Size and surface area affect capacitance, which affects power dissipation and EMI. So minimizing surface area (capacitance) and current loop areas is always critical to a successful design. Inter-winding capacitance adds to this aforementioned capacitance and its power dissipation contribution, so minimum inter-winding capacitance is paramount. Leakage inductance between windings is basically a measure of the coupling between the windings. Windings that are better coupled to each other have lower leakage inductance and better signal fidelity. Leakage inductance between windings causes ringing in voltage waveforms and slower/delayed voltage switching edges. However, as is typical in Class-D design, the desired characteristics are at odds with each other. Leakage inductance is decreased by tighter coupling between windings. But tighter coupling between windings increases inter-winding capacitance. Therefore, achieving the conflicting goals of low inter-winding capacitance *and* low leakage inductance at the same time requires careful transformer design. Manufacturers of gate drive transformers are skilled at doing exactly this. That is why it is beneficial to use a gate drive transformer as the DCDC converter transformer in this application. It has exactly the attributes critical for this design application. The SN6507 transformer driver will drive the primary of the transformer in a push-pull topology. Therefore, it requires a center-tapped primary. Gate drive transformers, however, are available predominantly with single primary windings. They are available with single and dual secondary windings, and most times can be driven in reverse. That is to say, the secondary windings can be configured in this application to be the driven (primary) windings, and can be wired as center-tapped windings (see Figure 7.6). Since it is the inter-winding capacitance and leakage inductance between the primary and secondary windings that are critical here, these parameters are reciprocal and independent of the orientation of the windings. The transformer manufacturer should always be consulted, to ensure that a selected gate drive transformer can be driven in reverse, as described here.

Creating Multiple Floating Bias Supplies

Another added convenience of using a gate drive supply as the DCDC transformer in this application is that more than one floating bias supply can be generated by a single drive circuit. For example, the single gate driver shown in Figure 7.7 can use a gate drive transformer with two separate secondary windings to create two separate isolated bias supplies. This could be used to generate an isolated, floating bias supply referenced to the switch node (V_S) for the upper mosfet, and a separate, isolated bias supply referenced to the negative power supply rail for the lower mosfet gate driver bias, *and* the primary side can be referenced to circuit ground (0V), and be powered from a circuit-ground-referenced voltage rail (instead of the −HV-referenced −HV+12V rail). Again, this topology of gate drive transformer, having a single primary winding and dual, isolated secondary windings, is very common and available. In addition, the voltage ratings between the primary and secondary windings, and between the secondary windings themselves, will be appropriate for this application, because these are exactly the voltage ranges that the gate drive transformer will have been designed for in its

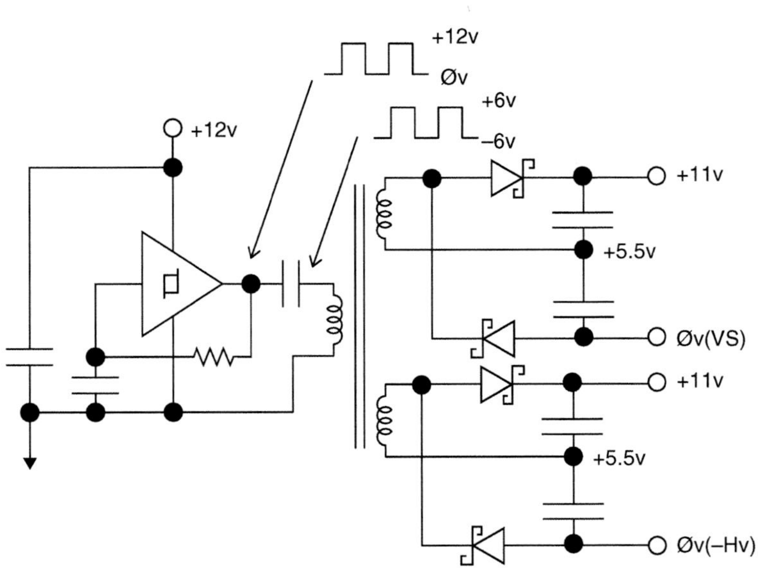

Figure 7.7 Multi-Output (Dual Output) Floating Bias Supply Constructed Using an Inverting Gate Driver IC and a Gate Drive Transformer with Single Primary and Dual Secondary Windings

intended application as a gate drive transformer. The fact that it can now easily be adapted as a DCDC transformer is a bonus.

It should be appreciated that using floating gate drive supplies instead of bootstrap-generated supplies completely eliminates the problems of excessive V_S undershoot and associated overcharging of the bootstrap supply. The downside is the extra component cost and space associated with these small DCDC circuits.

Another important note is that in multi-channel designs, *each* channel's upper mosfet requires a separate and independent gate drive supply referenced to *its* switching node. In designs like the reference design that use a single bias supply (−HV+12V) referenced to the negative voltage rail, multi-channel designs *can* use this single −HV+12V bias supply to power the gate drivers in multiple channels.

Bibliography

EPC. (2016). Development board EPC9066 quick start guide. El Segundo, CA: Efficient Power Conversion.

Fairchild Semiconductor (Onsemi). (2007). Datasheet, FAN3100 Single 2A high-speed, low-side gate driver. Rev. 1.0.4. Aurora, CO: ON Semiconductor.

Infineon (International Rectifier). (2013, October 24). Datasheet, IRS2092

Protected digital audio amplifier (p. 13 = input/output pin equivalent circuit diagrams). El Segundo, CA: International Rectifier.

Pulse Electronics. (2020, January). Datasheet. SPM2007_61, SMT Gate drive transformers (includes P0544NL). San Diego, CA: Pulse Electronics.

Skyworks Solutions Inc. (2021, October 18). High-side bootstrap design using ISOdrivers in power delivery systems. Application Note AN486. Rev. 0.3. Irvine, CA: Skyworks Solutions Inc.

Texas Instruments. (2019). Datasheet, LMG1210 200-V, 1.5-A, 3-A half-bridge MOSFET and GaN FET driver with adjustable dead time for applications up to 50 MHz, SNOSD12D. Originally published November 2018; revised January 2019. Dallas, TX: Texas Instruments.

Texas Instruments. (2022). Datasheet, SN6507 Low-emissions, 36-V push-pull transformer driver with duty cycle control for isolated power supplies, SLLSFM0A. Originally published March 2022; revised June 2022. Dallas, TX: Texas Instruments.

Chapter Eight

Half-Bridge and Full-Bridge Switching Power Stages

Modulation Types for Full-Bridge Arrangement

The great majority of Class-D amplifiers use either a half-bridge arrangement of the switching power transistors, or a full-bridge (also called "H-bridge") arrangement (more about this in Chapter 20). The half-bridge uses two transistors and the full-bridge uses four, assuming none of the transistors have been paralleled with additional transistors.

Note: Paralleling of power transistors is sometimes used as a way of increasing the output power capability of the switching stage, by allowing it to handle higher current levels. When this is implemented in a full-bridge (BTL) arrangement, it is called Parallel BTL, or PBTL.

Characteristics of Half-Bridge and Full-Bridge Circuits

Figures 8.1 and 8.2 depict generic arrangements for half-bridge and full-bridge power transistors, as could be used in a Class-D amplifier power stage.

The designators "U1" and "L1" are used as shorthand for "upper" and "lower" transistors. Generic "switch symbols" are used to represent the switching transistors, because for this initial discussion it is not important to know the type of power transistor used (e.g. bipolar or mosfet or GaN). Each transistor has a control or drive terminal, with first letter "C" in the figures. This control signal determines whether each individual transistor switch is ON (closed) or OFF (open).

DOI: 10.4324/9781003327943-8

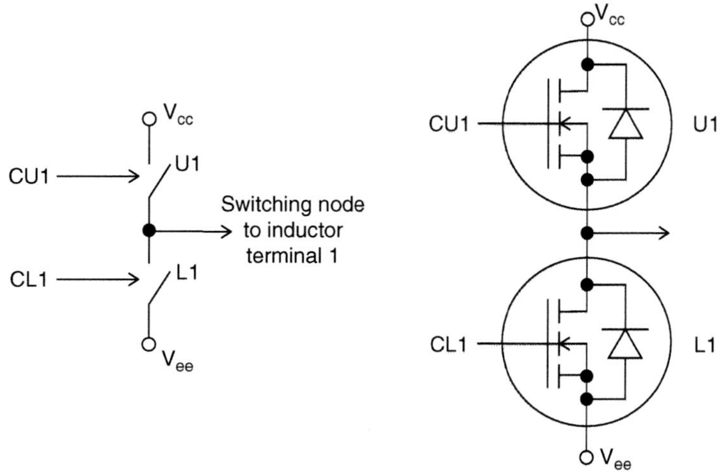

Figure 8.1 Power Switch Arrangement in a Half-Bridge Switching Power Stage

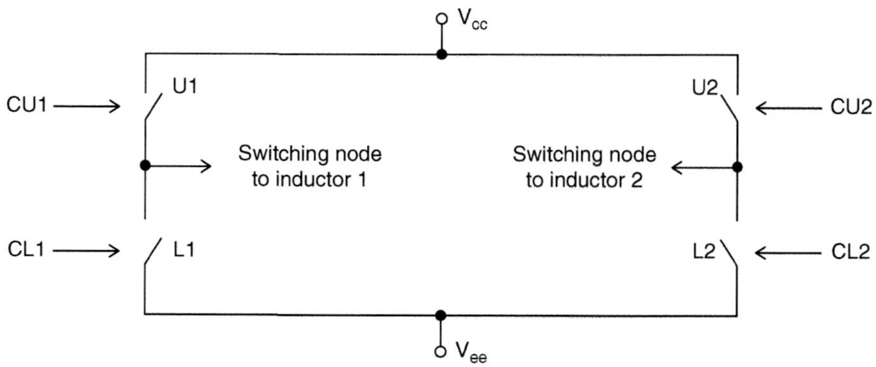

Figure 8.2 Power Switch Arrangement in a Full-Bridge Switching Power Stage

Power Supply Rail Connections

In Figures 8.1 and 8.2, the supply rails are labeled as "Vcc" and "Vee". These power supplies can be either bipolar, meaning a positive supply rail voltage for Vcc and a negative rail voltage, of equal value, for Vee, or unipolar, meaning a positive supply rail for Vcc and 0V (or "ground") for Vee. In either case, it is critical to realize that there is a direct path from Vcc to Vee, or from Vcc to ground, if both transistor switches happen to be ON at the same time. This represents

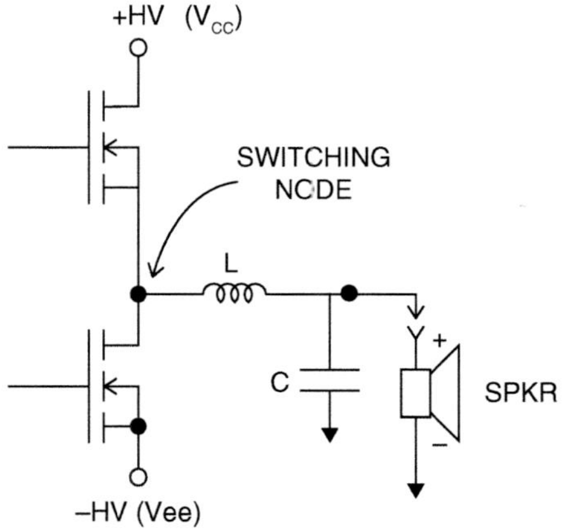

Figure 8.3 Half-Bridge Switching Stage Driving an L-C Demodulation Filter and Speaker

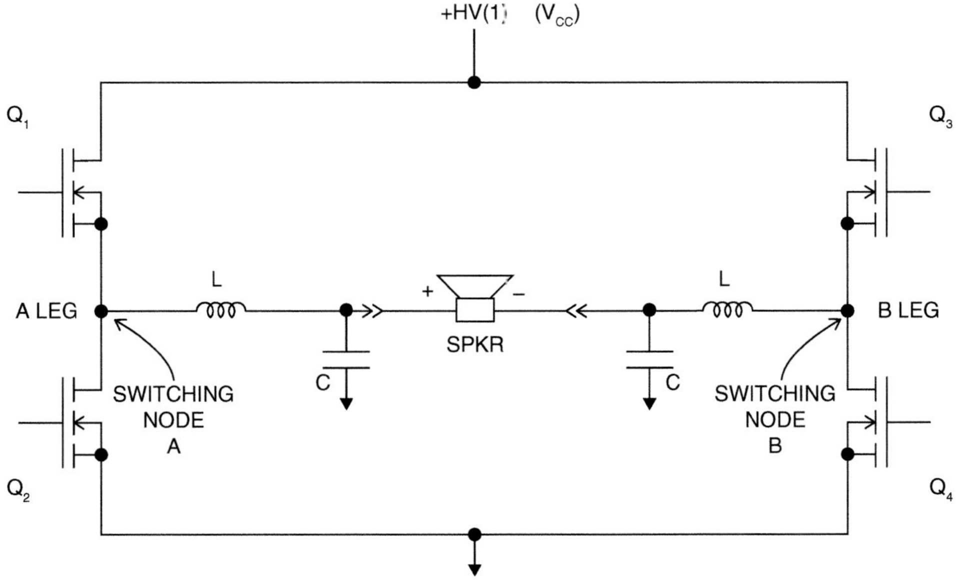

Figure 8.4 Full-Bridge Switching Stage with A- and B-Legs

a direct, high-current, very-low-resistance path from rail to rail, or from rail to ground, which for practical "real-world" transistors will cause failures, and therefore must be avoided. This type of conduction from rail to rail or from rail to ground is called cross-conduction.

Figures 8.3 and 8.4 show the half-bridge and full-bridge switching stages with the output L-C filters added.

Power Switch Operation, in Detail

If a deep understanding of Class-D audio amplifiers is desired, it is essential to understand in detail the switching action of the power stage. Although the power stage may appear to be a straightforward switching circuit, there is actually a great deal more going on than meets the eye. In this section, the switching behavior of the power stage will be analyzed in detail. By understanding the subtleties of these circuits, the designer will be better prepared to make the design decisions and many tradeoffs necessary to achieve the desired amplifier performance objectives. For this initial discussion, the half-bridge configuration will be used, and component values, reference designations, and voltages used in the reference amplifier will apply.

In Figure 8.5, the power switching stage is at idle, and assumed to already be in the steady-state condition (start-up and shutdown will be analyzed separately in other chapters). "Idle" is defined as the condition where the input signal to the amplifier is 0V, and the power stage is switching at its fixed frequency of 500KHz (period = T = 2μsec) and producing (ideally) a 0V output signal. There will of course be a very small DC offset voltage due to various offsets in the amplifier circuitry (discussed in other sections).

In order for a half-bridge power stage to produce 0V output, with positive and negative rails of equal voltage magnitude, the ON-times for the upper and lower switching transistors must be equal, and therefore the duty cycle is 50%. So, to a first approximation, the upper transistor is on for T/2 = 1μsec, and the lower transistor is also on for 1μsec. Since the output voltage is 0V, there will be +48V impressed across inductor L1A during the ON-time, so the change in inductor current (di) is expressed as:

$$di = (dt \times V_L)/L \qquad \text{EQ. 8.1}$$

Substituting circuit values yields:

$$di = (1\mu sec \times 48V)/22\mu H = 2.18A$$

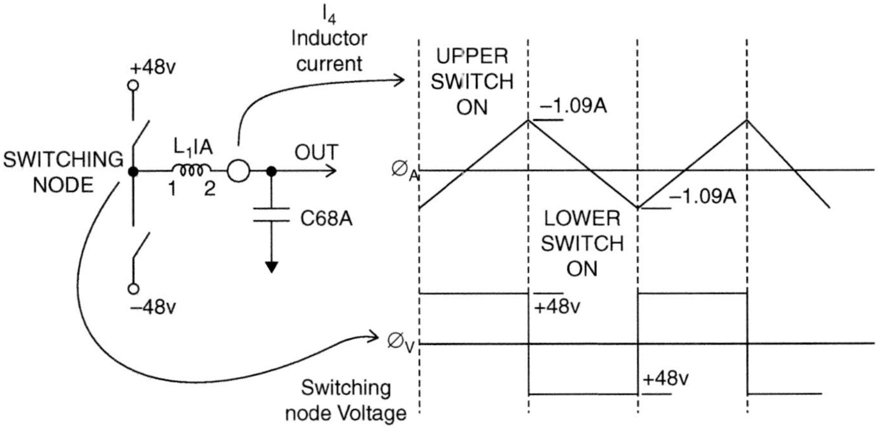

Figure 8.5 Half-Bridge Switching Stage with the Switching Node Voltage Waveform and Inductor Current Flowing through the Inductor of the L-C Demodulation Filter

Since the inductor current *is* the output current, and the output current is 0A, the *average* value of the inductor current must be 0A. This means that during the ON-time interval of the upper transistor, the inductor current will ramp from a starting value of −1.09A to an ending value of +1.09A. Likewise, when the lower transistor is ON, the inductor current will ramp from a starting value of +1.09A to an ending value of −1.09A.

The circuit node where the *source* terminal of the upper transistor (FET) connects with the *drain* terminal of the lower transistor (FET), as well as the first terminal of the inductor, is called the "switching node" in Class-D parlance, and is one of the most critical nodes in the design, because it has the highest-amplitude switching voltage waveforms in the design, with the highest dv/dt as well. This node is particularly important in the chapters on layout and EMI (Chapters 15 and 16), because of these same signal characteristics (amplitude and slew rate), which makes it a particularly potent radiator of electromagnetic energy and EMI. As depicted in Figure 8.5, the voltage at the switching node at idle is a high-amplitude square wave, switching from positive rail voltage to negative rail voltage.

This waveform is applied to the L-C filter (also called a demodulation filter), which behaves as a low-pass filter. The resultant waveform at the output of the L-C filter is a relatively small quasi-sinusoidal waveform, at the fixed switching frequency. The ripple voltage amplitude at the output is termed residual output voltage ripple. In Class-D audio amplifier design, it is critical to understand the interrelationships between various design parameters. For the half-bridge power stage just discussed, the following relationships should be appreciated (from EQ. 8.1):

- *decreasing* the fixed switching frequency, or
- *decreasing* the inductance value of L, or
- *increasing* the rail voltages

will *increase* the current through the inductor, and therefore the residual output voltage ripple.

Now, we will zoom in to take a closer look at a switching transition. This is analyzed in Figure 8.6.

This figure illustrates what is happening when the upper mosfet switches off and the lower mosfet switches on. As stated above, when the upper mosfet is ON, there is +48V at the switching node, applied to terminal 1 of inductor L1A. It is critical to maintain *non-overlapped* switching of the upper and lower switches

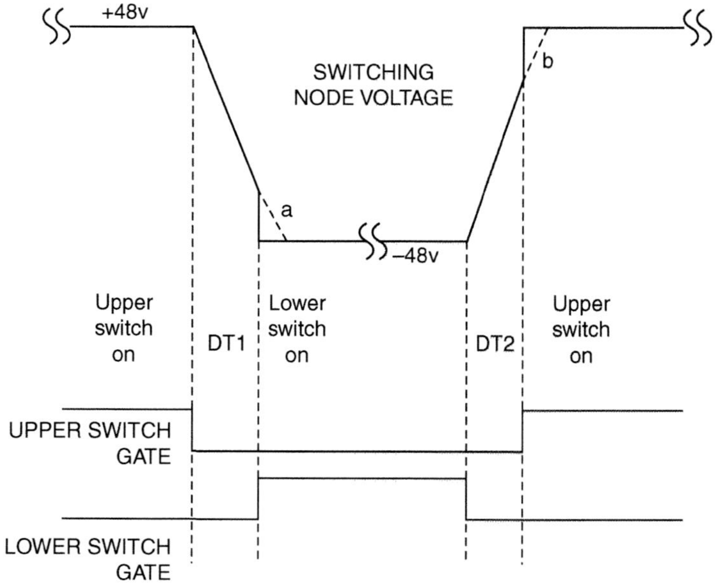

Figure 8.6 Idealized Waveform Diagram, Depicting the Switching Node Voltage Waveform and Gate Drive Waveforms for the Power Switches in a Half-Bridge Switching Power Stage

to avoid cross-conduction; therefore, an intentional *deadtime* (DT) is inserted into the gate drive waveform timing for the upper and lower switches. During the deadtime, *both switches* are OFF to prevent cross-conduction. Note what is happening at the switching node during DT1. At the instant the upper switch turns off, there is 1.09A flowing through the inductor from terminal 1 to terminal 2 (direction is very important for these detailed discussions). An inductor will always seek to maintain current at its present level (i.e. it will resist any change in current); therefore, this 1.09A of current will continue to flow, even though the drain-source channel of the upper fet is OFF. This current will be sourced from the various capacitances present at the switching node. These capacitances are, principally, the output capacitances of *both* mosfet switches. Figure 8.7 depicts the equivalent circuit during DT1.

The inductor looks like a constant current sink of 1.09A, and the total capacitance at the switching node will be C_U plus C_L (the output capacitances of the upper and lower mosfet switches, respectively). The output capacitance of a mosfet changes with V_{DS} (voltage drain-to-source), so some manufacturers will provide an equivalent (or effective) capacitance value that takes this variation into account. The datasheet for the mosfet used in the reference schematic (IRF6775) (Infineon (International Rectifier), 2014) lists an effective value of 175pF. So the total capacitance at the switching node will be 2 × 175pF = 350pF for the reference design.

Note: There is an easy way to indirectly measure the total capacitance at the switching node in a Class-D amplifier. With the amplifier enabled, and the power

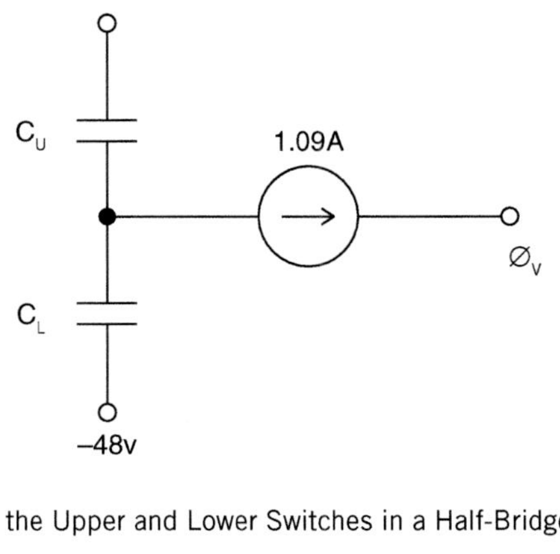

Figure 8.7 Capacitances of the Upper and Lower Switches in a Half-Bridge Switching Stage

stage switching in its "idle" state (i.e. audio output voltage = 0V), the power stage is then disabled (both transistors are switched off simultaneously). This will cause the switching node to ring. The ring frequency will be:

$$F_{RING} = 1/(2\pi SQRT(LC)) \qquad \text{EQ. 8.2}$$

Rearranging to solve for C:

$$C_{SWITCHING\text{-}NODE} = (1/2\pi F_{RING})^2/L$$

where L is the Class-D inductor value.

For the reference amplifier, assuming a total switch node capacitance of 350pF, and Class-D inductor value of 22μH, the measured value of F_{RING} will be 1.81MHz.

It is possible to calculate the voltage change at the switching node (dv) during the deadtime DT1:

$$dv = (dt \times i_L)/C \qquad \text{EQ. 8.3}$$

Substituting circuit values, and rearranging terms, it can be seen that the 1.09A peak inductor current will be able to slew the 350pF of effective capacitance at the switching node the full 96V (from +48V to −48V) in 30.7nsec. So, theoretically, if the deadtime setting (DT1) were greater than 30.8nsec, the switching node will have switched *passively*, on its own, from +48V to −48V without requiring either mosfet to actively switch it. At the end of this passive switching interval, the lower mosfet can switch ON, with its drain-to-source voltage already at 0V. This is called "zero-voltage switching" and is highly efficient, theoretically incurring zero switching power loss.

Referring again to the tradeoffs and interrelationships that are always present in Class-D design, the decision of setting the deadtime interval is critical and has many different ramifications, affecting many performance aspects of the Class-D amplifier. For example, the insertion of deadtime in Figure 8.6 has made the rising and falling edges of the switching waveform at the switching node non-ideal. Instead of having infinitely fast, vertical rising and falling edges, the square wave voltage at the switching node will have a ramping voltage with a finite dv/dt, or a piecewise linear shape, as depicted in Figure 8.6. These non-ideal aspects

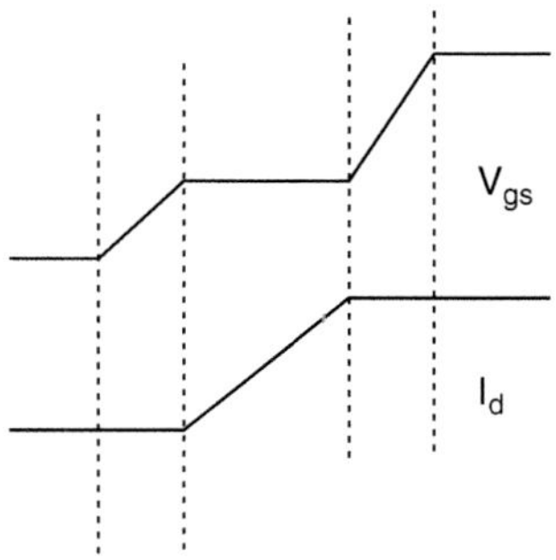

Figure 8.8 Idealized Waveform Diagram Showing Mosfet Switch I_D versus V_{GS} during Mosfet Turn-on Event

of the voltage waveshape at the switching node will be integrated by the output L-C filter and contribute to voltage error, which increases THD. Deadtime will have even more of an error-producing effect under other switching conditions, which will be discussed shortly.

- *Decreasing* deadtime will improve THD
- *Decreasing* deadtime will increase power dissipation in the mosfets.

It may seem at first that setting the deadtime should be a simple undertaking. However, it is complicated by the fact that mosfets do not switch on or off instantaneously, but instead have a drain current (I_D) waveform to V_{GS} voltage waveform relationship, as depicted in Figure 8.8.

This was discussed in greater detail in Chapter 7 (Bias Supply for Upper Gate Driver). For the purposes of this discussion, it will suffice to say that it is not always optimal to extend the deadtime so that there is *zero* cross-conduction current. This would indeed provide the lowest idle condition switching losses; however, it may result in unacceptably large values of THD and power loss under different loading conditions (to be explained shortly). Therefore, setting the optimal value of deadtime requires taking into account many different factors, principal among which are:

- THD
- IDLE power loss
- FULL LOAD power loss

- Voltage headroom (maximum effective duty cycle)
- Reverse recovery characteristics of the mosfet body diode.

To understand the last three factors above, an analysis of the same half-bridge switching circuit under a high-power condition is necessary. For this example, we will analyze the case where the amplifier is producing 100W into its minimum rated load impedance (4 ohms).

100W into a 4-ohm load:

- 20VRMS out
- 5ARMS out
- 28.28Vpeak out
- 7.07Apeak out.

Using EQ. 3.4 from Chapter 3 (Comparator and PWM Circuits):

$$Vout = V(2D - 1) \qquad \text{EQ. 8.4}$$

Duty cycle (D) = 0.7946 in order to produce +28.28V out, using +/−48V supplies.

The upper FET is on for (0.7946 × 2μsec) = 1.589μsec, and the lower FET is on for (1 − 0.7946) × 2μsec = 0.4108μsec.

Figure 8.9 depicts this condition.

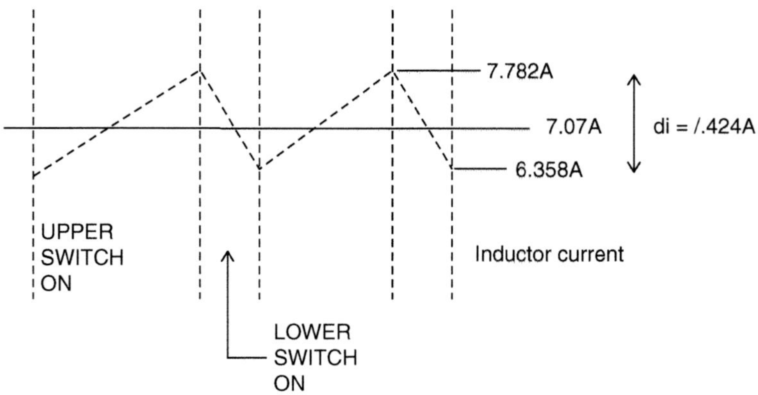

Figure 8.9 Inductor Current Waveform Detail, under the Conditions of 28.28V out across a 4-Ohm Load, Using Component Values of the Reference Design

Since the output terminal of the inductor (terminal 2) is now at +28.28V, there will be only (48V − 28.28V) = 19.72V across the inductor when the upper switch is ON. When the lower switch is on, there will be (48V + 28.28V) = 76.28V across the inductor. This difference in inductor voltage will cause very different di/dt ramp rates of inductor current, as illustrated in Figure 8.9.

Figure 8.10 zooms in at the switching transition when the upper mosfet is switching OFF and the lower mosfet is switching ON at this peak positive output current condition.

At the instant the upper mosfet switches off, there is 7.782A flowing through the inductor from terminal 1 to terminal 2. Even though the drain–source channel of the upper mosfet is off, this current will continue to flow, discharging the 350pF effective total output capacitance of the upper and lower mosfets, as depicted in Figure 8.11.

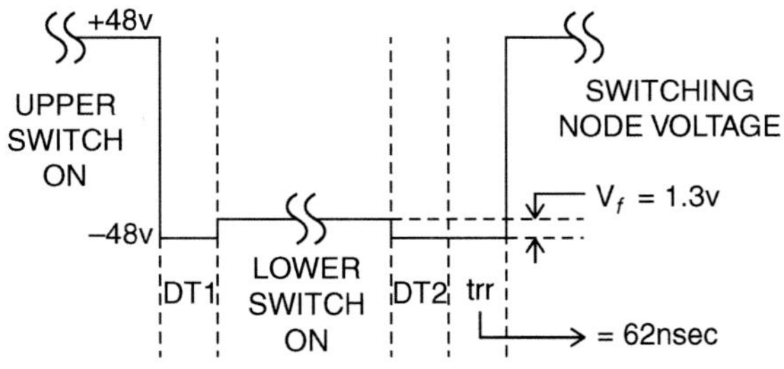

Figure 8.10 Idealized Waveform Diagram of the Switching Node Waveform, Detailing Deadtime Intervals and Reverse Recovery Time (trr) Intervals

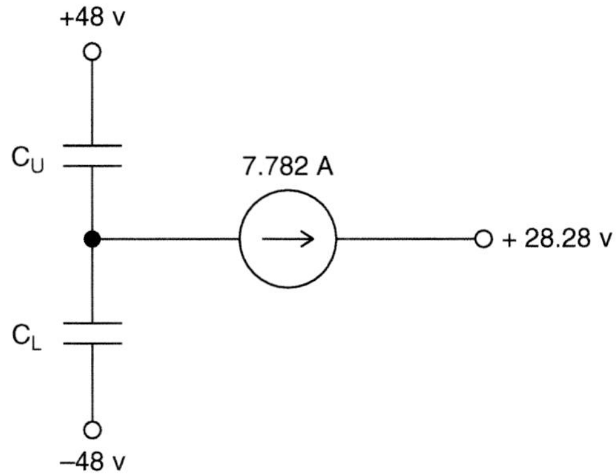

Figure 8.11 Idealized Schematic Diagram Showing Capacitances of the Upper and Lower Mosfet Switches in a Half-Bridge Power Stage under a Condition of High Current

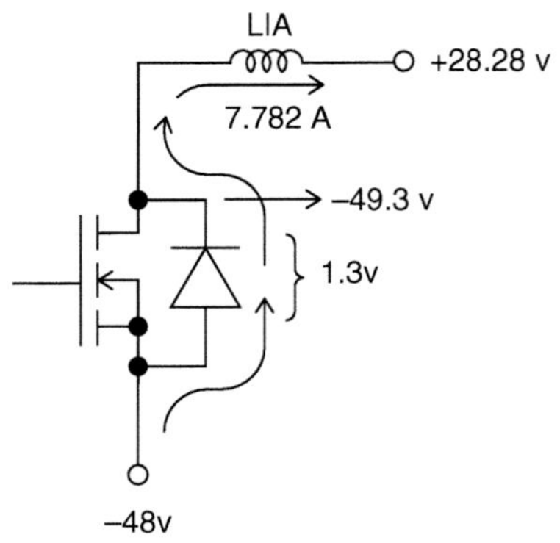

Figure 8.12 Condition of Body Diode Conduction in the Lower Mosfet of a Half-Bridge Switching Stage, When Sourcing High Current through the L-C Demodulation Filter

Using EQ. 8.3 above, with a current of 7.782A, this 96V = dv transition of the switching node voltage will take only 4.32nsec. Obviously, the falling edge of the switching node voltage will be dramatically faster than it was in the idle condition. Now, during DT1, neither mosfet is on, so the 7.782A will flow through the body diode of the lower mosfet. Figure 8.12 shows this, and illustrates why the switch node voltage will be *more negative* than the negative rail during the interval DT1.

The datasheet for the IRF6775 lists a body diode forward voltage of 1.3V @ 5.3A. The actual body diode current during this interval will be 7.782A, so the forward voltage of the lower mosfet's body diode will actually be slightly greater than 1.3V. At the end of DT1, the lower mosfet will be switched ON, and the inductor current will switch from the body diode to the drain–source channel of the lower mosfet. The voltage drop across the lower mosfet will now be determined by the ON-resistance of the drain–source channel (R_{DS}ON). For the IRF6775 this will be 47 milliohm typical, resulting in a voltage drop of 366mV at a current of 7.782A, as seen in Figure 8.13.

Note that the switch node voltage will still be *more negative* than the negative rail voltage due to the direction of current flow through the drain–source channel of the lower mosfet. During the DT2 interval, the lower mosfet is again switched off. By this time, the current has ramped down to 6.358A as shown in Figure 8.9, but the direction is such that the body diode of the lower mosfet will again need to conduct during DT2, causing the switch node voltage to again become more negative, dropping back to −49.3V. At the end of DT2,

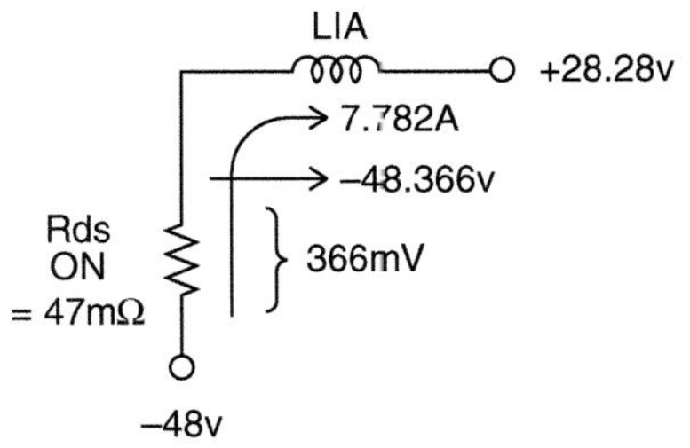

Figure 8.13 Effects of Lower Mosfet R_{DS}ON in a Half-Bridge Switching Stage, When Sourcing High Current through the L-C Demodulation Filter

the upper mosfet will be driven ON. At this time, 6.358A of current is still flowing through the body diode of the lower mosfet. This current must be swept out of the body diode before it can switch off and recover to its fully blocking (non-conductive) state. This is the reverse recovery interval of the body diode. For the IRF6775, this is specified as 62nsec under a specific set of conditions. The datasheet conditions are different than the application circuit conditions, but close enough for this analysis. The switch node will remain at the −49.3V level for the full reverse recovery interval. It will only be at the end of this interval that the body diode will switch off, allowing the switch node voltage to slew to +48V. It is at this instant in the switching cycle where the reverse recovery characteristics of the body diodes of the mosfets are critical. The body diodes of mosfets can differ dramatically in terms of these characteristics. Some can have a soft recovery, whereas others can be snappy, creating very different noise spectra. These recovery characteristics can have a critical influence on switching noise and EMI signature, and can affect switching power loss. It is important to note that during the reverse recovery interval, the upper mosfet has the full supply voltage across it, and close to maximum load current through it simultaneously. These conditions create the highest possible instantaneous power dissipation in the upper mosfet, and for this reason, this interval needs to be made as short as possible. This is best done by careful selection of the Class-D mosfets, and illustrates how critical the reverse recovery specifications of the body diodes are in that selection process.

From the discussion above, it should be evident that the switch node voltage waveshape will be quite different at full load than it was at idle. The actual duty cycle of the switch node (applied to the L-C filter) at high current will be different than the duty cycle programmed by the PWM comparator, due to the switching anomalies analyzed above. Further, the transition between the zero-voltage switching waveshape at idle and low currents, compared to the trr-influenced waveshape at high current, can be quite abrupt, happening at a particular range of output currents. This, then, illustrates a severe non-linearity in the open-loop switching characteristics of the amplifier, which will need to be corrected by negative feedback, and which translates into an important contributor to THD. (Firth, Quek, and Texas Instruments (2009) elaborates on how open-loop designs do not correct for this distortion. Further, it makes the point that digital purists seeking to eschew the use of D/A converters in the signal path ignore the fact that the power switching stage, followed by the L-C demodulation filter, in fact comprises a large signal D/A converter, operating with significant distortion.)

Please note that the illustrations given above were for switching transitions of the upper and lower mosfets under a specific set of conditions. These transitions are reciprocal, in that the same effects occur for the opposite transitions, i.e. at the negative output voltage and current signal peaks. The same principles apply.

Full-Bridge Switching Stages

Full-bridge switching stages have several advantages over half-bridge stages. These are:

1) Twice the voltage available across the speaker, and therefore four times the output power for the same rail voltages
2) Freedom to use a single-rail power supply, without having to use large AC coupling capacitors in the audio output path
3) Freedom from supply rail pumping (to be discussed)
4) Redundant overcurrent protection (either leg can switch off current to the speaker)
5) Less susceptible to PWM-start-up "pops and clicks"
6) More modulation schemes available (AD and BD modulation)
7) Greater power supply rejection ratio (PSRR) (qualified below).

These benefits are explained below:

1) Suppose an amplifier with a half-bridge topology utilizes +/−48VDC rails (as the reference amplifier does). Further, suppose that the maximum modulation

index (duty cycle) allows a maximum audio output voltage of +/−40V peak (28.28VRMS) into an 8-ohm speaker, since the − terminal of the speaker is connected to ground. This means this amplifier will be capable of providing 100W into this 8-ohm speaker load. Now suppose the same power supply rails are used, but the amplifier has a full-bridge power stage. Further, suppose the same maximum modulation index and peak output voltages apply, and AD modulation is being used (explained below). Now, when the A-leg of the full-bridge is supplying +40V peak to the + terminal of the speaker, the B-leg is providing −40V peak to the − terminal of the speaker. Therefore, 80V peak is being applied to the speaker + to −. The opposite output polarity is also produced, when the A-leg produces −40V and the B-leg produces +40V. This applies −80V peak to the speaker + to −. Therefore the 8-ohm speaker load "sees" +/−80V peak = 56.57VRMS or 400W. A critical point to understand about this full-bridge connection is that the load appears to each leg as one half of its actual impedance. To prove this, examine the operating conditions of the A-leg. When it is producing +40V (and the B-leg is producing −40V) it is sourcing 10A to the speaker. When it is producing −40V (and the B-leg is producing +40V) it is sinking 10A from the speaker.

$$Z_{LOAD} = \Delta V_{A\text{-}LEG}/\Delta I_{A\text{-}LEG} = 80V/20A = 4 \text{ ohms}$$

This is critical to keep in mind, because the current-handling ability of the power switches must be correctly specified for the operating conditions. If a 4-ohm speaker is to be driven by a full-bridge stage, it will look to this stage like a 2-ohm load.

2) The full-bridge switching stage allows a single supply rail to be used. This reduces cost and component count in the power supply. It also allows the great benefit of eliminating supply rail pumping (benefit #3 above) because power and charge are supplied by the same single power supply rail and storage capacitors during *both* the positive and negative audio signal peaks. When a single power supply rail is used, and no audio signal is being produced, *both* terminals of the speaker will be at a potential equal to one-half of the supply rail voltage. Since *both* speaker terminals are at this voltage, no current will flow through the speaker. This, however, would *not* be true for a half-bridge configuration using a single power supply rail. In that case, since the negative terminal of the speaker is connected to ground, an AC coupling capacitor will need to be used between the amplifier output (biased at an idle, no-signal voltage of one-half of the supply voltage) and the positive terminal of the

speaker. This capacitor will need to have a large value in order to allow for low-frequency response, and must handle the full audio output current. This can be a polarized aluminum electrolytic capacitor (more readily available than non-polarized types) because there is always a fixed polarity bias voltage across this capacitor of one half of the supply voltage.

If a −3dB maximum frequency response rolloff at 20Hz is required, then the value of the AC coupling capacitor can be calculated as follows:

$$C = 1/2\pi R(20Hz) \qquad \text{EQ. 8.5}$$

where R is the speaker impedance.

For an 8-ohm speaker this dictates a coupling capacitor = 995μF (1000μF is a standard value). For a 4-ohm speaker, 2000μF would be needed (2200μF is a standard value). The minimum voltage rating for this capacitor would seemingly need to be one-half of the supply voltage. However, for a conservative design, it would be preferable to specify a capacitor voltage rating equal to or exceeding the supply voltage range, in case the amplifier is driven to the power supply rail for an extended period of time (in the case of a DC fault, for example).

3) Supply rail pumping eliminated.
4) Since there are two half-bridge legs comprising the full-bridge stage, with one leg sourcing current and one leg sinking current at all times, *either* leg can interrupt current flow to the speaker. This allows for redundancy in overcurrent protection. If *either* leg shuts down due to a detected overcurrent event, it stops the current flow in both legs.
5) When Class-D power stages are first enabled, they can produce audible "pop and click" artifacts in attached speakers. Many schemes have been implemented specifically to reduce these PWM start-up artifacts. Some have patent protection. One issue that makes PWM start-up more complicated than it might at first seem is the following. In order for the upper gate driver to switch on the upper power switch in each leg, the supply capacitor (boot capacitor) for the upper gate driver needs to be charged. Further, the simplest and lowest component count method of charging the upper gate driver boot capacitor is the bootstrap method (discussed in Chapter 7), where this capacitor is charged when the lower mosfet turns on and pulls the switching node to ground (for a single supply implementation) or to the negative rail (for a bipolar supply

implementation). This necessarily means that the lower mosfet must be switched on first, and for a defined length of time, long enough to charge the boot capacitor. In addition to this, there is a further complicating detail in that the upper gate driver may not instantly be operational as soon as its supply voltage exceeds its UVLO threshold. Certain drivers have a turn-on time specification. This is the time delay from when the rail voltage rises into operational range, to when the driver can deliver a valid output state. In the reference amplifier design, for example, this specification is called "device start-up time", and is listed as 7μsec maximum. Therefore, the upper gate driver cannot be relied upon to produce a valid output state until this time has elapsed. *Note:* This same time delay applies to the lower gate driver as well, but that does not present a problem, because it is biased from the −HV+12V rail, which is up and regulated long before the gate driver is enabled. A solution to this problem is to drive the bottom mosfet ON for a defined time period long enough to charge the boot capacitor. Then wait the minimum 7μsec, then start normal PWM drive under control of the feedback loop. Note that during the 7μsec start-up delay time, the upper gate driver is inactive, but has quiescent current that will need to be supplied by the boot capacitor. This quiescent current is 3mA maximum. So during the 7μsec start-up delay time, the voltage on the boot capacitor ($V_{C\text{-}BOOT}$) will decrease by:

$$dv = (I_{QUIESCENT})(7\mu sec)/C_{BOOT} = (3mA)(7\mu sec)/0.1\mu F = 210mV$$

This extra voltage drop in C_{BOOT} must be taken into account when designing the bootstrap circuit, in order to ensure that the supply voltage for the upper gate driver (once it has risen above the rising UVLO threshold) remains above the falling UVLO threshold throughout the start-up delay time interval, as well as the first on-time interval of the upper mosfet. C_{BOOT} will be recharged again during the next lower mosfet on-time.

The above discussion is meant to familiarize the designer with the details involved in PWM start-up. It can now be appreciated that these start-up pulses and delay intervals can produce audible "click and pop" artifacts in attached speakers. An advantage that the full-bridge output stage has over the half-bridge is that both legs of the full-bridge can be controlled to produce these start-up pulses and delay times simultaneously. This forces the output artifacts of each leg to be common-mode, i.e. they appear on both terminals of the speaker simultaneously, and therefore no differential signal voltage or current is present to drive the speaker, rendering the artifacts inaudible.

6) The full-bridge output stage topology allows more modulation schemes than the half-bridge topology. It can be appreciated that the half-bridge power stage has only *two operational power states*. The switching node can either be pulled to the positive voltage rail (when the upper mosfet is on), or be pulled to the negative voltage rail (when the lower mosfet is on). It will be high impedance when *both* mosfets are off, but that is not an "operational" state. And the state where both mosfets are on simultaneously is destructive and never allowed. When a second half-bridge leg is introduced, in the full-bridge topology, there now are four operational states, determined by the on/off states of the four mosfets in the two legs (labeled "A" and "B" in Figure 8.4). This gives rise to two different modulation schemes termed "AD" modulation and "BD" modulation (Texas Instruments, 2016). The states of the switching node in each leg are listed in Table 8.1. For convenience, the logic "1" state is substituted for +HV.

The second and third bold rows in Table 8.1 are the two states that are available with AD modulation. All four states listed in Table 8.1 (all four rows) are available in BD modulation. The block diagrams in Figures 8.14 and 8.16 respectively show how AD and BD modulation can be realized in a PWM modulator. Figures 8.15 and 8.17 show the corresponding timing diagrams, as well as Common-Mode (CM) and Differential-Mode (DM) voltages produced by AD and BD modulation, respectively. AD modulation can be implemented by simply driving the B-Leg mosfets 180 degrees out-of-phase from the A-Leg mosfets. This is implemented with the inverter in the B-Leg.

BD modulation as shown in the block diagram (Figure 8.16) is accomplished by adding a second comparator, and inverting the audio signal to that comparator. The negative input of each comparator is connected to the PWM triangular ramp signal.

Table 8.1 The Four Possible Voltage States of the A-Leg and B-Leg Switching Nodes, and the Differential and Common-Mode Speaker Voltages They Produce in a Full-Bridge Switching Stage

A-Leg Switching Node	*B-Leg Switching Node*	*SPKR Voltage Differential Mode*	*SPKR Voltage Common Mode*
0	0	0	0
0	**1**	**−1**	**+0.5**
1	**0**	**+1**	**+0.5**
1	1	0	1

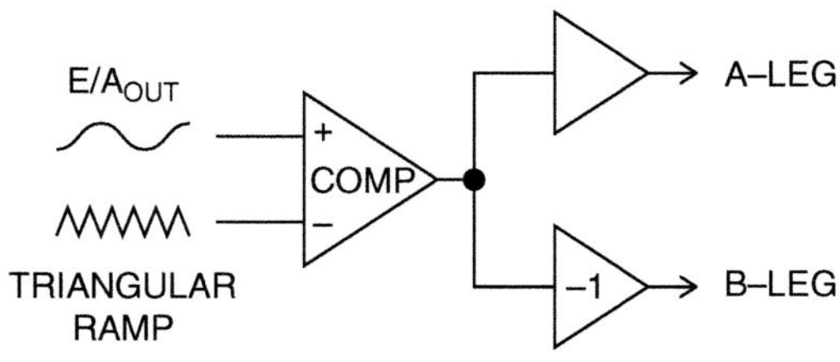

Figure 8.14 AD Modulation Block Diagram

These differences in the AD and BD modulation techniques illustrated above are responsible for important operational and functional differences.

AD modulation always has the maximum differential voltage applied across the L-C filters and SPKR load (connected between the switching nodes of the two legs). This results in maximum peak-to-peak switching frequency ripple voltage and current. BD modulation, in contrast, has *half* the DM voltage applied across the L-C filters and SPKR load. This difference has greater ramifications in filterless designs, where the switch node voltages appear on connecting cables to the SPKR. As will be discussed in Chapter 15 on EMI, CM EMI from cables causes more problems than DM, and AD modulation causes no significant CM switching voltages, whereas BD modulation results in significant CM voltage on the load cables. An important advantage to BD modulation, especially in filterless designs, is that the differential-mode voltage across the SPKR load is halved. This is important, because in "filterless" designs, although there is no discrete L-C filter, there *is* the inductance of the speaker that acts as the ripple current filter, so halving the voltage applied to this speaker inductance halves the ripple current through it, which has power savings benefits.

Another very important benefit of BD modulation is that the addition of the second PWM comparator means that there are now twice as many samples taken within a switching period. This means the sampling frequency has doubled. Thinking back to our rule of thumb concerning maximum loop closure frequency versus switching frequency, this means that the negative feedback loop can be closed at twice the frequency of an equivalent AD modulation scheme switching at the same switching frequency. This frequency-doubling effect can be seen in the *timing diagrams* above for the BD modulation scheme. Increasing the loop bandwidth means greater available loop gain and error correction at 20KHz, with commensurate improvements in THD and noise at higher audio frequencies.

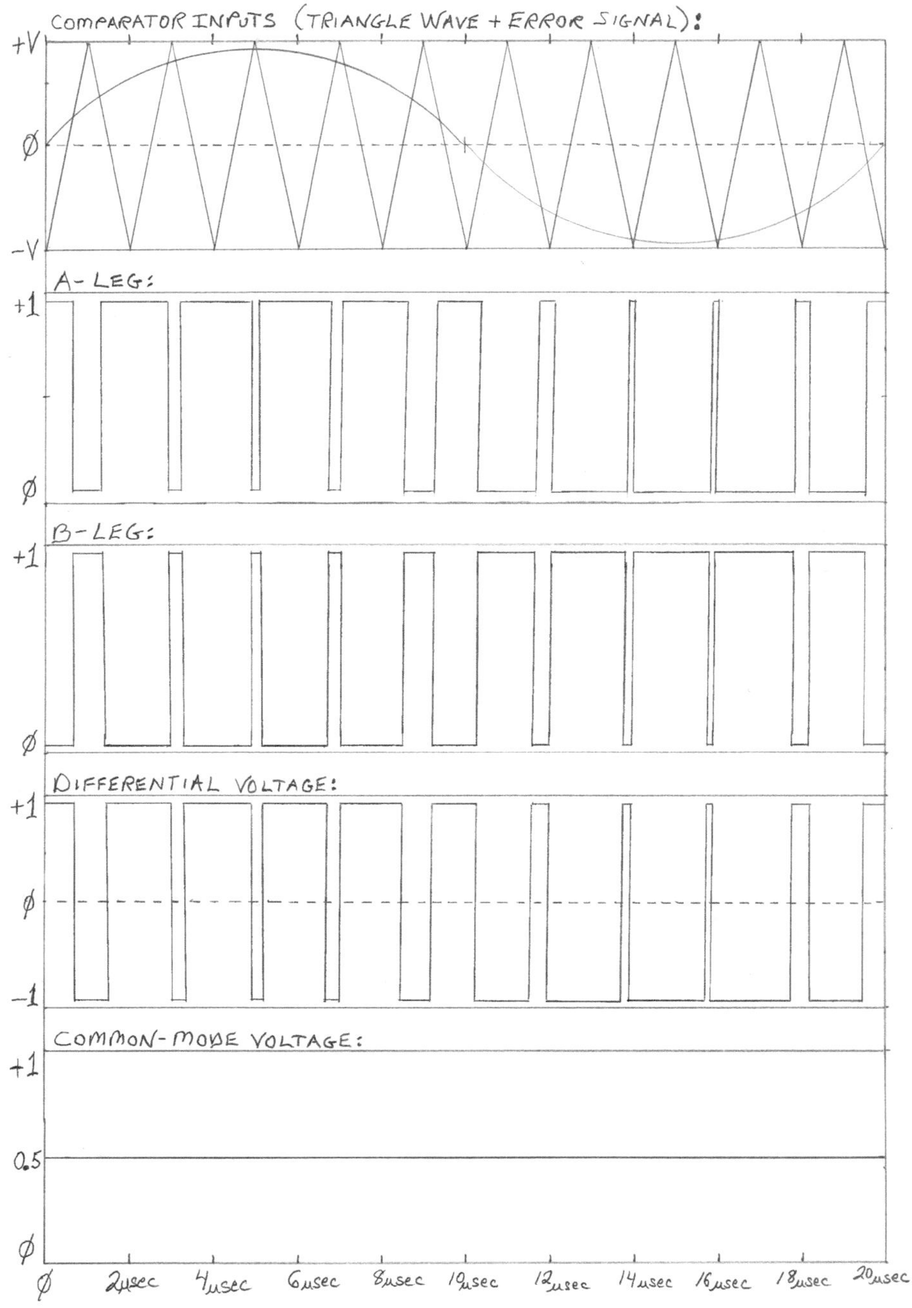

Figure 8.15 AD Modulation Timing Diagram

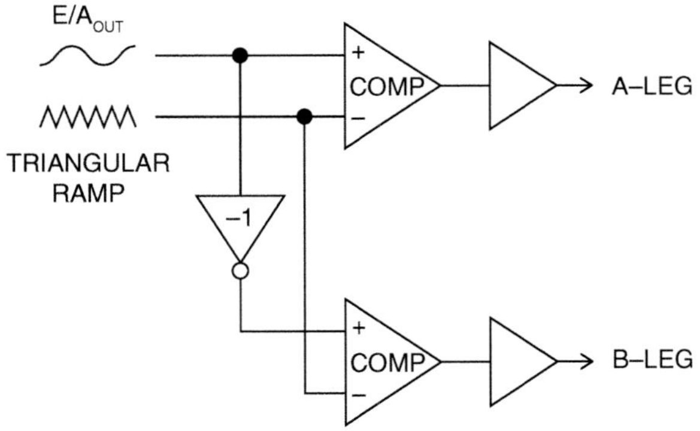

Figure 8.16 BD Modulation Block Diagram

7) Concerning PSRR benefits for full-bridge power stage topologies, compared to half-bridge topologies, the conventional belief is that full-bridge power stages have inherently high PSRR because power supply voltage deviations appear as CM signals at the load, and do not contribute to the THD and noise of the differential-mode signal (the audio output signal) (Ge, Chang, & Shu, 2008; Ferstenberg, 2013). Furthermore, this belief is used as the justification for open-loop Class-D designs using full-bridge topologies. This belief is correct; however, a more careful examination of the Class-D amplifier reminds us that the *open-loop* gain of the power stage is directly proportional to the supply voltage. Therefore, any modulation of the supply rail voltage appears as a modulation of the open-loop gain. Ge et al. (2008) make the point that typical PSRR measurements on open-loop Class-D amps are done by introducing perturbations into the power supply rail, under no-audio signal conditions. Under these conditions, the gain modulation would not be detected, because there is no audio signal present on which the gain modulation could be detected. These tests only look for perturbations on the power supply rail that propagate to the output. However, the gain modulation effect *could* be detected at the output, if an audio signal were injected while the power supply rail perturbation was being applied. What would be seen in that case is an *intermodulation distortion* effect (IMD), whereby the gain change caused by the power supply perturbation causes distortion in the audio signal. The remedy for this effect, of course, is the addition of negative feedback correction and closed loop control. Ge et al. (2016) also mentions the importance of component value matching in feedback loop components, as a means of achieving high PSRR.

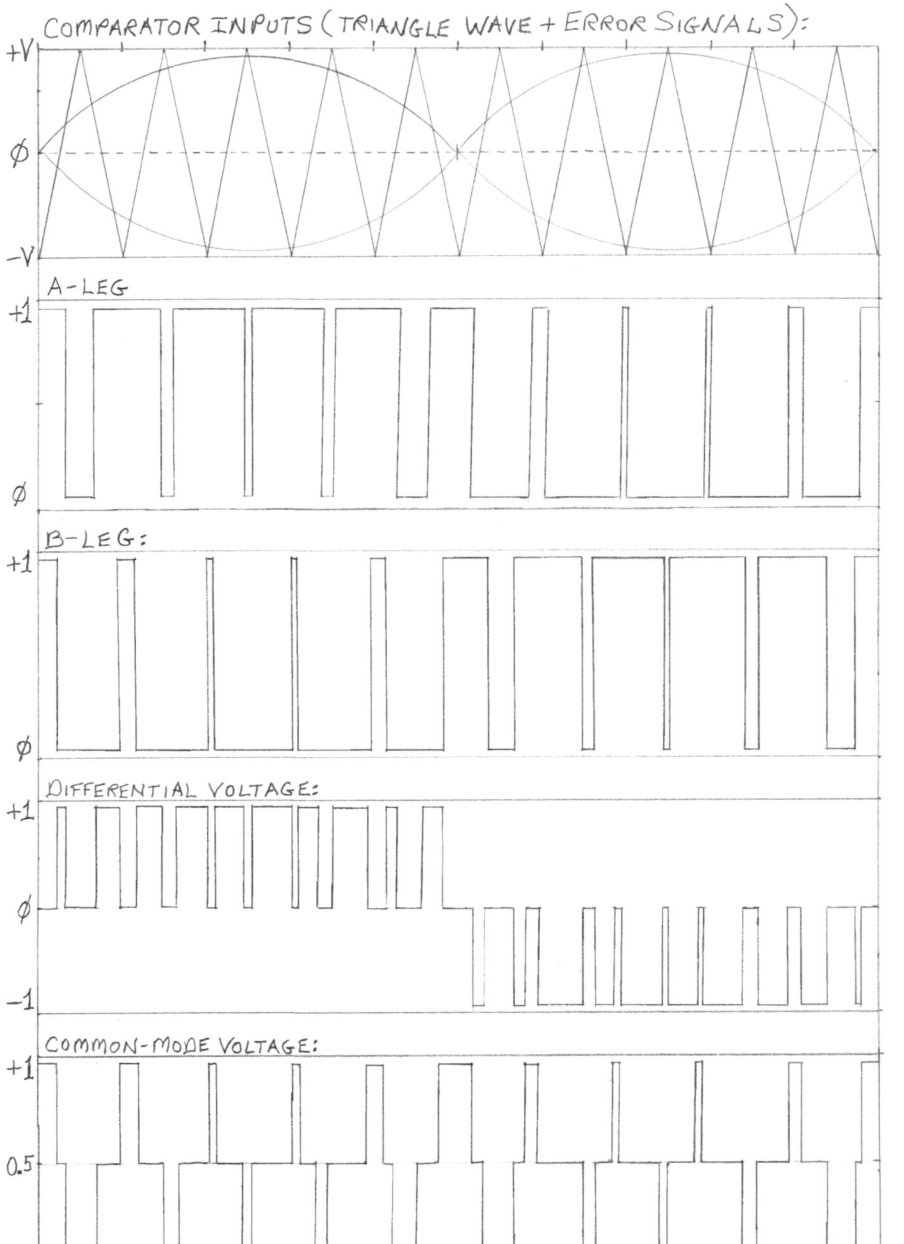

Figure 8.17 BD Modulation Timing Diagram

So far, the benefits and advantages of full-bridge topologies have been discussed. The disadvantages are component count (in the amplifier circuitry), cost (of added amplifier components), and (potentially) efficiency. The full-bridge topology requires twice as many power switches (though at one-half the voltage rating) and twice as many filter inductors. If the BD modulation scheme is realized, it also requires twice as many high-speed PWM comparators. Factors affecting efficiency comparisons between half- and full-bridge topologies are switching loss (twice as many switching transistors, but switching at half the voltage), and core and conduction loss in twice as many inductors. Inductor current is the same as half-bridge (for the same output power), but core loss for each inductor of the full-bridge can be less because the voltage swing across each inductor is halved. So, a careful assessment of all these loss components would need to be performed, if an accurate comparison of losses between half- and full-bridge topologies were to be made for a given application.

Bibliography

Clifton, R. (2018, September). Pre-filter parallel bridge-tied load (PBTL). Texas Instruments Application Report SLAA822. Dallas, TX: Texas Instruments.

Ferstenberg, I. (2013, July 17). Bridging Class-D amplifiers. Application Note AN-1198. El Segundo, CA: International Rectifier (now Infineon).

Firth, M., Quek, Y. B., & Texas Instruments. (2009). The real story about closed-, open-loop Class D amps. *EE Times-Asia*, eetasia.com.

Ge, T., Chang, J. S., & Shu, W. (2008). PSRR of bridge-tied load PWM Class D amps. In: *2008 IEEE International Symposium on Circuits and Systems (ISCAS)* (pp. 284–287). New York: IEEE Press.

Ge, T., He, H., Zhou, J., Kang, Y., & Chang, J. S. (2016). An investigation of THD of a BTL Class D amplifier. In: *2016 IEEE International Symposium on Circuits and Systems (ISCAS)* (pp. 470–473). New York: IEEE Press.

Infineon (International Rectifier). (2014, February 26). Datasheet, IRF6775MTRPbF Digital audio mosfet. El Segundo, CA: International Rectifier.

Skyworks Solutions Inc. (2021, July 31). Datasheet, Si824x Class D audio driver with precision dead-time generator. Rev. 1.0. Irvine, CA: Skyworks Solutions Inc.

Texas Instruments. (2016). LC filter design. Texas Instruments Application Report SLAA701. Originally published October 2016, revised November 2016. Dallas, TX: Texas Instruments.

Chapter Nine

L-C Demodulation Filters

L-C Filter Characteristics and Equations

The Class-D, L-C filter is positioned in-series with the speaker output, and therefore represents a low-pass filter, decreasing amplifier control over the speaker with increasing frequency (Texas Instruments, 2016). The main purpose of this filter, however, is to attenuate the high-frequency PWM switching signal produced at the Class-D power switching stage, leaving the audio frequency (20Hz–20KHz) signal voltage unattenuated at the filter's output. How well the filter accomplishes this task depends on a number of factors, which will be discussed in this chapter. As discussed in previous chapters, the audio signal amplitude information is converted to a modulated pulse-width (PWM) signal by the modulator section of the Class-D amplifier, and amplified by the power switching stage. The output of the power switching stage is a rail-to-rail, PWM voltage waveform. The "demodulator" function of the L-C filter is so named because the low-pass function of the L-C filter removes (attenuates) the high-frequency components of the switching voltage waveform and integrates (or averages) this time-domain voltage waveform. This integration or averaging of the PWM waveform effectively demodulates it, leaving the desired, amplified audio signal at the filter's output, with some (undesired) residual modulation voltage ripple.

In the reference design, the L-C filter consists of L1A and C68A (= C68-1A + C68-2A + C68-3A). When the attached speaker load is included, this filter is classified as a second-order LRC low-pass filter, where the "R" term is provided by the speaker load. *Note:* For the sake of simplicity and clarity in this initial discussion

DOI: 10.4324/9781003327943-9

of L-C demodulation filters, the *notch filter* added in the reference design (133A and L6A) is *excluded* from this analysis. The justification for this is that the high Q of this notch filter makes its phase shift and attenuation negligible below about 400KHz.

The chosen corner frequency for the filter (f_0) is a compromise between minimizing interaction with the speaker (over the audio band) and reducing high-frequency switching ripple (at the (much higher) switching frequency). The former goal advocates for a higher F_0, the latter for a lower F_0. This is yet another compromise to be negotiated in Class-D audio amplifier design, where conflicting design goals must be balanced in order to achieve the optimal results.

Typical values for these components in present-day Class-D amplifiers are:

L = 10μH to 47μH

C = 0.33μF to 1μF

In the reference design, the values chosen are:

L = 22μH (L1A on reference schematic)

C = 0.47μF (C68A = C68-1A + C68-2A + C68-3A on reference schematic)

These values result in an $F_0 = 1/(2\pi\sqrt{LC}) = 49.49$KHz. This is also termed the "cutoff frequency" of the filter.

$$F_0 = 1/(2\pi\sqrt{LC}) \qquad \text{EQ. 9.1}$$

Other helpful equations for designing L-C filters are EQ. 9.2 through EQ. 9.4.

Note: "R" in these equations is the load resistor at the output of the L-C filter. In a Class-D audio amplifier, this is the load impedance of the speaker, usually represented as a resistance, for simplicity.

Q = Quality Factor (see EQ. 9.2). EQ. 9.3 and EQ. 9.4 calculate L and C for the specific case of Q = 0.707. For that special case, the filter is critically damped and has a Butterworth response (Texas Instruments, 2016). It is important to also note that those equations require that you enter the desired f_0 as well.

$$Q = R \times \text{SQRT}(C/L) \qquad \text{EQ. 9.2}$$

For the special case of Q = 0.707 (critically damped), EQ. 9.3 and EQ. 9.4 apply.

$$L = (R \times SQRT(2)) \, / \, (2\pi F_0) \qquad \text{EQ. 9.3}$$

$$C = 1 \, / \, ((2\pi F_0)\, R \times SQRT(2)) \qquad \text{EQ. 9.4}$$

It is quite informative to run an AC analysis (Bode plot) of the L-C filter alone in LTspice. Since the characteristics of the L-C filter are so crucial to the design and performance of the Class-D amplifier, it is important to get a feel for what factors affect its response. In particular, it will be seen from the simulations that the response of the filter will vary widely with loading. In particular, the unloaded case (which includes no parasitic parameters for the L and C components) will show significant peaking at the resonant frequency (F_0). This cannot be ignored, because operating the amplifier with no speaker loads connected is a very common practice, and therefore the amplifier must be stable in this "no-load" condition. *Note:* One exception for this case might be powered speakers, where the speaker load is permanently connected to an internal Class-D amplifier. However, even in that case, the amplifier should be designed to be stable in the event that the speaker becomes disconnected, due to a broken or disconnected wire connection to the speaker voice coil or crossover network. Another very interesting observation is that including parasitic parameters in the simulation will again alter the response. These characteristic parasitic parameters are:

- For the inductor:
  - DCR (DC resistance)
  - Shunt capacitance
- For the capacitor:
  - ESR (equivalent series resistance)
  - ESL (equivalent series inductance).

The LTspice sweeps below illustrate this.

These plots not only show the expected filter response in the audio frequency range, but also show the response over the loop bandwidth (BW). This information (gain and phase of the filter response over the loop BW) is critical in designing the feedback loop and compensation, to achieve feedback loop stability and optimal audio performance in a closed-loop Class-D amplifier design. This is discussed further in Chapter 12 (Feedback and Compensation).

Figure 9.1 shows the response of an ideal 22μH/0.47μF L-C filter (with no parasitics included), when loaded with 2-ohm, 4-ohm, 8-ohm, and 10-Kohm loads. *Note:* It is always a good idea to have at least *some* resistive loading on the Class-D output. This will provide a path for leakage currents. When *both* upper and lower mosfets are disabled, the Class-D output will be high impedance, and the voltage at the output will be determined by the leakage currents of the "OFF" mosfets and any attached loads. A 10-Kohm load will keep the output voltage below +/−1V, with leakage currents as high as +/−100uA. When adding such a load to a Class-D output, the power dissipation of this load must be considered, in order to maintain high efficiency. For the reference amplifier, this 10-Kohm load will dissipate a maximum of $28.28V^2/10K = 80mW$, when the amplifier is producing its full rated output voltage.

Figure 9.2 shows the LTspice circuit used to generate the Bode plot in Figure 9.1.

There are several important observations to be made from the Bode plot in Figure 9.1:

1) There is greater than 30dB of peaking at the cutoff frequency (49.49KHz) in the lightly loaded (10 Kohm) case (uppermost curves).
2) The *phase* response of the lightly loaded case transitions to −180 degrees of phase shift extremely quickly, whereas the phase transitions of the 2-ohm, 4-ohm, and 8-ohm loaded cases show much more gradual phase transition.
3) The frequency response in the audible range (20Hz–20KHz) is relatively *flat* for loading of 4 ohm, 8 ohm, and 10 Kohm, with about 4.5dB of rolloff at 20KHz for the 2-ohm loaded case.
4) In the frequency range of the targeted unity gain crossover frequency for the feedback loop (approximately 100KHz), the gains for the 4-ohm, 8-ohm, and 10-Kohm loaded cases only differ by about 4dB, making stable compensation and loop closure easier.
5) The attenuation at the switching frequency (which will determine the amplitude of the residual modulation voltage ripple) is approximately −40dB for all loading cases. This will be explained in more detail.

Some references recommend designing the L-C filter for a Q of 0.707. This provides critical damping with a Butterworth response. This is possible when operating with a fixed load. But most general-purpose Class-D audio amplifiers

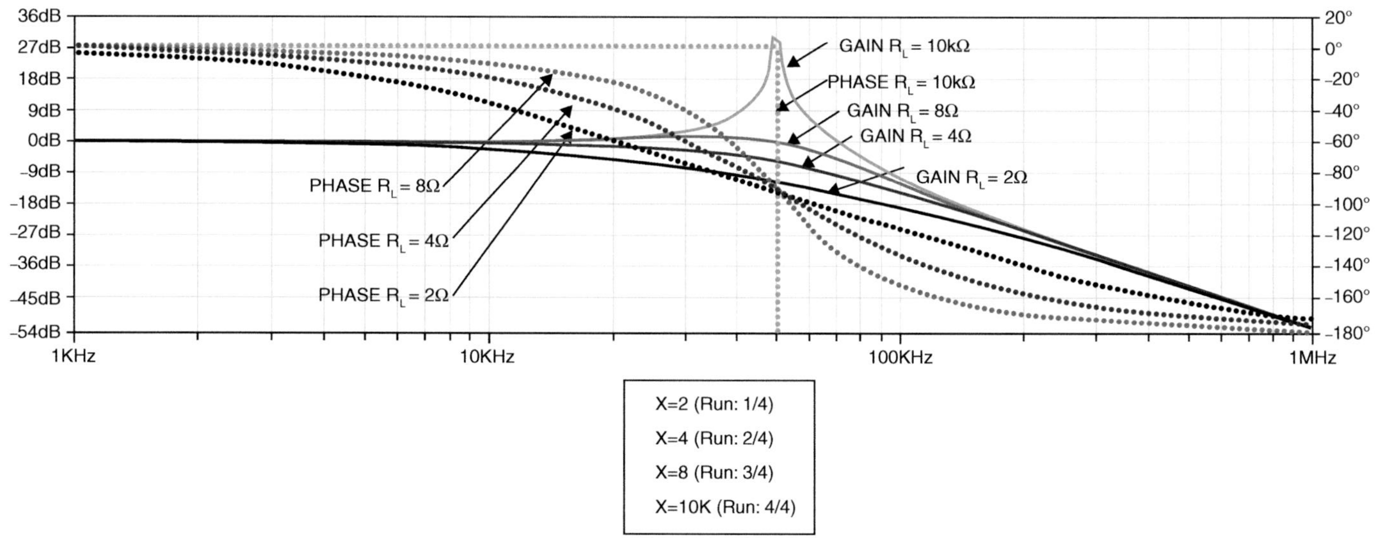

Figure 9.1 Bode Plot in LTspice for an L-C Filter with Loads of 2 Ohms, 4 Ohms, 8 Ohms, and 10 Kohms
Solid traces are *gain* curves; dashed traces are *phase* curves.
Traces, from top to bottom: 10 Kohms; 8 ohms; 4 ohms; 2 ohms.

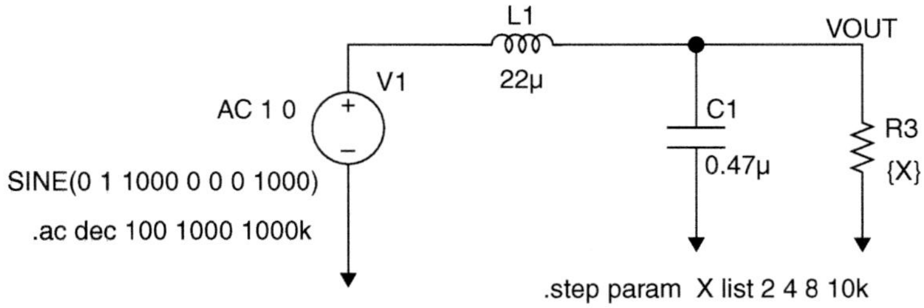

Figure 9.2 LTspice Schematic Used to Generate the Bode Plot in Figure 9.1

Table 9.1 Q versus Load Resistance for the L-C Demodulation Filter of Figure 9.2

Q	*R*
0.292	2 ohm
0.585	4 ohm
0.707	4.837 ohm
1.169	8 ohm
1462	10 Kohm

are designed to drive low-impedance speaker loads that range (typically) between 4 and 8 ohms impedance. In this case, the Q will vary 2:1 (using EQ. 9.2). Table 9.1 shows the range of Q for the 22µH/0.47µF L-C filter, with various loads.

Residual Output Ripple Voltage at the Class-D Switching Frequency

The Bode plot in Figure 9.1 also shows the expected attenuation at the switching frequency. This is critical for determining the amplitude of the switching frequency voltage ripple which will be present at the amplifier output. This parameter is sometimes referred to as "residual modulation voltage". When Class-D amplifiers were first introduced, it was shocking to some Class-AB amplifier designers to see relatively high-amplitude residual modulation voltage at the output of an audio amplifier. These designers were accustomed to analog (non-switching) amplifiers, with perhaps only millivolts of noise and ripple at the output of the amplifier. In contrast, most Class-D amplifiers produce hundreds of millivolts to volts of ripple at the output, at the switching frequency. This ripple is usually of no consequence to attached speakers, because the inductance of the speaker components (woofer/midrange/tweeter drivers and crossover networks) presents a relatively high impedance at the switching frequency, and therefore little current flows at this high frequency, and little power is dissipated in the voice coils of the speakers as a consequence of this residual modulation.

The Bode plot in Figure 9.1 shows about 40dB of attenuation at the 500KHz switching frequency of the reference amplifier. A rough approximation therefore will predict that the +/−48V square-wave switching waveform present at the input side (left side) of the L-C filter (switching node) will produce +/−480mV peak residual modulation (40dB of attenuation = /100) at the output load (right side). This is only an approximation, because this assumes a sinusoidal input voltage of +/−48V peak. Whereas, the actual input voltage is a square wave, and thus has higher-frequency harmonics added to the 500KHz switching fundamental frequency.

Note: The special case of *filterless Class-D amplifiers* will be discussed later in this chapter.

Benefits and Disadvantages of Zobel Networks

In the reference amplifier, a *Zobel* network (also called a Bocherot cell) is added at the output of the Class-D amplifier. This provides several benefits, and some disadvantages, that must be considered and weighed.

Benefits:

1) Provides resistive loading of the amplifier at high frequency, where an attached speaker load may become inductive or capacitive (reactive).
2) "Damps" the unloaded response of the L-C filter, making compensation of the closed-loop amplifier (with feedback taken *after* the L-C filter) manageable.

Disadvantages:

1) Dissipates power, with increasing dissipation as frequency increases. Power dissipation at the maximum audio frequency must be calculated and considered.

Zobel Resistor Power Dissipation

The power dissipation of the Zobel network can be easily simulated in LTspice. It can also be calculated as follows.

First, calculate the total impedance of the series R-C Zobel network at 20KHz, according to EQ. 9.5.

$$Z_{TOTAL} = SQRT\ (R^2 + Z_c^2) \qquad \text{EQ. 9.5}$$

where Z_c is the impedance of the Zobel capacitor = $1/2\pi FC$.

Note: "F" = Frequency, and 20KHz is normally used to calculate the worst-case power dissipation of the Zobel resistor (R) in a Class-D audio amplifier application.

Next, calculate the maximum RMS current that will flow through the Zobel network. This is calculated by dividing the maximum RMS output voltage of the amplifier, by the Zobel impedance (at 20KHz) = Z_{TOTAL} calculated in EQ. 9.5.

For the reference amplifier, producing 100W into an 8-ohm load, the maximum output voltage will be 28.28VRMS.

The resulting Zobel current is then squared and multiplied by the Zobel resistance to determine the power dissipation in the Zobel resistor.

Tables 9.2 and 9.3 list power dissipation values for the Zobel resistor versus Zobel capacitance values.

Table 9.2 Power Dissipation of Zobel Resistor (@ 20KHz), Using R_{ZOBEL} = 4 Ohms

C_{ZOBEL}	$f_o = 1/2\pi R_{ZOBEL} C_{ZOBEL}$	*Pd in R_{ZOBEL} (@ 20KHz)*	*Zobel current (@ 20KHz)*
0.1μF	397.88KHz	0.503W	354.61mA
0.22μF	180.86KHz	2.41W	776.21mA
0.33μF	120.57KHz	5.35W	1.16A
0.47μF	84.66KHz	10.55W	1.62A
0.68μF	58.51KHz	20.87W	2.28A

Table 9.3 Power Dissipation of Zobel Resistor (@ 20KHz), Using R_{ZOBEL} = 8 Ohms

C_{ZOBEL}	$f_o = 1/2\pi R_{ZOBEL} C_{ZOBEL}$	*Pd in R_{ZOBEL}(@ 20KHz)*	*Zobel current (@ 20KHz)*
0.1μF	198.94KHz	0.999W	353.38mA
0.22μF	90.43KHz	4.65W	1.08A
0.33μF	60.29KHz	9.89W	1.11A
0.47μF	42.33KHz	18.20W	1.51A
0.68μF	29.26KHz	31.76W	1.99A

Figures 9.3 and 9.4 characterize the 22µH / 0.47µF L-C filter of the reference design, with the addition of a Zobel network. The capacitor (C2) of the Zobel network is *stepped* in *value* in the LTspice simulation. The resistor (R1) of the Zobel network is *fixed* at 4 ohms. The legend shows that from top to bottom, the sweeps are for: C2 = 0.1µF, 0.22µF, 0.33µF, 0.47µF, 0.68µF.

Note that the capacitance value is presented alternately in units of nanofarads (nF) and microfarads (µF).

Figures 9.5 and 9.6 characterize the 22µH / 0.47µF L-C filter, with the addition of a Zobel network. The capacitor (C2) of the Zobel network is *stepped* in *value* in the LTspice simulation. The resistor (R1) of the Zobel network is *fixed* at 8 ohms. The legend shows that from top to bottom, the sweeps are for: C2 = 0.1µF, 0.22µF, 0.33µF, 0.47µF, 0.68µF.

Non-Linearities

High linearity (in inductance and capacitance) is required over the full audio frequency range and over the full power range for which the amplifier is designed. Magnetic core hysteresis and saturation effects in the inductor cause non-linearities (with current) in ferrite and iron core inductors used in Class-D, L-C filters. Capacitance changing with applied voltage is a characteristic of some capacitor dielectrics, and causes a non-linear (with voltage) response in the L-C filter capacitor.

Inductor Types

Typical inductor types used in Class-D amplifiers are toroidal (with cores made of powdered iron or gapped ferrite, or distributed-gap core materials); and PQ, RM gapped ferrite cores.

The Class-D inductor needs to handle the full output current of the Class-D amplifier, as well as fault currents, without significant saturation. Saturation refers to the decrease in inductance that occurs due to decreasing permeability of the core material with applied current. A more detailed explanation of saturation is warranted and follows. But first, a basic understanding of magnetic component terms and theory must necessarily be introduced. When current flows through turns of a coil of wire, wound around a magnetic core, magnetic flux is induced in the core, with an amplitude proportional to the current × number-of-wire-turns

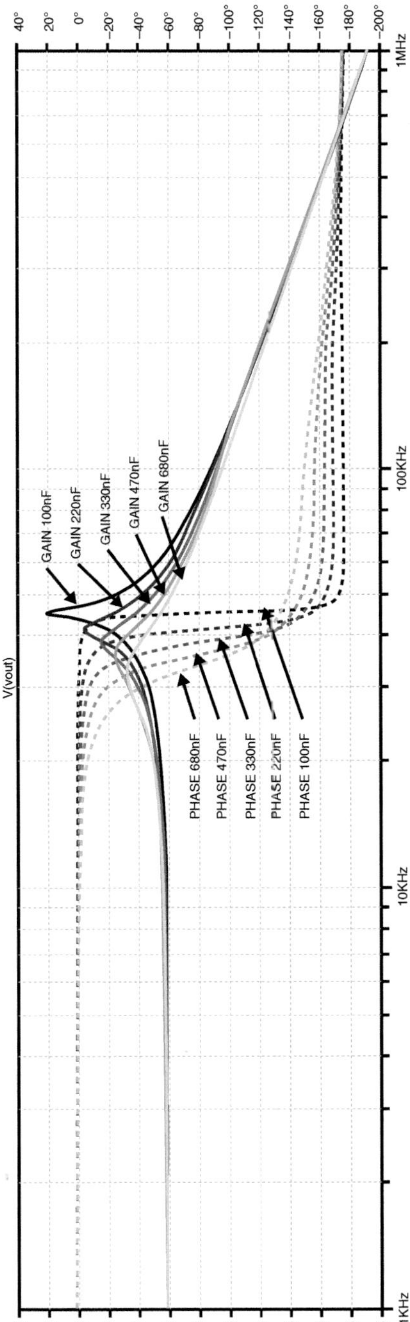

Figure 9.3 Bode Plot Run in LTspice for an Unloaded L-C Filter with Only a Zobel Network Connected at its Output (Zobel Resistor Fixed at 4 Ohms)

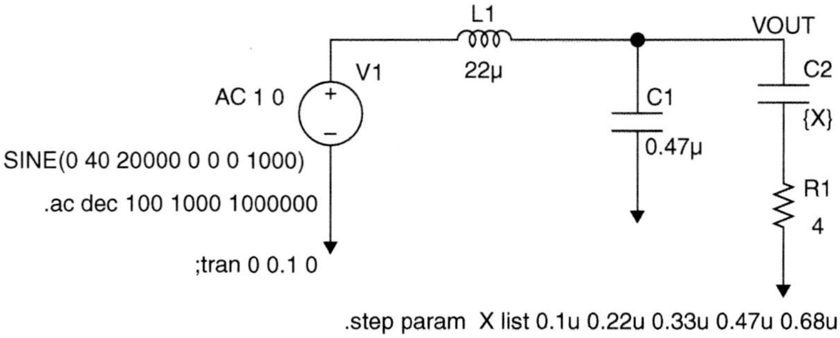

Figure 9.4 LTspice Schematic Used to Generate the Bode Plot in Figure 9.3

product (abbreviated as ampere-turns). Every core has a maximum flux capacity which is termed flux density. Permeability is a term that denotes how readily magnetic flux can be made to flow in a core material, compared to air (an air core). For example, if an inductor were constructed using a number of turns (X) around a core, then a core material having a permeability of 1000 would have 1000 times the magnetic flux flowing, as compared to the same core made of air (or non-magnetic material such as phenolic), as long as the number of turns and current remain the same. Another way of understanding this is that the core acts as a magnetic flux concentrator (concentrating magnetic flux lines within the core material, preferentially, compared to the surrounding air).

Detailed Design of the Class-D Inductor

It will be instructive to work through the design of an inductor for a Class-D audio amplifier. In this section, the design of the inductor used in the reference design will be discussed in detail. This inductor uses a toroidal core composed of a low-permeability powdered iron. The reasons that support this core selection are:

1) Toroidal core shape exhibits very low leakage flux, therefore magnetic field emissions from this inductor (that could interfere with adjacent circuitry) are low. In contrast, *gapped core* structures exhibit a phenomenon called "fringing flux", which is flux that bulges out from the gaps in the magnetic path, and may interfere with adjacent circuitry.
2) The toroidal core allows for multiple wire turns on a single-layer winding, with a distinct "start-end" gap. The utilization of a single-layer winding ensures

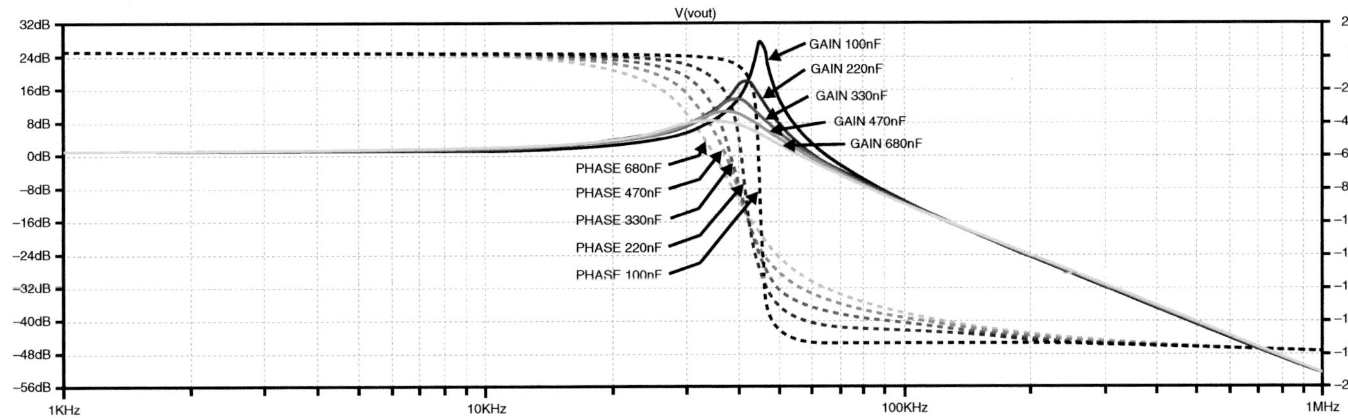

Figure 9.5 Bode Plot Run in LTspice for an Unloaded L-C Filter with Only a Zobel Network Connected at its Output (Zobel Resistor Fixed at 8 Ohms)

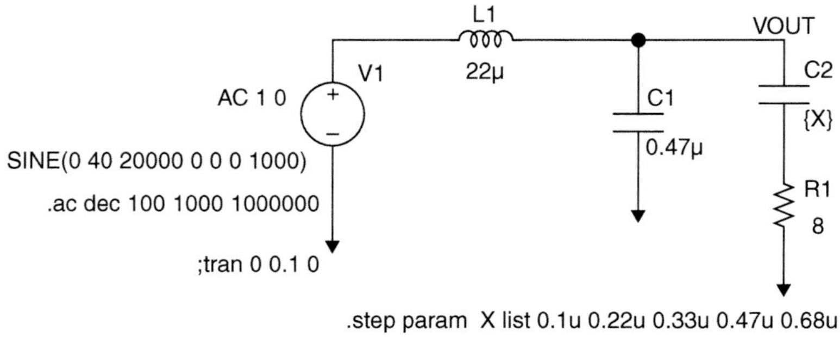

Figure 9.6 LTspice Schematic Used to Generate the Bode Plot in Figure 9.5

low end-to-end shunt capacitance across the inductor. This shunt capacitance will lower the Q of the part, and increase high-frequency noise in the residual ripple at the output of the amplifier. The utilization of a "start-end" gap also helps to reduce this shunt capacitance. This gap keeps the "start" turn and the "end" turn of the winding separated by a distinct physical distance. The winding area available on a toroidal core allows for the required inductance to be achieved using multiple wire turns in a single layer on a low-permeability core. This allows the core to handle higher current levels without saturating. This will be explained further in the design discussion which follows.

3) Powdered iron cores are very forgiving when it comes to saturation effects. Specifically, they have a "soft-saturation" characteristic. This means the permeability of the core (and therefore the inductance of the wound inductor) decrease *gradually* with current, as opposed to abruptly for a core type with a hard-saturation characteristic (gapped ferrite core, for example). The curves in Coilcraft (2023) and Sagami (n.d.) illustrate this.

The curve in Coilcraft (2023) illustrates the saturation characteristics of a commercially available toroidal inductor for Class-D, using a powdered iron, low-permeability core. It can be seen that the inductance rolls off very gradually at high currents (soft saturation). In fact, at 36.0A, the inductance has only dropped by 10% from its no-load current value. For this part, the RMS (heating) current rating is limited by the wire size and DCR. The part temperature will rise 40 degrees C, at a current of 7.2A, due to the wire size used. As will be seen in the detailed design example in this chapter, the design constraint of having a single-layer winding restricts the diameter of the wire that can be used, when trying to achieve a certain number of wire turns in a single (non-overlapping) winding around the toroidal core. An inductor design of this type is very forgiving of transient high currents, as will occur during overcurrent events, such as momentary short circuiting of the Class-D output. It allows a larger headroom (or tolerance of delays) in the protection response, in the design of the overcurrent sensing circuitry.

The characteristic in the figure in Sagami (n.d.) illustrates a hard-saturation effect. The inductance can be seen to rolloff sharply with current. This is the characteristic of a high-permeability material (such as ferrite) with a discrete gap (Sagami, n.d.). The rolloff characteristics of different inductor values are displayed in Sagami (n.d.).

The first step in the design of the Class-D inductor is to choose the required inductance value. This will be based on the following parameters:

1) Class-D switching frequency
2) Desired amplitude of residual modulation ripple at the switching frequency
3) Desired resonant frequency of the L-C filter
4) Speed (propagation delay) of overcurrent protection circuitry.

The reference design has a switching frequency of 500KHz. The attenuation of the L-C filter *after the resonant frequency* can be approximated at 12dB/octave (double pole rolloff). Values (VRMS) of "residual modulation *voltage*" present at Class-D amplifier outputs vary widely. Typical values range from 500mVRMS to 2VRMS. Values in this range do not cause any voice coil heating problems with speakers designed to handle moderate power (>25W). For the reference amplifier, a maximum residual ripple voltage of 1VRMS at 500KHz is a design goal. This is the amplitude of the signal at the *output* of the L-C filter. The signal at the *input* of the L-C filter will be the rail-to-rail switching voltage of the Class-D power stage. Therefore, the input to the L-C filter will be +/−48Vpeak. This is equivalent to 48VRMS. The required attenuation is 1VRMS/48VRMS = −33.62dB. Assuming 12dB/octave, double pole rolloff, this will require approximately 3 octaves of rolloff.

Working backwards…

- 500KHz
- 250KHz (1 octave)

- 125KHz (2 octaves)
- **62.5KHz (3 octaves).**

Therefore, the resonant frequency of the L-C filter should be set to 62.5KHz or less. Note that this is just an approximation, because the calculation assumes sinusoidal waveforms, and the input signal in the example above is a square wave, and not a sinusoid. So, it will have higher-frequency harmonics, in addition to its 500KHz sinusoidal fundamental frequency. However, its fundamental will be attenuated by 33.62dB as previously calculated.

For the reference design, a value of 22μH (with 0.47μF filter capacitor) produces a resonant frequency of 49.49KHz. It also has a Q value in an acceptable range for speaker loads between 4 and 8 ohms (see Table 9.1). A 100W Class-D amplifier will need to produce 28.28VRMS/3.54ARMS output into an 8-ohm load, and it will need to produce 20VRMS/5ARMS output into a 4-ohm load. Thus, the 8-ohm load defines the maximum output voltage requirement, and the 4-ohm load defines the maximum output current requirement. In addition to these RMS values above, the peak values must be calculated.

28.28VRMS = +/−40Vpeak

5.0ARMS = +/−7.07Apeak

In Chapter 14 on Protection Circuits, Figure 14.5 illustrates how quickly the inductor current can ramp up during a short-circuit fault. Using that information, and assuming the overcurrent protection circuits have been designed to switch OFF the Class-D power stage in less than 4μsec, an overcurrent trip threshold of +/−8Apeak can be used. This threshold provides some margin above the +/−7.07Apeak current required by 4-ohm loads, accounts for highly reactive speakers, and will reduce nuisance tripping on program content peaks. It can be seen from Figure 14.5 that the inductor current can ramp up to 15.93A in 4μsec. So, the *design goals* for the output inductor for the reference amplifier will be:

- 22μH nominal (+/−20% is a typical tolerance for Class-D inductors)
- 5ARMS (heating limit due to I^2R wire losses) for 40 degree C temperature rise
- >16A saturation current (inductance dropping less than 30% from no-load inductance value).

Magnet Wire Size (AWG)

Using the Micrometals (2007) single-layer winding table, it is noted that 20 AWG magnet wire has a temperature rise of 40 degrees C at an RMS current of 5.70A, so this is adequate to meet the above requirement of 40 degrees C rise at 5ARMS.

Core Selection

A low-permeability powdered iron core is typically used for Class-D toroidal inductors. Micrometals -2 material is a typical choice, and will be used in the reference design. An excellent reference is the Micrometals Power Conversion catalog (Micrometals, 2007). This catalog conveniently includes data on single-layer windings (p. 64). Some trial and error is required in order to arrive at a toroid size that will accommodate the number of turns (in a single layer) that will provide the desired 22μH inductance value.

For example, in an effort to minimize the size of the part, initially a 0.942" diameter core is selected (T94 size). This core, in -2 material (T94-2), has an A_L value of 8.4nH/N^2. To calculate the number of wire turns required on this core to achieve an inductance of 22μH, simply divide the desired inductance by the A_L value, and take the square root of the result.

Note: Magnetic core vendors typically provide A_L values in one of two ways: inductance per 1000 turns, or normalized to a single turn with units of nH/N^2, where N is the number of turns used.

$$\text{Number of Turns Required (N)} = \text{SQRT (Desired Inductance / } A_{L\ (\text{expressed in nH/N}^2)}) \quad \text{EQ. 9.6}$$

$$N = \text{SQRT } (22\mu H / 8.4nH/N^2)$$

$$N = 51 \text{ turns}$$

Referring to the Micrometals (2007) single-layer winding table, it is noted that this core can accommodate 45 turns of 20 AWG wire. This is not adequate; 45 turns on this T94 core will only achieve an inductance of $8.4nH \times (45)^2 = 17.01\mu H$.

The choices now are either to decrease the wire AWG, or to increase the core size. The goal of a conservative design impels us to choose the next larger core size, rather than decrease the wire AWG and incur increased I^2R heating losses. The next larger core size is T106. In the -2 material, this core has an A_L of 13.5nH/N^2. Using this core, an inductance of 22μH can be achieved using just 40.37 turns. This core can accommodate 46 turns of 20 AWG wire in a single layer. Since fractional turns cannot be used, 41 turns will be used, for an inductance of:

$$13.5\text{nH/N}^2\ (41\ \text{turns})^2 = 22.69\mu\text{H}$$

Note: It is good practice *not* to use the full number of turns allowed by the single-layer winding table. The reason is that it is advisable to have a "gap" or space between the starting (first) wire turn and the end (last) wire turn ("start-end gap"). These start and end wires should not touch, because they have the maximum differential voltage that is impressed across the inductor, and having them touch creates a capacitive coupling between these two circuit nodes. It will allow switching noise from the high amplitude, high dv/dt switching node, to shunt the inductance of the part and couple capacitively to the output node, increasing output noise and degrading electromagnetic compatibility (EMC) performance. A good rule of thumb is to try to allow for a 0.1" minimum gap between the start and end turns on the wound toroid (see Figure 9.7).

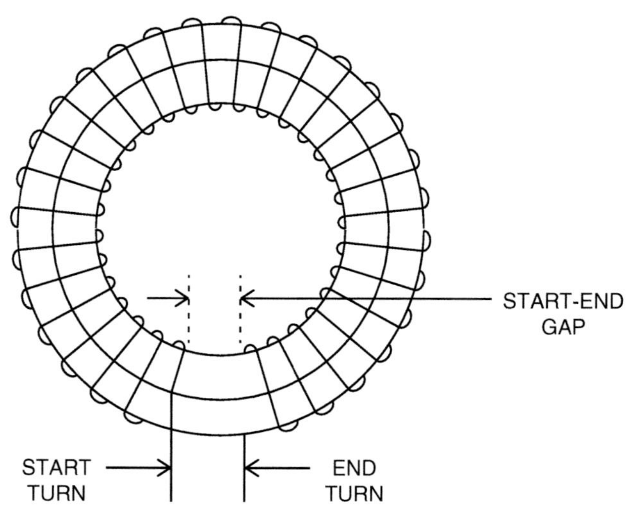

Figure 9.7 Toroidal Inductor, Using a Single-Layer Winding with a "Start-End" Gap

Checking Flux Density and Saturation

Using the T106-2 core, and consulting the Micrometals catalog (Micrometals, 2007, p. 26), a graph of "Percent Initial Permeability versus DC Magnetizing Force" is reviewed. As mentioned previously, the -2 material is low-permeability material, and has a very gradual saturation characteristic. In fact, this graph only displays a maximum saturation level of −12% at 350 Oersteds (DC magnetizing force). Using the equation associated with this graph (EQ. 9.7):

$$H = 0.4\,\pi\, N\, I \,/\, l \qquad \text{EQ. 9.7}$$

where H = DC magnetizing force (in Oersteds), N = number of turns, I = current in amps, and l = magnetic path length (in cm).

Substituting the design values for the above variables, and looking up the core data for the T106 core, it is found that l = magnetic path length = 6.49cm. Now the equation is rearranged to find the *current value* that will generate 350 Oersteds of DC magnetizing force and 12% saturation.

$$I = H\, l \,/\, 0.4\,\pi\, N = (350\ \text{Oersteds})(6.49\text{cm})/(0.4\pi)(41) = 44\text{A}$$

This 44A saturation current value indicates that this inductor will be extremely linear in terms of inductance over the nominal operation currents (+/−7.07Apeak), and will be extremely tolerant of overcurrent fault conditions.

Core Loss Calculation

In order to calculate core loss, the *flux swing* that the core is subjected to must be calculated. It is changing (AC) flux within the core that causes the majority of core losses in this application. This type of core loss is determined by the flux density swing and the frequency. Core loss increases with increasing flux density and also with increased operating frequency. This is another parameter in Class-D design that is influenced by the choice of switching frequency. Increasing the switching frequency of the Class-D amp will result in increased core losses for the Class-D filter inductor. This must be considered for efficiency reasons.

Note regarding magnetic unit quantities: There are two systems of units for magnetic component quantities. These are the SI (or international) units, and cgs (Gaussian or emu) units. Both systems are still in use today, although there is a tendency to adopt SI units moving forward. However, many magnetic components vendors still provide component data using cgs units. In this discussion, cgs units will be used. These units may be converted, using conversion factors and tables.

For the core loss calculation, the operating condition that will be used is the IDLE condition. This is because the flux density swing in the core that is *most* responsible for core loss will be the high-frequency (500KHz) component, and *not* the much lower audio frequency component. So even though, at full output amplitude, the inductor current will be swinging from +7Apeak to −7Apeak, this will only be occurring at 20KHz, where core losses are much lower than they are at 500KHz. To show this, EQ. 9.8 will be used, along with the graph of core loss vs flux density provided in Micrometals (2007, p. 31):

$$\text{Bmax} = \text{L Ipeak} / \text{N Ae} \times 10^{-8} \qquad \text{EQ. 9.8}$$

where:

- Bmax is the maximum peak flux density (in Gauss)
- L is the inductance
- Ipeak is the peak inductor current for the operating condition
- N is the number of wire turns
- Ae is the core window area (in cm^2) (from core data for the T106-2 toroid in Micrometals (2007)).

Using the IDLE switching condition, and the calculated value of 22µH, it has already been shown that the current waveform in the inductor will be a triangular signal, ramping from −1.09A to +1.09A at −500KHz (see Figure 9.8).

In the equations in Figure 9.8, di is the total change in inductor current from −Ipk to +Ipk, and dt is the time for which the respective switching voltage is applied. In this case, for a 500KHz switching frequency at IDLE, the period (T) is 2µsec, and at 50% duty cycle (the IDLE condition), each switched voltage (+48V or −48V) is applied to one terminal of the inductor for 1µsec. The other end of the inductor is at 0VDC (no output signal). Therefore, V_L, the voltage across the inductor, is 48V.

$$di = (48V)(1\mu sec) / 22\mu H = 2.18A$$

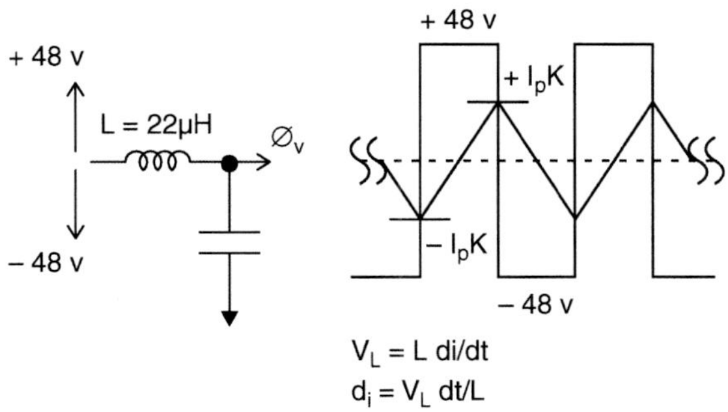

Figure 9.8 L-C Filter with Associated Voltage and Current Waveforms at IDLE (i.e. Amplifier Switching at 500KHz, at 50% Duty Cycle, and with No Output Audio Signal Present)

The inductor current will swing from −1.09A to +1.09A. Therefore, Ipeak is 1.09A.

$$\text{Bmax} = \text{Bpk} = (22\mu\text{H})(1.09\text{A}) \,/\, (41 \text{ turns})(0.659 \times 10^{-8}) = 88.75 \text{ Gauss}$$

Next, using this calculated value of Bmax, the graph of Core Loss (mW/cm^3) versus Bpk = Peak AC Flux Density (Gauss) in Micrometals (2007) is consulted (p. 31). It shows that at 88.75 Gauss, a vertical line drawn up from this value on the X-axis will intersect the slanted line for 500KHz operation, at approximately 45–50mW/cm^3. A core loss value of 50mW/cm^3 will be used.

Next, the *volume* of the core in cm^3 is found from the core data for the T106-2 core. It is 4.28cm^3. Therefore, the total core loss will be 50mW/cm^3 × 4.28cm^3 = 214mW.

The next step is to combine this core loss with a calculation of copper (Cu) loss from I^2R heating of the 20 AWG wire winding.

To do this, the total resistance of the wire must be calculated. The Micrometals single-layer winding table has a column labeled MLT. This stands for Mean-Length-of-Turn in units of cm/turn. For the T106 size toroid, MLT = 4.49cm. Therefore, 41 turns will require 41 × 4.49cm = 184.1cm of 20 AWG wire.

The wire table also lists the resistivity of 20 AWG wire as 0.330 milliohm/cm.

$$\text{Total wire resistance} = 184.1\text{cm} \times 0.330 \text{ milliohms/cm} = 60.75 \text{ milliohms}$$

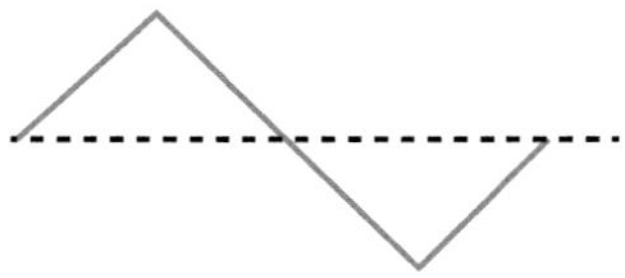

Figure 9.9 Triangular Ramp Waveform and Equations for Calculating the RMS, AVERAGE, and Peak-to-Peak (P-P) Values for a Triangular Ramp Waveform

Next, the RMS value of the triangular current waveform of the inductor current must be calculated.

For a triangle wave:

$$I_{RMS} = 0.577 \times \text{Ipeak} \qquad \text{EQ. 9.9}$$

$$= (0.577)(1.09\text{A}) = 0.629\text{A}_{RMS}$$

Figure 9.9 lists equations for calculating the RMS, average, and peak-to-peak (P-P) values for a triangular ramp waveform.

$$\text{Cu loss} = I^2R = (0.629\text{A}^2)(60.75 \text{ milliohms}) = 24.0\text{mW} \qquad \text{EQ. 9.10}$$

Therefore, Total loss = Core loss + Copper loss = 214mW + 24mW = 238mW EQ. 9.11

The single-layer winding chart (Micrometals, 2007) also lists (on the right side) data for temperature rise versus total power dissipation. For the T106 core size, a power dissipation of 492mW will result in a 10 degree C rise. Therefore, the calculated power dissipation of 238mW will result in less than a 10 degree C rise for this component.

It will be informative to also check the power dissipation due to audio frequency current. As mentioned above, core loss increases with increased frequency of flux change. The calculation for IDLE core loss above was based on 500KHz switching frequency. For the audio frequency calculation, 20KHz will present the highest core loss situation. In the case of a 4-ohm load, the load current will be 5ARMS for 100W = P_{OUT}. Figure 9.10 shows a detailed sketch of inductor current at the $100W_{OUT}$ / 4 ohm / $5A_{RMS\text{-}OUT}$ / $28V_{PK\text{-}OUT}$/$20V_{RMS\text{-}OUT}$ condition.

Under this full-load condition, Cu losses will be:

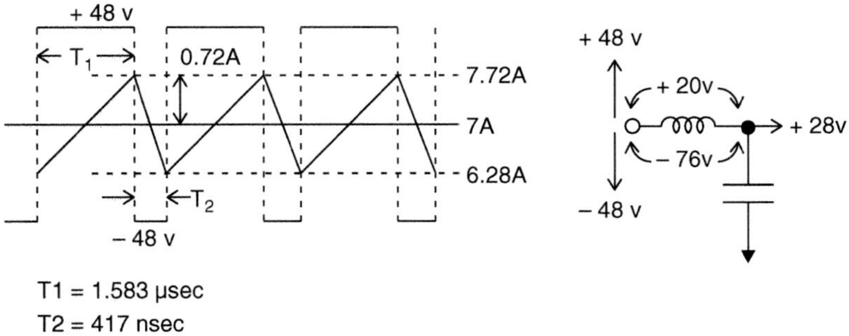

Figure 9.10 Inductor Current Waveform When the Amplifier is Producing its Maximum Output Current

$(5A^2)(60.75 \text{ milliohms}) = 1.519W$

Core losses will be based on the peak flux generated by the 20KHz audio frequency current of 7Apeak. *Note:* The 7.72Apeak current in Figure 9.10 is the real-time peak inductor current generated at the 500KHz switching frequency during this operating condition. This current value is important for determining the overcurrent trip threshold current, but the audio frequency peak current of 7A should be used for this 20KHz core loss calculation.

$$\begin{aligned} Bmax\text{-}_{20KHz} &= L\ Ipeak\ /\ N\ Ae \times 10^{-8} = (22\mu H)(7A)\ /\ (41)(0.659 \times 10^{-8}) \\ &= 570 \text{ Gauss} \end{aligned}$$

A vertical line drawn up from the X-axis at 570 Gauss will intersect the slanted line of 25KHz at approximately 100mW/cm^3. (Note there is no power dissipation line for 20KHz. Using the 25KHz line will overestimate the losses, so is a more conservative value.)

At 100mW/cm^3, the T106 core will dissipate:
(100mW/cm^3)(4.28cm^3) = 425mW.

$$\text{Total power dissipation} = Pd_{CU} + P_{CORE} = 1.518W + 425mW = 1.943W$$

The total power dissipation data on the single-layer winding table show a temperature rise of 25 degrees C at a power dissipation of 1.47W, and a temperature rise of 40 degrees C at a power dissipation of 2.59W. Therefore, an extrapolation between these two data points yields a temperature rise of approximately 31.3 degrees C at a power dissipation of 1.943W.

Also note that operation at full output power at 4 ohms at a frequency of 20KHz is *only* a lab test condition and will not occur with any musical program material in real-world usage. The designer needs to decide if certain worst-case lab tests are valid test conditions which an amplifier should tolerate. An advantage of designing to meet these worst-case test conditions is the realization of a very conservative and reliable design. The disadvantage is that the design may be too conservative, and more expensive, large, heavy, etc. than it would otherwise need to be, if designed specifically for real-world use-conditions.

Since this T106-2 inductor design has so much margin, and low temperature rise, it will be illustrative at this point to circle back and see if this inductor design could be made smaller by using the next smaller core size (T94-2) with 22 AWG magnet wire instead of 20 AWG magnet wire. This serves to illustrate how the design process may proceed in arriving at an optimal design.

Following the steps described above, 51 turns of wire are required on the T94-2 core. This core will accommodate 58 turns of 22 AWG wire in a single-layer winding. So, 51 turns should fit in a single-layer winding with an adequate start-end gap. The MLT (mean-length-of-turn) for wire on the T94 core is 3.44cm/turn. So, 51 turns will be 175.44cm. The resistivity of 22 AWG magnet wire is 0.53 milliohms/cm. So, 175.44cm will equal 92.98 milliohms.

Cu_{LOSS} @ 500KHz (*IDLE condition*) = $(0.629A_{RMS}{}^2)$(92.98 milliohms) = 36.8 mW

Cu_{LOSS} @ 20KHz ($5A_{RMS}$) = $(5A^2)$(92.98 milliohms) = 2.47W

Saturation current for T94-2 core: (12% saturation)

$$I = H\,l / 0.4\,\pi\,N = (350\text{ Oersteds})(5.97\text{cm}) / (0.4\,\pi)(51\text{ turns}) = 32.6\text{A}$$

So this core, though smaller than the T106-2, still has plenty of margin against saturation.

Core loss @ IDLE (500KHz @ 1.09Apeak):

$$Bmax\text{-}_{IDLE} = L\,Ipeak / N\,Ae \times 10^{-8} = (22\mu H)(1.09A) / (51)(0.362 \times 10^{-8}) = 129.9\text{ Gauss}$$

129.9 Gauss peak at 500KHz loss line = $100mW/cm^3$

$Pd_{CORE\text{-}IDLE} = (100mW/cm^3)(2.16cm^3) = 216mW$

Total Pd @ IDLE = 36.8mW + 216mW = 252.8mW

Core loss @ 20KHz, 7Apeak:

$$Bmax\text{-}_{20KHz} = L\ Ipeak\ /\ N\ Ae \times 10^{-8} = (22\mu H)(7A)\ /\ (51)(0.362 \times 10^{-8})$$
$$= 834 \text{ Gauss}$$

834 Gauss peak at 25KHz loss line = $2500mW/cm^3$

$Pd_{CORE\text{-}20KHz} = (2500mW/cm^3)(2.16cm^3) = 5.4W$

Total Pd @ 20KHz/100W/4 ohms = 2.47W +5.4W = **7.87W**

Note: This power dissipation will result in a core temperature rise far exceeding 40 degrees C.

This extremely high power dissipation at the 20KHz full power, 4-ohm test case is concerning. It could again be argued that this is a lab test condition only, which could be intentionally limited in duration, but some type of protection would most likely need to be incorporated into the design to protect against sustained high-frequency operation, to protect the inductor from excessive temperature rise, if this smaller T94-2 core were used. This will be discussed in Chapter 14 on Protection Circuits. For the reference design, the more conservative T106-2 core design is chosen, yielding very low temperature rise under all conditions.

Notch Filters

One of the big benefits of a Class-D amplifier operating with a fixed switching frequency is that a passive notch filter can be added to the output L-C filter. This notch filter is designed for extremely high attenuation, specifically at the switching frequency (high Q).

Figure 9.11 shows an example of a passive notch filter added to a fixed-frequency L-C filter output, to attenuate switching frequency ripple.

$$F_{NOTCH} = F_{SWITCHING} = 1/2\pi(SQRT\ (L_2\ C_2)) \qquad \text{EQ. 9.12}$$

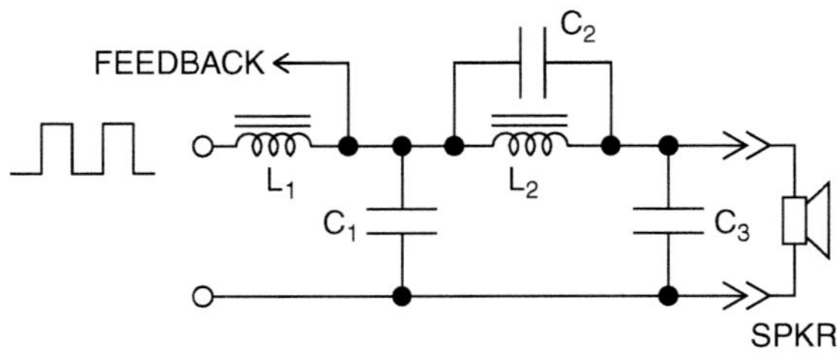

Figure 9.11 Passive Notch Filter Added to a Fixed-Frequency L-C Filter Output

$$C_2 = (1/2\pi F_{SWITCHING})^2 / L_2 \qquad \text{EQ. 9.13}$$

To design a notch filter to attenuate the 500KHz switching frequency ripple of the reference design amplifier, it is easiest to start by selecting a value for L_2 in EQ. 9.12 = L_{NOTCH}. In the reference design, L_{NOTCH} is L6A. Typically, a value much smaller than the Class-D inductor value (L1A = 22μH) is selected. Another reason that L_2 is selected first is that it is desirable for the inductance value of L_2 to be as precise as possible, in order for the *notch* frequency to match the (crystal-oscillator-controlled) switching frequency. A high-current inductor with tight tolerance is not as readily available, or available in as wide a range of inductance values, as are precision film or NP0/C0G capacitors. For this example, and for the reference design, a value of 705.6nH is chosen. This seemingly strange value is a result of using a readily available small powdered-iron toroidal core (with 7% permeability tolerance = Micrometals T50-2) with 12 turns of 18 AWG magnet wire evenly spaced around the circumference of the core. With the value of L_{NOTCH} determined, EQ. 9.12 is used to calculate C_2 = C_{NOTCH} for a 500KHz *notch* frequency. For this example, C_{NOTCH} is determined to be 0.144μF. This very specific, non-standard value of C_{NOTCH} is made up of the *sum* of C133-1A + C133-2A + C133-3A in the reference design. The nominal value achieved for C_{NOTCH} therefore is 0.141μF. The closer the actual values for C_{NOTCH} and L_{NOTCH} are to the calculated values, the "deeper" will be the notch attenuation, and the greater the reduction in switching frequency residual ripple amplitude.

Figure 9.11 shows the feedback pickoff point *after* the first L-C filter (*before* the notch filter). This pickoff location is *sometimes* desirable when any extra phase shift caused by the notch filter cannot be tolerated by the feedback loop, because this phase shift can decrease the phase margin of the closed loop and possibly cause instability and oscillation. However, in the reference design, the feedback pickoff point is *after* the notch filter. The loop compensation used (Chapter 12) is minimally affected by the phase shift of the notch filter, because the Q of the notch filter is high, so there is very little phase shift at the frequency where the loop closes (unity-gain frequency). This feedback pickoff location also provides the benefit that the switching frequency ripple present on the feedback signal will be very low (at 500KHz), so the input voltage to the PWM comparator (driven by the error amplifier output) has minimal 500KHz ripple on it. If 500KHz ripple *is* present at the PWM comparator input, it will cause a non-linear modulation of the PWM timing (duty-cycle modulation) over the 0%–100% duty cycle modulation range of the PWM comparator. This feedback pickoff location also provides the added benefit that voltage drops across the notch filter are compensated for by the feedback loop. This decreases the effective output impedance of the amplifier, and therefore increases damping factor.

Capacitor Types

Capacitor types typically used in the L-C demodulation filter application for Class-D are:

- Metalized polypropylene (film)
- Metalized polyester (film)
- Ceramic.

These types are all characterized as non-polarized, low-ESR capacitors, available in a wide range of voltages. These different capacitor types have different characteristics.

ESR

ESR is a key specification for this application. Though film and ceramic cap datasheets do not usually state ESR directly, they *do* provide useful graphs of AC voltage versus frequency. These are based on the internal heating that results, and maximum hotspot internal temperature rise (KEMET, 2024).

The graph of allowable AC voltage (RMS) versus frequency in KEMET (2024) shows the maximum RMS voltage that can be applied versus frequency, under the worst-case operating temperature condition of 85 degrees C local air temperature around the capacitor *or* trace temperature.

The same reference (KEMET, 2024) includes a graph of maximum RMS current that the capacitor can handle, versus frequency, under the worst-case operating temperature condition of 85 degrees C local air temperature around the capacitor *or* trace temperature.

Looking at the 0.33μF value, it can handle approximately 2.5VRMS at 500KHz, using the graph in KEMET (2024). The *impedance* of a 0.33μF capacitor at 500KHz is $1/2\pi fC = 1/2\pi(500\text{KHz})(0.33\mu\text{F}) = 965$ milliohms. A voltage of 2.5VRMS would induce a current flow of 2.5V/0.965 ohms = $2.59\text{A}_{\text{RMS}}$. Checking the current versus frequency graph in KEMET (2024), it can be seen that at 500KHz, the allowable current is approximately 2.5A_{RMS}, which demonstrates agreement between the data provided in these two graphs. In the reference design, a capacitor value of 0.47μF will be used, consisting of the sum of the capacitances of C68-1A, C68-2A, and C68-3A. The graphs in KEMET (2024) do not show data for that specific capacitance value. However, it will lie between the curves for 0.33μF and 1μF. Even the 1μF cap can tolerate more than 1V_{RMS} at 500KHz. Therefore, the residual modulation voltage amplitude of 1V at 500KHz will be well within the specification limit of a 0.47μF or 0.33μF capacitor in this capacitor series.

In addition to the capacitor's voltage handling at the switching frequency (500KHz for the reference amplifier), it is also important to check the capacitor's voltage limitation at 20KHz. Referring again to the AC voltage versus frequency graph in KEMET (2024), it can be seen that a 1μF value can handle 60VRMS at 20KHz. Therefore, the 0.47μF value chosen for the reference amplifier will handle even more than 60VRMS, and is well suited for the reference design application of 28.28VRMS maximum (100W @ 8 ohms).

With regard to ESR and its heating effects, ceramic caps perform best, followed by metalized polypropylene and then metalized polyester. The film dielectrics (metalized polypropylene and metalized polyester) have stable capacitance over voltage, whereas certain ceramic dielectrics do not. For example, X7R and X5R ceramic dielectrics are characterized by having a voltage coefficient of capacitance, whereby capacitance will change with applied voltage. This phenomenon can be particularly pronounced in some dielectrics, and can vary from manufacturer to manufacturer. This characteristic is particularly undesirable in a Class-D filter application, as it will produce non-linearities in the filter response as the amplitude of the audio signal varies. Another undesirable characteristic of some ceramic dielectrics is piezoelectric effects. This means that the dielectric actually undergoes a physical change as applied voltage varies. This physical change causes audible vibration effects when these are mounted to a circuit board. For these reasons (voltage coefficient of capacitance and piezoelectric effects), the most desirable ceramic dielectric for this Class-D application is NP0 or C0G. The drawback of this more stable dielectric is that its volumetric efficiency is decreased as compared to other ceramic dielectrics. In other words, they will be physically larger for the same capacitance and voltage value than other dielectrics, and will only be available up to certain capacitance values, depending on voltage. This may require paralleling of multiple capacitors in order to achieve the desired capacitance. They are readily available in surface-mounted device (SMD) packages, which helps to reduce board size. This is in contrast to the film types, which are not readily available in SMD packages, and are therefore physically much larger through-hole devices, taking up considerable PCB area.

Multi-layer ceramic capacitors (MLCCs) using the X7R and X5R dielectric *are* suitable for decoupling applications in Class-D amplifiers, as long as the designer is cognizant of the reduced capacitance value at the applied circuit voltages. However, another characteristic to keep in mind is temperature rating. X5R capacitors are typically only specified to 85 degrees C, whereas X7R capacitors are typically specified to 125 degrees C. As Class-D designs evolve, the trend is to use an increasing number of SMDs, and to use the PCB itself as the heatsink, or at least the heat conducting medium. This means increased PCB surface temperatures. Therefore, MLCCs which are placed close to power SMDs could easily experience temperatures exceeding 85 degrees C.

Film Capacitors

Film capacitors used in Class-D amplifier designs (L-C filter application) are typically one of two types. These are metalized polyester film and metalized polypropylene film. Metalized polypropylene films handle higher levels of AC current, and come in either single-metalized or double-metalized varieties. The double-metalized variety handles the highest currents. The graphs of AC voltage and AC current versus frequency, in the manufacturers' datasheets, should be consulted when selecting these capacitors for L-C filter application. These film capacitors are available in a wide range of voltages, suitable for Class-D amplifier application. Their capacitance stability with varying voltage and temperature is excellent, and for that reason, high-linearity filters (resulting in low THD) can be designed using these film capacitors.

Filterless Class-D Amplifiers

The role of the L-C demodulation filter function, discussed thus far in this chapter, is to extract the audio frequency content from a high-frequency PWM signal. This extraction can, however, be accomplished in other ways. Speakers are principally characterized by their nominal DC resistance (typically 4–8 ohms for low-impedance loudspeakers driven by amplifiers of the type discussed in this book, or 16–32 ohms for headphones). Though characterized by their nominal resistance, a loudspeaker is actually more complicated. In fact, a typical electromagnetic loudspeaker is constructed of a voice coil suspended in a strong magnetic field. This construction gives it an inductive characteristic, and a defined frequency response or bandwidth over which it is responsive. It can be appreciated that a speaker cone represents a moving mass, with inertia, and therefore there are limits to how fast it can accelerate and vibrate given the power it is driven with, and its physical constraints. All this is to say that the speaker already presents a low-pass function in and of itself. It will respond to the audio content in the driving signal and will be unresponsive to the high-frequency signal content due to its inductive nature.

Figure 9.12 is an electrical circuit model for a typical electrodynamic speaker. It can be seen that there are inductive components, as well as an L-C low-pass arrangement of circuit elements, which will limit high-frequency response.

Texas Instruments (2013) also makes the interesting observation that human hearing itself presents a low-pass characteristic, also being unresponsive to the high-frequency signal content. In other words, even if the speaker *were* responsive to the high-frequency signal content and reproduced it faithfully, the finite response of human hearing would prevent the high-frequency signal content from being heard. Both of these low-pass functions (loudspeaker response and human hearing response) underlie the justification for filterless Class-D designs. In these designs, the discrete L-C filter is eliminated, and the filtering of the high-frequency PWM signal is provided by the loudspeaker itself.

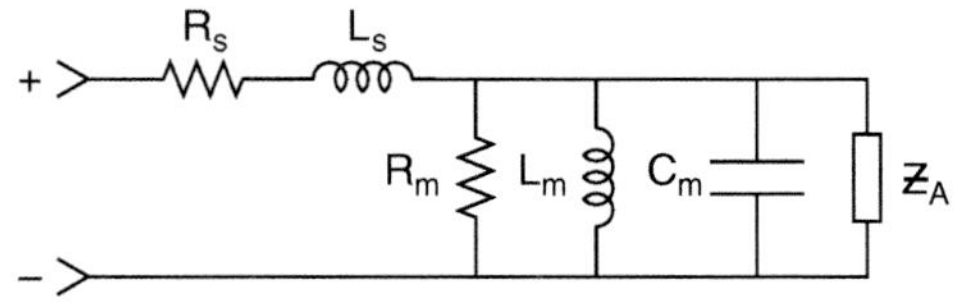

Figure 9.12 Schematic Diagram Representation of an Electrodynamic Speaker

Because of the large signal swings and high dv/dt at the outputs of filterless Class-D amplifiers, they have greater potential for producing undesired radiated emissions. Because of this constraint, most designs that employ filterless Class-D amplifiers are low-power (and therefore low-voltage and low-current) designs, and include the speaker and amplifier in close proximity within the same enclosure, ensuring that the connection length or wire length between the Class-D amplifier output and the speaker are as short as possible. As will be discussed in Chapter 15, the principal source of radiated emissions in any Class-D design is the attached cables. Minimizing the length of interconnecting cables, or eliminating them completely, is particularly beneficial for reducing radiated emissions in filterless Class-D amplifier designs.

Filterless designs also benefit from novel and innovative switching schemes (as described in Texas Instruments, 2013). Filterless designs are typically low-voltage designs and therefore benefit from using a BTL (Bridge-Tied Load, or full-bridge) arrangement of the power mosfets in the switching power stage. Conventional BTL designs have the two half-bridges switching 180 degrees out of phase. This means at idle, when the output of one half-bridge is high, the output of the other half-bridge will be low. The switching duty cycle will be 50%, and the average differential output voltage will be 0V. This arrangement generates the maximum instantaneous differential voltage across the load, and a triangular ripple current through an inductive load, even though the average output current and voltage are zero. An audio frequency differential voltage is produced across the speaker when the duty cycle changes from 50%. Even though the two half-bridge outputs of the BTL switching stage still remain 180 degrees out of phase at all times, by changing the duty cycle, the average (audio frequency) voltage across the load can be changed, varying from the positive rail voltage (at one duty cycle extreme) to the negative rail voltage (or ground), for a single supply configuration.

Figure 9.13 is a BTL power switching stage for the filterless Class-D amplifier producing the waveforms in Figures 9.14 and 9.15 below.

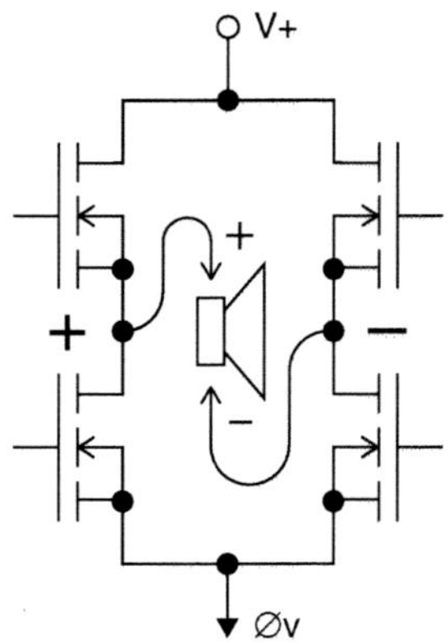

Figure 9.13 Schematic Diagram of a BTL Power Switching Stage for the Filterless Class-D Amplifier Producing the Waveforms in Figures 9.14 and 9.15

Waveforms 1 and 2 of Figure 9.14 are the respective outputs of each half of the full-bridge and connected to the + and − terminals of the speaker, at the 50% (IDLE) duty cycle condition (no audio output).

Waveform 3 is the *differential* voltage seen across the speaker terminals (+ terminal with respect to negative terminal), at the 50% (IDLE) duty cycle condition (no audio output).

Waveforms 4, 5, and 6 are waveforms at the same locations as 1, 2, and 3 above, except under the condition where there is a *positive* voltage (across the speaker terminals) audio output condition.

Waveforms 7, 8, and 9 are waveforms at the same locations as 1, 2, and 3 above, except under the condition where there is a *negative* voltage (across the speaker terminals) audio output condition.

It can be seen from waveforms 1–9 that the voltage swing across the terminals of the speaker is always maximum and equal to 2 × V. This means that there is a triangular ripple current flowing through the (inductive) speaker and attached interconnecting cables at the (IDLE) 50% duty cycle condition, even though there is no audio output.

Texas Instruments (2013) also describes an innovative switching arrangement. In this arrangement, the two half-bridges switch *in-phase* at idle (see Figure 9.15). With this arrangement, the instantaneous voltage across the speaker load is always zero at IDLE (50% duty cycle condition). This reduces ripple current through the inductive

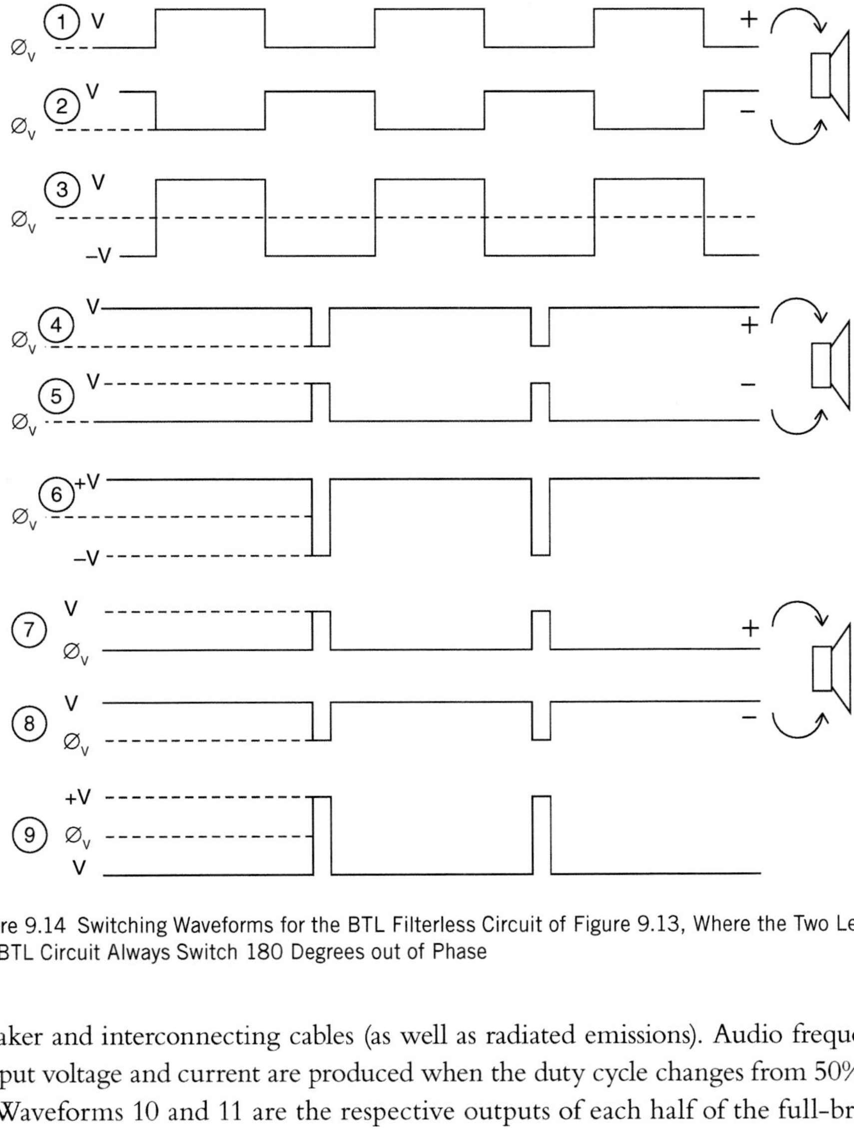

Figure 9.14 Switching Waveforms for the BTL Filterless Circuit of Figure 9.13, Where the Two Legs of the BTL Circuit Always Switch 180 Degrees out of Phase

speaker and interconnecting cables (as well as radiated emissions). Audio frequency output voltage and current are produced when the duty cycle changes from 50%.

Waveforms 10 and 11 are the respective outputs of each half of the full-bridge and connected to the + and − terminals of the speaker, at the 50% (IDLE) duty cycle condition (no audio output).

Waveform 12 is the *differential* voltage seen across the speaker terminals (+ terminal with respect to negative terminal), at the 50% (IDLE) duty cycle condition (no audio output).

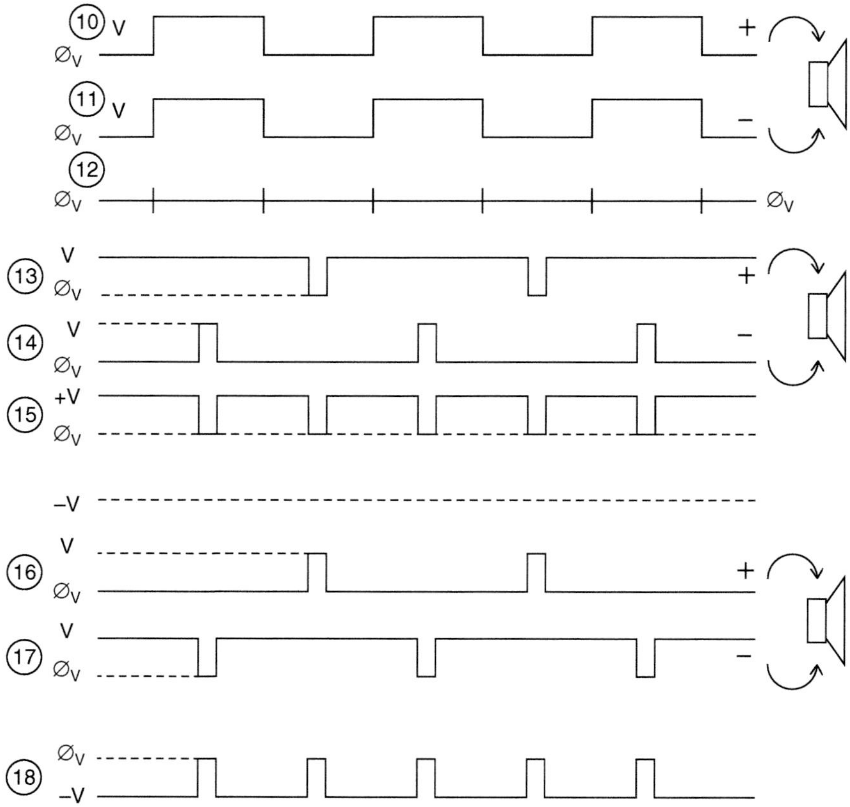

Figure 9.15 Switching Waveforms for the BTL Filterless Circuit of Figure 9.13, Where the Two Legs Switch in a Manner so as to Produce Zero Ripple Current through the Speaker, during the IDLE (No-Signal) Condition

Waveforms 13, 14, and 15 are waveforms at the same locations as 10, 11, and 12 above, except under the condition where there is a *positive* voltage (across the speaker terminals) audio output condition.

Waveforms 16, 17, and 18 are waveforms at the same locations as 10, 11, and 12 above, except under the condition where there is a *negative* voltage (across the speaker terminals) audio output condition.

It can be seen from waveforms 10–18 that the voltage swing across the terminals of the speaker is always limited to V, as opposed to 2 × V for the conventional switching case described for waveforms 1–9 previously in Figure 9.14. This means that there is *zero* ripple current flowing through the (inductive) speaker and attached interconnecting cables at the (IDLE) 50% duty cycle condition. Likewise, even when producing positive or negative audio output across the speaker terminals, the differential voltage swing across the speaker terminals is always limited in amplitude to V, as opposed to 2 × V.

Before leaving this chapter on L-C demodulation filters and filterless designs, a cautionary note is warranted. There are situations where excessive output voltage (at high frequency) can damage output capacitors and attached speakers. There are two typical cases. The first is when an unloaded, high Q, L-C filter produces an output voltage exceeding the voltage limit of the L-C filter capacitor(s). This can occur if the L-C filter is unloaded, or has an inadequate (or absent) Zobel network, or when feedback is lost on a feedback amplifier design. Adding output overvoltage sensing circuitry to the design can help protect against this scenario. The other method of protection for this first case is to add protection diodes to the output, that clamp the output voltage to the supply rails. Caution must be taken to ensure that the current rating of these diodes is adequate, as significant current will flow through these diodes when the high Q, underdamped filter is allowed to resonate. The second case is when the L-C filter capacitor is absent, or has a connection failure. In this case, the residual modulation amplitude at the output of the Class-D amplifier will be much greater than expected. Any attached loudspeaker will see this much higher residual voltage, and tweeters can be overheated and damaged. This is akin to attaching a loudspeaker to a filterless Class-D amplifier having an output signal swing much higher than the loudspeaker can safely handle. For these two reasons presented, the construction of the L-C filter should be robust, so that its function and integrity remain reliable over the lifetime of the amplifier.

Bibliography

Coilcraft. (2023). Datasheet, Power filter inductor ED0006-AL. Document 565-1. Revised April 10, 2023. Cary, IL: Coilcraft Inc.

Hupp, G. (2005, October). Measuring Class-D amplifiers for audio speaker overstress testing. Texas Instruments Application Report, SLOA116. Dallas, TX: Texas Instruments.

KEMET. (2024, January 2). Datasheet, Polypropylene pulse/high frequency capacitors R75, single metallized polypropylene film, radial, DC and pulse applications (automotive grade), F3106_R75. Fort Lauderdale, FL: KEMET Corporation.

Materion. (2019, June). Magnetic fields and unit systems. Materion Technical Tidbits (126). Mayfield Heights, OH: Materion Performance Alloys.

Maxim Integrated (Analog Devices). (2002, April 30). Class D audio amplifier output filter optimization. Application Note 624. San Jose, CA: Maxim Integrated Products, Inc.

Micrometals. (2007, February). Power conversion & line filter applications. Catalog, Issue L. Anaheim, CA: Micrometals Inc.

Sagami. (n.d.) Datasheet, Power inductors for Class-D amplifier, 7G Series. Kanagawa: Sagami Elec Co.

Texas Instruments. (2013). Filterless Class-D amplifiers. Texas Instruments Application Report, AN-1497, SNAA034A. Originally published May 2006, revised May 2013. Dallas, TX: Texas Instruments.

Texas Instruments. (2016). LC filter design. Texas Instruments Application Report, SLAA701A. Originally published October 2016, revised November 2016. Dallas, TX: Texas Instruments.

van der Hulst, P., Veltman, A., & Groenenberg, R. (2002). An asynchronous switching high-end power amplifier. Paper presented at the 112th AES Convention, Munich, Germany, May 10–13, 2002. New York: Audio Engineering Society.

Chapter Ten

Sources of Distortion and Noise

Overview

Often in this book, the concept of using negative feedback as a tool to reduce distortion in Class-D amplifiers has been repeated. It is easy and convenient to begin to think of feedback as a panacea, or cure-all for any and all distortion inherent in a design. A designer, if not vigilant, can begin to become complacent and less careful in minimizing open-loop distortion mechanisms, in the thought that these will, after all, be corrected by feedback. This mindset must be avoided. The wise design approach is to seek out and be cognizant of all sources of open-loop distortion in a Class-D design. Then, each should be minimized as much as possible before negative feedback is applied. Nyboe, Risbo, and Andreani (2005) makes the point that Class-AB amplifiers have had a long history. Over that time, the sources of open-loop distortion have been uncovered, and various techniques have evolved in minimizing these. The further point is made that this approach is even more important in Class-D design, where it is more difficult to add high feedback loop gain, needed to reduce this distortion (especially at higher frequencies), while still maintaining loop stability. The authors of that reference recommend a thorough understanding of Class-D open-loop distortion mechanisms, just as has evolved for Class-AB designs. For example, suppose a Class-D amplifier has an open-loop distortion (THD) of 1% at 20KHz, and it is desired for this amplifier to achieve a distortion level of 0.01% at 20KHz. This is a 100:1 reduction in THD (40dB). Further, suppose that the maximum feedback gain that can be applied, while still maintaining adequate gain and phase margin over all loading conditions, is 20dB

DOI: 10.4324/9781003327943-10

(10:1). One approach to this problem is to try to find and understand the sources of the open-loop distortion and minimize it. If the open-loop distortion can be reduced by 20dB, then the 20dB available feedback gain will be sufficient to meet the closed-loop performance target of 0.01% THD at 20KHz.

EQ. 10.1 establishes the mathematical definition of THD:

$$\mathrm{THD}\ (\%) = \frac{\mathrm{SQRT}((V_{out\,@\,2f})^2 + (V_{out\,@\,3f})^2 + (V_{out\,@\,4f})^2 + \ldots)}{V_{out\,@}} \times 100\% \qquad \text{EQ 10.1}$$

where "f" is the fundamental test frequency, and 2f, 3f, 4f, etc. are the harmonics of the test frequency.

Nyboe et al. (2005) analyzes open-loop distortion mechanisms of the Class-D output stage *only*, using time-domain analysis – the idea being that the PWM signal fed to the output stage has been generated by highly linear, low-noise circuits, which, relative to the output stage, will have contributed very little THD.

In actuality, however, the linearity of the triangle ramp waveform generated by the triangle wave generator in fixed-frequency PWM Class-D amplifiers is an extremely important source of THD. Great care should be taken in the design of that circuit block to ensure optimal linearity and the lowest possible noise.

Power Switching Stage Distortion

The power switching stage of a Class-D amplifier, it can be argued, is the most distinguishing circuit block in differentiating the Class-D topology from other amplifier topologies, and more heavily influences performance in terms of efficiency, distortion, power output capability, etc. than other circuit blocks. For this reason, the power switching stage will be analyzed from multiple perspectives. In this chapter, its effect on distortion is discussed. In the following chapter, its voltage and current switching waveforms are explained in detail, and its power losses broken down and analyzed.

The switching power output stage distortion is analyzed in the time domain, by separating the switching waveform into four distinct time intervals: $T_{FALLING\text{-}EDGE}$, T_{LOW}, $T_{RISING\text{-}EDGE}$, T_{HIGH}. Each interval is examined for distortion contribution. Nyboe et al. (2005) posit that time-domain analysis can be used to quantify root cause distortion mechanisms in the Class-D power stage in greater detail than THD graphs. Time-domain analysis is also more accessible in terms of visualizing actual switching waveforms on an oscilloscope and correlating non-idealities and non-linearities in those waveforms to levels of distortion. While extremely useful, fast Fourier transforms (FFTs) display distortion components at specific frequencies, but give no information as to the mechanisms producing those distortion components. Concerning switching deadtimes (necessary in any Class-D design to avoid cross-conduction of the power switches), these will contribute to THD, and the explanation is illuminating. If ideal switches and power rails are assumed, and the switching stage had zero deadtime and infinitely fast switch rise and fall times, the output PWM timing would perfectly match the input PWM timing, except for voltage scaling. There would be no THD introduced by this stage. The introduction of deadtime produces a timing mismatch between the PWM driving waveform and the output high-voltage PWM switching waveform. Since the actual output voltage at the switching node during these deadtime intervals is load dependent (as will be explained in this chapter), it has a large effect on THD.

When output current is low, such as when the amplifier audio output is close to zero volts, the amplifier can be said to be in "idle" or no-signal mode. For a fixed-frequency design, the power stage will be switching at a 50% duty cycle, or very close to it. In this mode, resonant or "soft-switching" is possible. In this mode, the current flowing through each mosfet switch at the moment the mosfet switches off, slews the output capacitance at the switching node towards the opposite rail. If this current is high enough, the switch-node capacitance low enough, and the deadtime long enough, the switch-node voltage can slew all the way to the opposite rail within the deadtime interval, and the mosfet switches will incur zero switching loss. However, if the opposite occurs, i.e. the current at switch-off is not high enough, or the switch-node capacitance not low enough, or the deadtime interval not long enough, the opposite mosfet will switch on *before* the voltage has resonantly slewed all the way to the opposite rail. In these cases, the mosfets *will* incur switching loss. As in so many areas of Class-D audio amplifier design, this example illustrates another area where compromise is required. Increasing the deadtime might allow for lossless mosfet switching, but adds distortion to the switch-node waveform. Remember, in this sense, switch-node waveform distortion is a measure of how well the switch-node waveform represents an amplified version of the input PWM stream from the PWM comparator. Figure 10.1 illustrates the soft-switching concept.

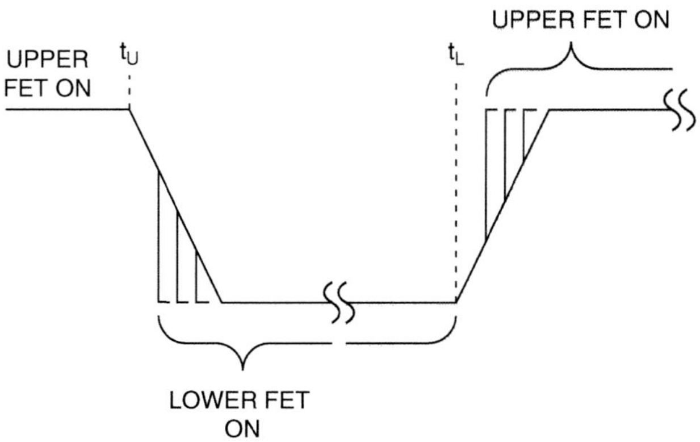

Figure 10.1 Continuum of Mosfet Switch Timing, from Hard-Switched to Increasingly Soft-Switched, as Deadtime is Increased

Figure 10.1 starts with the upper mosfet ON. At the moment the upper mosfet switches OFF (t_U), deadtime begins (both mosfets are off). The switching node voltage slews resonantly towards the opposite voltage rail. The three vertical line segments depict three different turn-on timings for the lower mosfet. If the switching node voltage is able to slew all the way to the opposite rail *before* the end of the deadtime, then completely lossless switching occurs. If, however, the lower mosfet switches ON before the end of the deadtime, a noticeable inflection will be seen in the switch-node waveform at the instant the opposite mosfet switches on, and there will be switching losses incurred.

When higher current flows in the switching stage, *hard-switching* will occur, and switching losses will be incurred (to be discussed in detail in the next chapter). A thorough understanding of the mechanics of hard-switching is critical for understanding Class-D switching stages.

Figures 10.2, 10.3, and 10.4 illustrate hard-switching.

Figure 10.2 illustrates the "falling-edge" switching interval. Just prior to t_0, the upper mosfet is on, and a large current is flowing from the positive supply rail, through drain-to-source, through inductor L1A towards the speaker load at Vout. This current causes a voltage drop across the upper mosfet from drain-to-source (V_{DS}ON) which is equal to the product of drain current × R_{DS}ON. In Figure 10.2 this voltage drop is labelled V_{ONu}. The time interval during which the switch-node voltage slews from the positive rail towards the negative rail is labelled $t_{SW\ OFF}$ = t_0 to t_1. The duration of the time interval is determined by how fast the upper

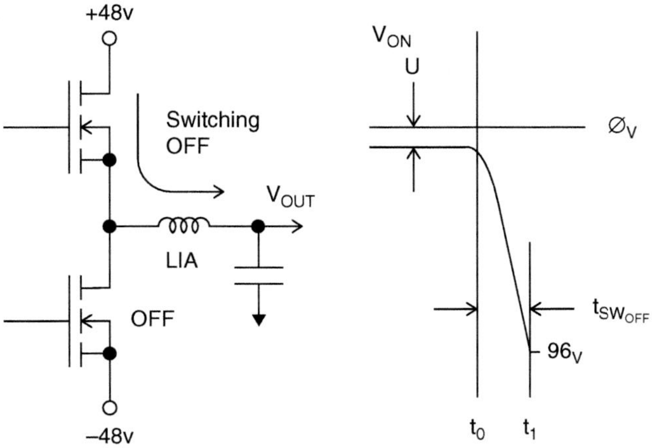

Figure 10.2 Falling Edge Transition of the Switching Node of a Half-Bridge Stage, as the Upper Mosfet Switch Turns off, during t_0–t_1

mosfet is switched off. At time t_1, the switch-node voltage has reached the opposite rail, and there is now the full total rail voltage (96V) across the upper mosfet.

Figure 10.3 illustrates what happens when the switch-node voltage reaches the negative rail at time t_1. During the deadtime interval (t_1–t_2), both mosfets are off, so the body diode of the lower mosfet acts as a "catch diode" and clamps the switch-node voltage to one diode drop below the negative rail. During this deadtime interval, current is flowing through the body diode of the lower mosfet, from source-to-drain, so the voltage at the switch node is actually one diode drop *more negative* than the negative rail voltage. At time t_2 deadtime ends and the lower mosfet switches ON. The voltage across the lower mosfet is now the product of the inductor current × R_{DS}ON of the lower mosfet. But, it is important to note that due to the current direction, the switch node (drain) is more negative than the negative supply rail (source). This condition remains through the entire ON-time of the lower mosfet.

Figure 10.4 depicts the next switching interval, where the lower mosfet switches OFF at t_3. This is the start of the next deadtime interval. Again, due to the current direction, current will transfer from the source-drain channel of the lower mosfet, to its body diode. Again, the switch node will be clamped to one diode drop more negative than the negative rail. At time t_4 the deadtime ends, and the upper mosfet begins to switch ON. This is an extremely critical interval for

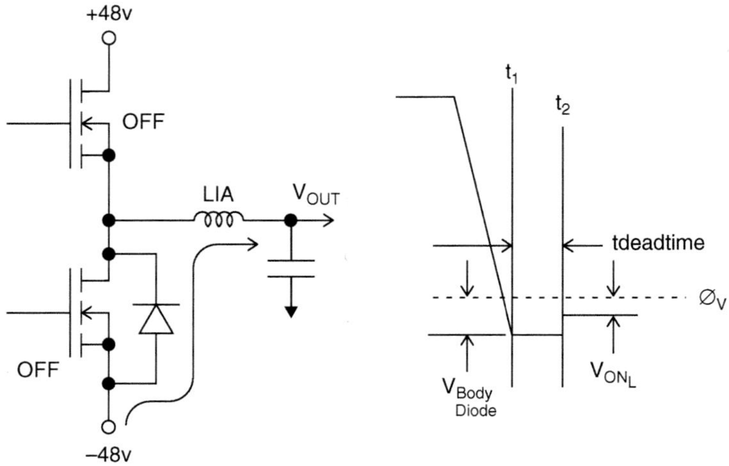

Figure 10.3 Falling Edge Transition of the Switching Node of a Half-Bridge Stage, as the Upper Mosfet Switch Turns off, Followed by the Deadtime Interval of t_1–t_2

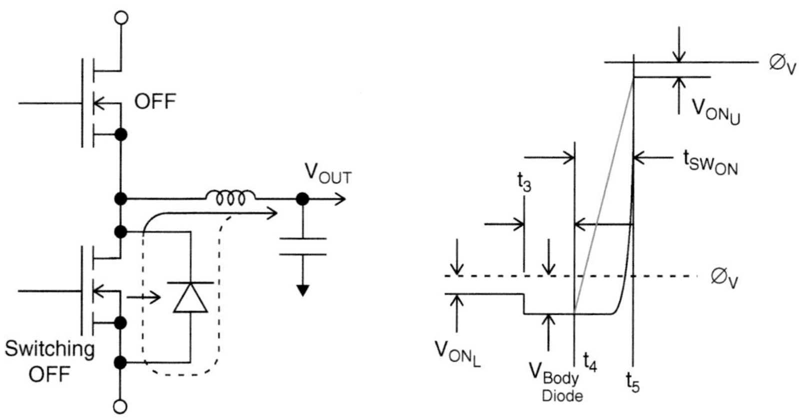

Figure 10.4 Switching Node of a Half-Bridge Stage, as the Lower Mosfet Switches off, Followed by Deadtime Interval t_3–t_4

the power stage, because current flow through the conducting p-n junction body diode of the lower mosfet must cease, before the switch-node voltage can begin to slew positive. This is the *reverse recovery* interval for the lower mosfet's body diode. The switching characteristics described here are quite dependent on the specific reverse-recovery characteristics of this body diode. The next chapter will deal with this in detail. For the purposes of this chapter (distortion effects), it is enough to say that there is a time delay (t_4–t_5) during which the upper mosfet supplies reverse recovery current and charge to the body diode of the lower mosfet to sweep out the minority carriers in its p-n junction, and make it switch off ("commutate"). This time varies between devices and is called the Reverse Recovery Time t_{rr}. Once reverse recovery of the lower mosfet's body diode has been completed, the switch-node voltage rises, and will ring, dependent on many factors (switch-node capacitance, stray inductance, switching speed, reverse recovery characteristics (snappy, soft, etc.)). At t_5 the ON-time for the upper mosfet continues, and the switch-node voltage will be the same as it was at the start of the sequence at t_0 (inductor current x R_{DS}ON of upper mosfet).

The above sequence examined the condition where high current is flowing out of L1A towards the speaker. The same concepts apply for the opposite case, where high current flows *from* the speaker, through L1A. The waveforms will be the same, but will apply to the opposite mosfet, i.e. it is the upper mosfet's body diode that will conduct.

Another source of THD is the mosfet body diode itself. The body diode voltage is load and temperature dependent, and thus affects the resultant switching output voltage waveform, contributing to THD. The mosfet body diode conducts when the current flows from source to drain, and the mosfet channel is OFF. The body diode will also conduct when the mosfet is ON, but only when the R_{DS}ON × Current product exceeds the forward voltage of the body diode. However, for most N-channel mosfets, this would only occur at very high levels of drain current. For example, if R_{DS}ON for the IRF6775 is 47mΩ, and the body diode starts to conduct at 0.6V, then the ON-state drain current for the mosfet would need to exceed 1V/47mΩ = 21.3A, before the body diode would begin to conduct while the mosfet channel is on. However, for this discussion, the operating condition of interest for the body diode is when current is flowing from *source*-to-*drain*, with the mosfet channel off. This occurs during the switching deadtimes. For the lower mosfet, this will occur when a positive current is flowing in the demodulation filter inductor (i.e. from switching node-to-speaker), and the upper mosfet has just switched off. For the upper mosfet, this will occur when a negative current is flowing in the filter inductor (i.e. from speaker to switching node), and the lower mosfet has just switched off. Under these conditions, the mosfet's source pin voltage will be positive with respect to its drain voltage, and current will flow in the mosfet's body diode, from source to drain. At the end of the deadtime, when the mosfet switches ON, current will transfer from the body diode to the channel, and the ON-state voltage will transition from the forward voltage drop of the body diode, to the $R_{DS}ON \times$ Current product. Therefore, the actual voltage magnitude (V_{SD}) will depend on the current amplitude and whether the mosfet is ON or OFF, and the $R_{DS}ON$ of the mosfet. For example, if 1A of current flows through the IRF6775 mosfet at 25 degrees C, when its $R_{DS}ON$ is 47 milliohm, the voltage V_{SD} will be −0.047V. If the mosfet is off, and the body diode is conducting, it will have a much higher V_{SD} equal to the forward voltage drop of the body diode at this 1A current level (perhaps 0.7V–1V). The body diode's forward voltage is both temperature and current dependent, decreasing with increasing temperature, and increasing with increasing current. In addition to this, the mosfet's $R_{DS}ON$ is also highly temperature dependent, increasing with temperature.

To summarize:

1) During the switch deadtime intervals, only the mosfet body diode can conduct. During deadtime, V_{SD} will be equal to the forward voltage drop of the body diode at the specific current level and temperature. For the IRF6775, this can be as high as 1.3V at 5.6A and 25 degrees C. During switch ON-time (with current flow from source-to-drain), V_{SD} will be the product of current × $R_{DS}ON$. Also, $R_{DS}ON$ has a temperature dependency as well. In mosfets, there is a positive temperature coefficient, so $R_{DS}ON$ will increase with increasing temperature. For the IRF6775, $R_{DS}ON$ will increase to 1.5 times its 25 degree C value when the junction temperature reaches 100 degrees C.
2) During the actual rising and falling voltage switching intervals, the timing and linearity of the switch-node voltage waveform also depend on many factors. These include the direction and amplitude of the inductor current, the magnitude of the total switch-node capacitance (which must be charged and discharged during the rising and falling voltage transition), as well as the turn-on and turn-off switching speed of the mosfet switches (which is controlled by the available gate drive source and sink current to each mosfet).

Other Sources of Noise and Distortion

Passive Components

Capacitors

Capacitors have some non-ideal characteristics that affect performance. These are:

- Voltage coefficient (of capacitance)
- Temperature coefficient (of capacitance)
- ESR (equivalent series resistance)
- ESL (equivalent series inductance)
- Piezoelectric effects.

Voltage coefficient of capacitance is particularly important for multi-layer ceramic capacitors (MLCCs). When engineers first encounter this, the usual response is disbelief. How can a 4.7µF capacitor have an effective capacitance of only 1.88µF at its rated voltage (KEMET, n.d.)? As can be seen in Figure 1 of KEMET (n.d.), this is in fact so, and is a characteristic of the dielectrics typically used for MLCCs in Class-D amplifier design (X7R and X5R are preferred). It therefore is important to design for this change in capacitance with applied voltage. The solutions are typically to use a larger nominal capacitance value, and/or to use an MLCC with a higher rated voltage. Or to use multiple capacitors in parallel. The X7R dielectric has better temperature stability and higher operating temperature than X5R, but trades off volumetric efficiency (maximum capacitance in a given case size). C0G/NP0 dielectrics have the best overall performance of all ceramic capacitor dielectrics, but volumetric efficiency is poor, so case sizes are much larger than for X7R and X5R at the same capacitance and voltage. Capacitances greater than 0.1µF are rarely used in C0O/NP0 in Class-D designs, because size and costs become prohibitive.

There is also an AC voltage effect on capacitance for MLCCs (as discussed in KEMET, n.d.). Capacitance will vary with the AC voltage impressed across the capacitor. It is important to consider this for AC signal-coupling applications. For an AC signal-coupling circuit, consisting of an R-C, the AC voltage across the coupling capacitor will increase as the frequency drops. At higher frequencies (above the corner frequency), the AC voltage across the capacitor will asymptote to 0VAC. For this reason, as well as the volumetric inefficiencies of large-value ceramic capacitors in the C0G/NP0 dielectric, ceramic capacitors are rarely used for low-frequency (down to 20Hz) AC signal-coupling applications in Class-D amplifiers.

It is important to note that aluminum electrolytic (AE), film, and C0G/NP0 dielectric ceramic capacitors do not suffer from the voltage coefficient phenomenon. Another circuit location where this effect can easily add to THD is in the capacitor used in the L-C demodulation filter. The linearity of that filter with voltage amplitude is critical. If the capacitance value changes with applied voltage, this will produce non-linearity and THD.

The temperature coefficient of capacitance is less critical for ceramic and film types, but does affect AE capacitors. In AE capacitors, ESR increases at low temperature, and capacitance dropoff with frequency occurs at lower frequencies as temperature drops (Cornell Dubilier, n.d.). This would be most important in the power supply rails, where the PSRR of the amplifier is critical. Increased ESR and decreased capacitance will result in greater power supply ripple and noise (especially at the higher switching frequencies used in switch-mode power supplies). The effect of this higher noise and ripple level on the output noise and distortion of the amplifier will depend on the PSRR of the design (which is strongly affected by negative feedback and the negative feedback factor).

ESL is mostly a packaging phenomenon. Leaded parts have higher parasitic inductances than surface-mounted (SMD) parts.

Film Capacitors

In addition to being available as either a metal foil or vacuum-deposited aluminum, film capacitors are also available as either a wound foil or stacked foils, or deposited films. Wound foils can introduce a non-linear inductive component which can have a sensitivity to stray fields (Muniz, 2019), which can be present in Class-D amplifiers. For this reason, stacked foils are preferred for Class-D applications.

Different capacitor dielectrics have different temperature sensitivities. Thus, the capacitance change with temperature will vary. It is always advisable to be mindful of the operational temperature range of any selected capacitor, and to consider heat sources that may be in close proximity to the capacitor during PCB layout and component placement. For film capacitors used in circuit areas that experience high-frequency AC currents, such as the output L-C filter, or the high power rail bypass capacitors, be mindful of RMS currents and internal heating.

Use the nomographs provided by the capacitor vendors to determine temperature rise.

AE Capacitors

For AE capacitors, consult the datasheet for ESR values. The internal power dissipation can be calculated knowing this ESR value and the capacitor current. Temperature is an important concern for AE capacitors, because it directly influences service life. AE capacitors are rated for service life (in hours) at specific operating temperatures. For Class-D applications, 105 degree C rated AE capacitors are recommended. Since operational life roughly *doubles* with every 10 degrees C reduction in operating temperature, life for a typical AE capacitor could be extrapolated as follows.

Example datasheet specification = 10,000 hours at 105 degrees C.

Extrapolating...

- 20,000 hours at 95 degrees C
- 40,000 hours at 85 degrees C
- 80,000 hours at 75 degrees C
- 160,000 hours at 65 degrees C.

One year = 8760 hours. So, this capacitor could be expected to achieve close to a 10-year operational life, if the amplifier were ON 24 hours a day, 7 days a week, at a local ambient temperature (in the immediate vicinity of the capacitor) of 75 degrees C. With normal ON/OFF cycling of the amplifier in normal use, it will last much longer than ten years. *Note* that it is the local ambient air temperature around the capacitor that matters here, as well as the PCB temperature, because the leads of a through-hole AE cap will be approximately the same temperature as the PCB, and these leads connect internally to the foils and dielectric. For an AE capacitor that is not dissipating significant internal power, its internal and external temperatures will be very close to the local ambient air temperature and/or PCB temperature, unless there is moving air (fan-cooling).

Resistors

Resistors exhibit distortion due to their voltage coefficients and temperature coefficients. Smaller resistors have less thermal mass and therefore change temperature more rapidly than larger resistors (physical package size). This effect can be seen up to about 200Hz (Muniz, 2019). Muniz (2019) advises limiting power in SMD resistors to less than 20mW, by employing series/parallel arrangements of multiple resistors.

For SMD resistors, thin film types are more accurate and exhibit lower temperature coefficients than thick film types, but are more expensive and less available. They also exhibit lower current noise. Wire wound resistors are not recommended for Class-D audio applications, because their inductive winding construction can pick up stray magnetic fields. A winding technique known as Ayrton-Perry reduces inductance, but these types still exhibit some inductance, and are therefore not recommended. *Note:* This should also be kept in mind when using "dummy-load" resistors for testing (in place of loudspeakers). The inductance of these load resistors can skew the measured results. Since high-power load resistors of this type (up to 2KW) are ubiquitous, it is almost impossible to not encounter these when testing a Class-D amplifier. It is always advisable to do a "sanity-check" by testing the amplifier with truly non-inductive loads as a comparison, to see if there is any load-effect in the measured response. The inductive load resistors can always be used for other tasks such as burn-in, or where optimum performance is not required.

Inductors

The inductor used in the L-C demodulation filter has been dealt with in detail in other chapters. However, with regard to distortion effects, it is linearity and saturation effects which will have the most influence. Low-permeability cores, such as Micrometals -2 material, will have very high linearity. Air core inductors have excellent linearity and freedom from saturation, but it is difficult to achieve any significant useful inductance value in a reasonable size, and there is no containment of the stray magnetic field, which can interfere with other circuits. Gapped structures (such as gapped ferrite cores) can have excellent linearity, as the inductance is determined by the air gap rather than the core material itself. However, care is needed to prevent stray fields from the gap from interfering with neighboring circuits. In a multi-channel design, where adjacent channels' inductors are in close proximity, such that their stray fields could interfere with each other, it is sometimes helpful to alternate their orientations by 90 degrees to cut down on this stray field interference. Another phenomenon caused by having a gap in a ferrite core is called fringing flux. This is flux that bulges out from the air gap and impinges on wire turns that are in close proximity to the gap. These fringing flux lines can cause eddy currents to flow in these wire turns, causing them to heat up. Also, this power loss decreases amplifier efficiency. The solution is to move wire turns away from the air gap, or to use a distributed gap instead of a single gap. Interestingly, the powdered core materials are essentially distributed gap materials.

Diodes

When using Transient Voltage Suppressors (TVSs), make sure in selecting for clamping voltage that there is sufficient voltage margin, to avoid distortion caused by partial TVS conduction at signal peaks.

Mosfet Body Diode

The effect of the mosfet body diode was previously discussed.

Comparator and Ramp Circuitry

The comparator ramp is particularly sensitive to noise and can cause jitter. Any noise spikes near the switching threshold can result in premature switching and therefore duty cycle errors.

Figure 10.5 illustrates noise spikes on the triangle ramp at the PWM comparator input. The ramp is applied to the comparator's negative input, and the positive comparator input is from the output of the error amplifier, and is at 0V in this example. In the absence of any noise, the comparator output should switch

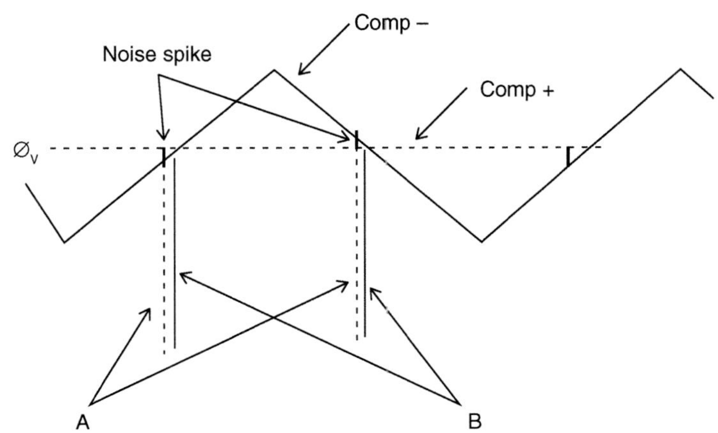

Figure 10.5 Effect of Noise Glitches on the Triangular Ramp Voltage Seen by One Input of the PWM Comparator

at the time points labeled "B". However, the noise spikes make the comparator switch prematurely at time points "A". One cause of these noise spikes could be the switching of mosfets in an adjacent Class-D amplifier channel, in a multi-channel design.

Of course, the linearity of the triangle ramp itself is crucial. Any nonlinearity in the ramp will contribute directly to distortion.

Muniz (2019) advises the use of a low-frequency, low-jitter clock and a *synchronous counter* for frequency division, to reduce jitter.

Gate Driver

Differences in the supply voltage for the upper and lower drivers can cause differences in propagation delays and switching times, which cause pulse-width distortions. These supply voltage differences are caused by the bootstrap circuitry (discussed in Chapter 7) as well as the actual switching conditions (current direction etc.).

Switching Power Stage

Both hard and soft switching conditions were described above. When the average inductor current exceeds the peak ripple current, the system transitions from soft-switching to hard-switching.

Pulse *amplitude* distortion is caused by bus-pumping, ringing, and high- or low-frequency supply ripple.

Summary

In summary, main sources of distortion in Class-D amplifiers are:

- Open loop:
  1) Deadtime
  2) Switching time, and shape of rising/falling edges
  3) ON-time voltage effects (R_{DS}ON and body-diode V_F)
  4) Noise on triangle ramp (for fixed-frequency PWM designs)
  5) Power supply rail noise and ripple
  6) Passive components.

- Closed loop (to be discussed in Chapter 12 on Feedback):
  1) Phase and duty cycle error caused by feedback (exacerbated by excessive gain at $F_{SWITCHING}$ and ½ $F_{SWITCHING}$). See Shu and Chang (2008) and Attwood (1983).

Bibliography

Attwood, B. E. (1983). Design parameters important for optimization of very high-fidelity PWM (Class D) audio amplifiers. *Journal of the Audio Engineering Society, 31*(11), 842–853.

Cornell Dubilier. (n.d.). Aluminum electrolytic capacitor application guide. Liberty, SC: Cornell Dubilier.

Esslinger, R., Gruhler, G., & Stewart, R. W. (2001). Digital audio power amplifiers using sigma delta modulation – linearity problems in the Class-D power stage. AES Convention Paper 5400. Paper presented at the 110th AES Convention, Amsterdam, the Netherlands, May 12–15, 2001.

KEMET. (n.d.). Tech Topic: VCC: Capacitance change vs voltage in ceramic capacitors. Fort Lauderdale, FL: KEMET Corporation.

Mosely, I. D., Mellor, P. H., & Bingham, C. M. (1999). Effect of dead time on harmonic distortion in Class-D audio power amplifiers. *Electronics Letters, 35*(12), 950–952.

Muniz, R. (2019). Noise and distortion mechanisms encountered in switching audio amplifier design. AES e-Brief 549, 147th Convention, October 16–19, 2019. New York: Audio Engineering Society.

Nielsen, K. (1998). Linearity and efficiency performance of switching audio power amplifier output stages – a fundamental analysis. Paper presented at the 105th AES Convention, San Francisco, CA, September 26–29, 1998.

Nyboe, F., Risbo, L., & Andreani, P. (2005). Time domain analysis of open loop distortion in Class-D amplifier output stages. Paper presented at the AES 27th International Conference, Copenhagen, Denmark, September 2–4, 2005.

Shu, W., & Chang, J. (2008). THD of closed-loop analog PWM Class-D amplifiers. *IEEE Transactions on Circuits and Systems, 55*(6), 1769–1777.

Shu, W., Chang, J., Ge, T., & Tan, M. T. (2006). Fourier series analysis of the nonlinearities in analog closed-loop PWM Class-D amplifiers. In: *Proceedings of the 2006 IEEE International Symposium on Circuits and Systems (ISCAS)*, May 21–24, 2006, Island of Kos, Greece (p. 4). New York: IEEE.

Tan, M. T., Chang, J. S., Chua, H. C., & Gwee, B. H. (2000). An investigation on the parameters affecting total harmonic distortion in Class-D amplifiers. Paper presented at ISCAS 2000 – IEEE International Symposium on Circuits and Systems, Geneva, Switzerland, May 28–31, 2000.

Vanderkooy, J. (1985). Comments on "Design parameters important for the optimization of very-high-fidelity PWM (Class-D) audio amplifiers". Letter to the Editor. *Journal of the Audio Engineering Society, 33*(10), 809–810.

Chapter Eleven

Class-D Power Stage Losses, a Deep Dive

Overview

High power conversion efficiency is the hallmark of Class-D amplifier design. It could be argued that it is the largest benefit achieved by adopting this class of amplifier instead of other classes. For this reason, it is justified to take a deeper dive into understanding the loss mechanisms in Class-D. A better understanding of these mechanisms is necessary to be able to make intelligent component selection decisions, and to optimize efficiency for a given application.

This discussion will focus on the losses in the power switches. These losses can be divided into the following areas:

- Conduction losses
- Switching losses
- Reverse recovery losses
- Capacitive losses
- Gate drive losses.

Conduction losses are the easiest to explain and understand. These losses result from the flow of current through a non-ideal power switch, having an on-state resistance. In the case of a silicon *mosfet* it is current flow through the $R_{DS}ON$ of the channel. In a Class-D amplifier, the audio frequency output current is essentially equal to the L-C filter inductor current.[1] At all times, this current flows through *either* the upper or lower mosfet in a *half-bridge* power stage topology, since these mosfets alternate ON/OFF switching. Furthermore, since *both* the upper

DOI: 10.4324/9781003327943-11

and lower mosfets have identical on-resistance ($R_{DS}ON$), the model for conduction losses can be represented by a single "ON" switch in series with the output inductor, and conducting the full output RMS current. Therefore, *conduction losses* can be approximated by the relation:

$$P_{CONDUCTION} = I_{OUT}^{2} \times R_{DS}ON$$

This will result in the *total conduction loss.*

For the reference design, using IRF6775 mosfets with an $R_{DS}ON$ of 47mΩ, the highest conduction loss will occur with a 4-ohm/100W load, where 5ARMS is the output current.

Total conduction losses for the mosfets will be: $(5)^2 \times 47\text{m}\Omega = 1.175\text{W}$

Each of the two IRF6775 mosfets will dissipate 1.175W/2 = 587.5mW

Switching losses are a bit more complicated, but the discussion below provides a useful simplification.

It is important to understand how the conduction modes for the power mosfets change as the inductor current waveform changes within a Class-D switching period T, where T = 2µsec for the reference amplifier. Recall that when the output current is zero, the current flowing through the inductor is only the residual switching frequency ripple current. It is a triangular RAMP waveform, ramping linearly up from a negative peak current to a positive peak current while the upper mosfet is on, and ramping down from a positive peak current to a negative peak current when the lower mosfet is on, as illustrated in Figure 11.1 using values from the reference design.

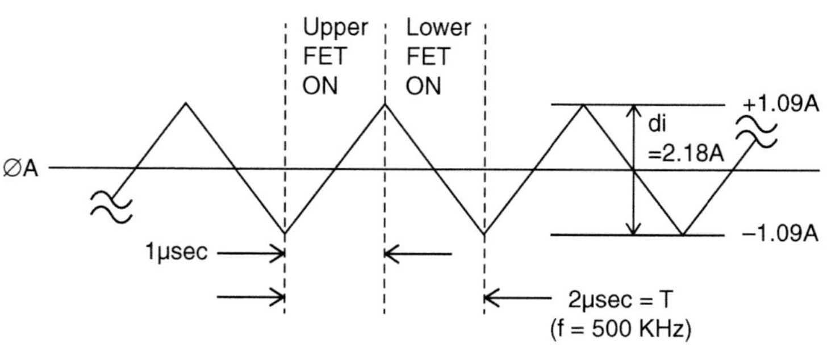

Figure 11.1 Inductor Ripple Current at Idle (the No Audio Signal Condition) for the Reference Amplifier

Using the inductor equation:

$$V_L = (L)(di/dt)$$

and rearranging:

$$di = (V_L)(dt) / L = (48V)(1\mu sec) / 22\mu H = 2.18A$$

This means that when the amplifier is at idle (producing Vout = 0V), the inductor current will be ramping symmetrically about 0A, from −1.09Apeak to +1.09Apeak. Since the current reverses polarity during the *ON-time* of each mosfet, *resonant switching* is possible. During resonant switching, the energy stored in the inductor is sufficient to passively slew the switching-node voltage from one rail to the opposite rail. This occurs when the ON mosfet switches off, and *before* the opposite mosfet switches on. This means that essentially lossless switching can occur at idle. However, this is dependent on deadtime. If deadtime is long, there can be sufficient time for the inductor current to slew the switch-node capacitance from one rail to the opposite rail. However, if deadtime is short, or switch-node capacitance is high, or if inductor peak current is low, the switch-node voltage may not be able to slew completely from one rail to the other *before* the opposite mosfet switches on and actively slews the switch node the remaining voltage to the opposite rail. This is illustrated in Figure 11.2.

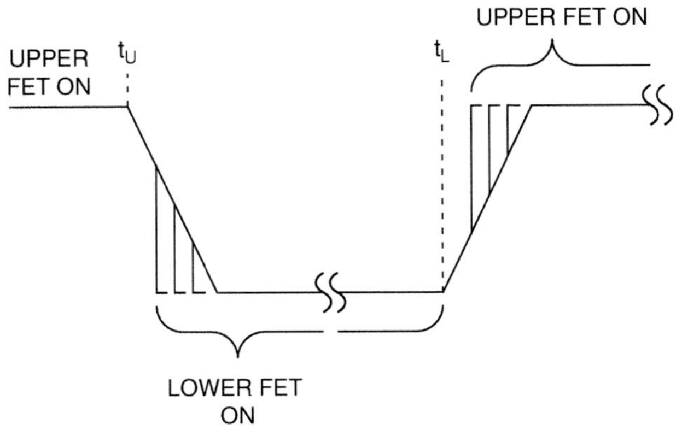

Figure 11.2 Resonant Switching Voltage of the Class-D Power Stage at Idle, with Different Deadtime Intervals

Figure 11.2 timing starts with the upper mosfet ON. The upper mosfet switches off at time t_U. At that instant, current is flowing out of the inductor, towards Vout, with a peak amplitude of +1.09A. Due to inductive action, the current will continue to flow in that same direction as the upper mosfet switches off. This current will force the switch node to slew towards the negative rail. The capacitance at the switch node therefore is passively discharged by this resonant switching action. The different vertical lines in Figure 11.2 illustrate different deadtimes, ranging from maximum deadtime, with fully resonant switching, to progressively shorter deadtimes and hard-switching. In the hard-switching cases, the opposite mosfet switch turns on and actively slews the switch node (and its capacitance) before the switch-node voltage has had time to reach the opposite rail. Those cases result in higher levels of quiescent power loss in the mosfets. At this idle condition, since the currents are low, the primary loss component is usually capacitive switching loss, which will be discussed in this chapter as well.

The next condition to be analyzed will be mosfet switching loss at peak output current. The maximum losses will occur with the maximum output current. For the reference design, capable of delivering 100W of output power, the peak current will occur with a 4-ohm load. This will result in output values of 20VRMS, +/−28.28Vpeak and 5ARMS, +/−7.07Apeak. Peak current will be rounded off to +/−7A for simplification.

Figures 11.3 and 11.4 illustrate an important concept for understanding Class-D switching behavior. It shows why the switching *mode* of the mosfets changes as the operating conditions change. Figure 11.3 illustrates the inductor current ramping up from zero. Recall that at IDLE (audio frequency output current = 0A), the inductor current (switching frequency current) ramps between +1.09Apeak and −1.09Apeak. However, as the inductor (and output) current increase away from 0A average, there will come a point where the inductor current, though still ramping

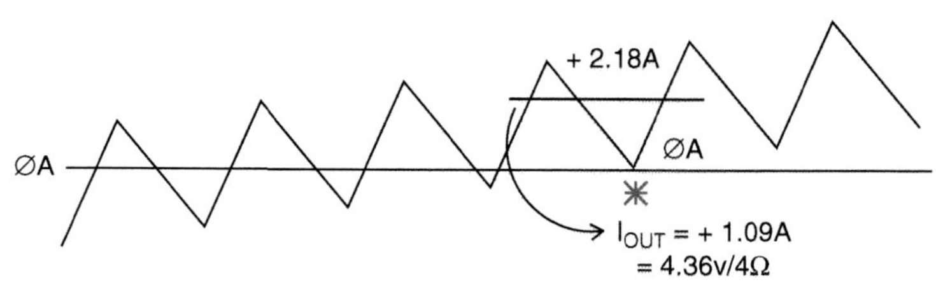

Figure 11.3 Inductor Ripple Current with Increasing Output Current

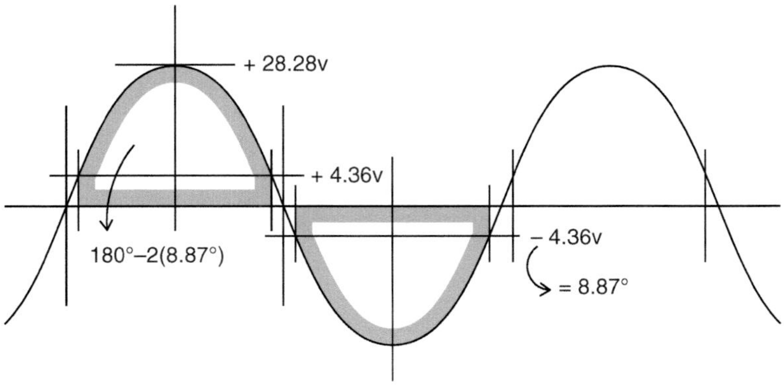

Figure 11.4 Time Intervals during which the Mosfets of the Class-D Switching Power Stage Operate in Hard-Switched Mode

in a triangular fashion, will remain in a single polarity (direction) over the full switching cycle. This is seen to occur at the output current level indicated by the asterisk (*). At this point (where Iout = 1.09A), the inductor and output current remain positive over the full Class-D switching cycle. This level of current will result in an instantaneous output voltage of +4.36V into 4 ohms. Next, assuming a sinusoidal output voltage where 28.28V is equal to $Vout_{peak}$, an output amplitude of +4.36V occurs at 8.87 degrees. This means that the resonant switching condition discussed previously will only occur between −8.87 and +8.87 degrees of the output waveform's 360-degree period. For the remainder of the output voltage waveform period (illustrated by the shaded areas in Figure 11.4), the mosfets will be operating in *hard-switched mode.*

Hard-Switched Mode

For this analysis, Figures 11.5, 11.6, and 11.7 will be used. The peak power condition will be analyzed, which occurs at $Iout_{PEAK}$ = 7A. It must be recognized that the conditions analyzed for the upper mosfet during the positive half-cycle of output current apply equally to the lower mosfet for the negative half-cycle of output current. Therefore, only a half-cycle of sinusoidal output voltage/current operation needs to be analyzed. Assume that, just prior to time t_0, 7A is flowing through the upper mosfet in a direction from the +48V rail, through

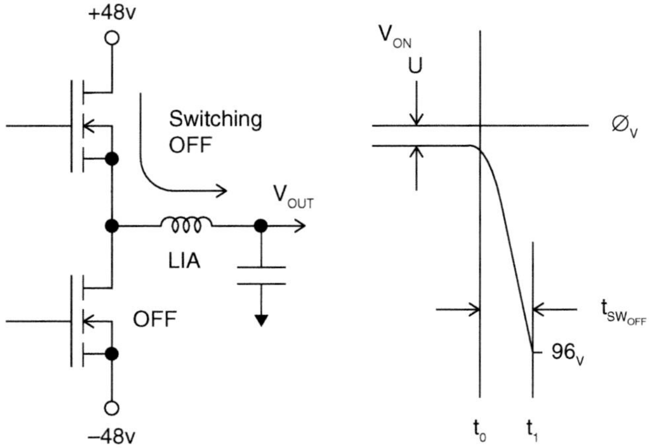

Figure 11.5 Switching Interval t_0–t_1, when the Upper Mosfet is Switching off, and the Switching Node Voltage is Falling

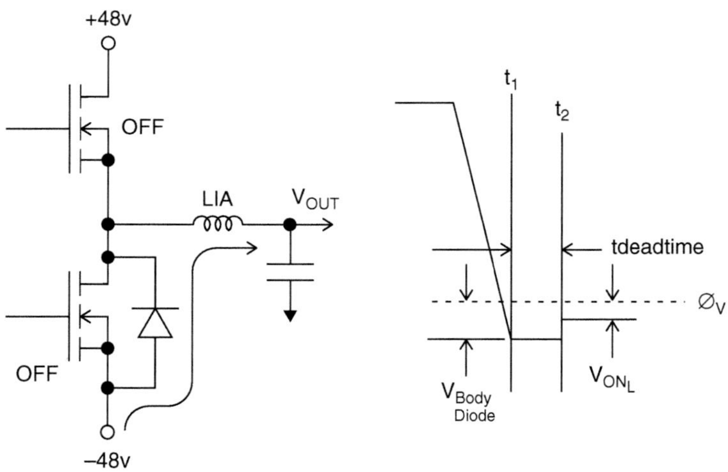

Figure 11.6 First Deadtime Interval (t_1–t_2), after the Upper Mosfet Has Switched off, and before the Lower Mosfet Has Switched on

drain-to-source, then through the inductor L1A, towards Vout. At time t_0 the upper mosfet begins to switch off. The 7A current will continue to flow through the upper mosfet over the entire switching time interval (t_{SWoff}).

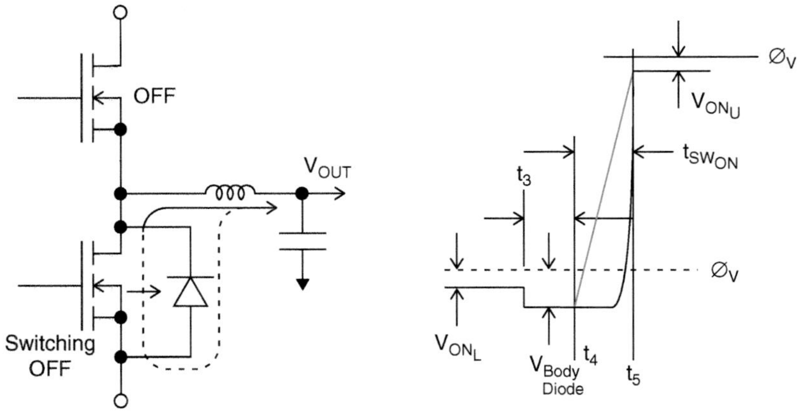

Figure 11.7 Second Deadtime Interval (t_3–t_4), Followed by the Reverse Recovery Interval (t_4–t_5)

The *average power dissipation* due to this specific switching interval (averaged over the Class-D switching period of 2μsec at the 500KHz switching frequency) is:

$$P_{D(t0-t1)}\ (Iout)(V_{DS\text{-}AVG})(t_{SW}/T) = (7A)(96V/2)(25nsec/2\mu sec) = 4.20W$$

Note:

1) Iout remains constant over this small time interval, because inductor current cannot change instantaneously.
2) V_{DS} for the upper mosfet changes from practically 0V (when it is ON) to the full differential rail voltage of 96V. So the average V_{DS} voltage is 96V/2 = 48V.
3) A switching time (t_{SW}) of 25nsec is assumed. This can be decreased or increased by design, depending on mosfet gate-drive conditions and component selection.

The next interval to be analyzed is the deadtime interval between the switch off of the upper mosfet and the switch on of the lower mosfet. During this interval (t_1–t_2 in Figure 11.6), the lower mosfet is not yet switched on, so the inductor current will flow through the *body diode* of the lower mosfet.

In this interval, the power dissipation (again averaged over a 2μsec Class-D switching period) is:

$$P_{D(t1-t2)} = (Iout)(Vsd)(t_{DEADTIME}/T) = (7A)(1.3V)(50nsec/2\mu sec) = 227.5mW$$

Note:

1) Deadtime is assumed to be 50nsec. This is adjustable.
2) Vsd is the forward voltage of the *body diode* of the lower mosfet. For the IRF6775, a voltage of 1.3V is used. This value is taken from the datasheet (Infineon (International Rectifier), 2014). It is at a temperature of 25 degrees C and a current of 5.6A. This voltage will increase with higher current, and will decrease with higher temperature.

At time t_2 the lower mosfet switches on and current transfers from its body diode to its channel. This will normally be seen (on an oscilloscope) as a decrease in Vsd voltage, since the voltage drop of the ON channel (Id × R_{DS}ON) will normally be less than the forward voltage drop of the body diode. For the IRF6775, this will be: (7A × 47mΩ = 329mV) compared to 1.3V for the body diode. It is important to note that during this interval, the drain voltage of the lower mosfet will be more *negative* than its source voltage, due to the current direction. This time interval (t_2–t_3) is a conduction time interval, contributing to *conduction losses*, which are not part of this *switching loss* analysis. *Note:* t_3 is shown in Figure 11.7.

The next switching interval begins at time t_3, when the lower mosfet switches OFF. This causes the drain current to transfer from the channel of the lower mosfet back to its body diode, where again, the losses are (averaged over the 2μsec switching period):

$$P_{D(t3-t4)} = (Iout)(Vsd)(t_{DEADTIME}/T) = (7A)(1.3V)(50nsec/2\mu sec) = 227.5mW$$

The next interval begins at t_4 and is arguably the most interesting (see Figure 11.7).

This is when mosfet body diode commutation occurs. The definition of commutation is when forward current flow is stopped in a free-wheeling diode. In this case, that free-wheeling diode is the body diode of the lower mosfet. The mechanism of body diode turn-off in this section will be discussed in greater detail later in this chapter. For now, we will say that at time t_4 the upper mosfet begins to switch on. It is important to realize that during this entire switching event, the current has been assumed to remain essentially constant at the 7A peak level due to the inductance of the L-C filter (remember we are ignoring the inductor's switching frequency ripple current component for this simplified calculation). During this time interval, 7A is flowing through the body diode of the lower mosfet. As the upper mosfet begins to switch on, and starts supplying some of this current to the inductor path, the current through the lower mosfet's body diode will decrease accordingly, *but* since *some* current is still flowing through this body diode, it is still biased at its forward voltage, therefore the switch node remains at $-48V - V_{SD} = -48V - 1.3V = -49.3V$ over this entire interval, until the current in the body diode has gone to 0A, and it can again support reverse voltage. This means that during the entire t_3–t_5 interval, as the current in the upper mosfet is ramping up from 0A to 7A, the voltage across the upper mosfet has remained at (48V + 49.3V) = 97.3V.

Therefore, the switching power losses during this interval (averaged over the 2μsec switching time period) are:

$$P_{D(t4-t5)} = (Iout/2)(V_{SUPPLY\text{-}TOTAL})(t_4-t_5)/T) = (7A/2)(97.3V)$$
$$(50nsec)/2\mu sec) = 8.51W$$

Notes:

1) It is assumed that the current in the upper mosfet ramps up from 0A to 7A during this interval, therefore the *average* current through the upper mosfet during this interval is Iout/2.
2) As discussed previously, the voltage across the upper mosfet remains constant at the full total supply voltage *plus* the forward-biased body-diode voltage of the lower mosfet.
3) The time interval is assumed to be 50nsec for this example, but is variable depending on gate-drive conditions (adjustable via component selection), and reverse recovery time t_{rr} of the lower mosfet's body diode.

Summing these switching losses yields: 4.20W + 227.5mW + 227.5mW + 8.51W = 13.17W.

This total power figure is the power dissipation at the peak output current level, averaged over the 2μsec switching period. In order to translate this figure into an average power dissipation, it must be averaged over a full period at the audio signal output frequency (for example, at a 1KHz test case). What makes this possible is that all the equations above are functions of Iout, and vary directly proportionally

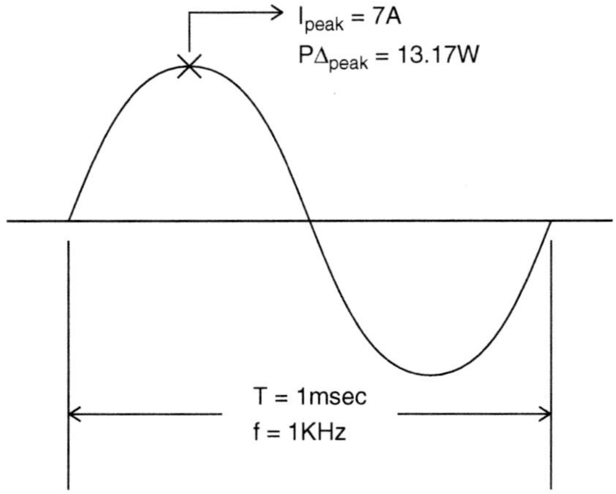

Figure 11.8 Peak Power Condition Relative to a Single Period of a Sine Wave at 1KHz

to Iout. Therefore, as Iout is sinusoidal, so shall be the graph of P_D over a 1KHz period. Switching power loss (*total* for both mosfets combined) will be the *average* value over the 1KHz output power waveform. The average of a sinusoidal function is 0.67 times the peak value. Figure 11.8 illustrates this, showing power dissipation over a 1msec period, for a 1KHz test tone at full output power (100W/4 ohms).

$$P_{Daverage} = (0.67)(P_{Dpeak}) = (0.67)(13.17W) = 8.82W$$

Body Diode Reverse Recovery (a Detailed Discussion)

It is important to understand reverse recovery in general, and the reverse recovery characteristics of the selected mosfets in particular, in order to design an efficient and reliable Class-D power stage. P-N junction diodes (to which the body diodes of silicon mosfets belong) have a reverse recovery characteristic. Basically this is due to the build-up of charge at the junction due to minority carrier conduction in the semiconductor. This charge builds up during the forward conduction time of the body diode and must be depleted during commutation, before the device can again withstand reverse voltage. This reverse recovery characteristic presents problems in Class-D power stages. These problems include: additional power dissipation in both the body diode of the mosfet experiencing commutation, as well as the opposite mosfet sourcing the forced commutation current, as well as potential avalanche or overvoltage, and EMI issues due to the high dv/dt, di/dt, and ringing that can be induced by reverse recovery waveforms. It is important to be aware of the voltage spike and ringing that occur during body-diode commutation in the Class-D power stage. An oscilloscope must be used to probe the switching node to observe the voltage waveform and to be sure that the maximum V_{DS} voltage of the mosfets is never exceeded. Typically, these ringing transients will last less than 50nsec, and some manufacturers' mosfet datasheets allow for avalanche (overvoltage breakdown operation). However, it is not good design practice for a reliable Class-D amplifier design to allow these transient overvoltages to occur. In addition to possible overvoltage damage, these ringing transients are likely to cause EMI issues, which will be addressed in Chapter 15. This ringing and overvoltage can be reduced by careful layout (reduced current loop inductances), slowing down mosfet switching times, or the selection of mosfets with different body-diode reverse recovery characteristics.

Reverse recovery current (Irr) is current that flows during the reverse recovery time (Trr) to deplete stored charge (Qrr). In the power loss examples above, the additional loss due to the reverse recovery of the body diode of the lower mosfet can be calculated as an additional current (Irr) that flows during the t_4–t_5 interval of Figure 11.7.

$$Irr = Qrr/(t_{4-5}) = 164nC/50nsec = 3.28A$$

Note: Qrr for IRF6775 = 164nC.

The true shape of the reverse recovery current waveform is triangular, reaching a negative peak (I_{RM} or I_{RR}) and then ramping back to 0A when the body diode is completely off, and can support its full blocking voltage. Additionally, this reverse recovery current flows through the upper mosfet, for some time while it sees its full V_{DS} equal to the total rail voltage, and then for a portion of time while its V_{DS} voltage is slewing to 0V. For simplification, and to overestimate these losses (for a conservative design approach), it will be assumed that this reverse recovery current flows while V_{DS} (for the upper mosfet) is maximum. Therefore, the power dissipation due to this reverse recovery effect (averaged over the 2µsec Class-D switching period) is:

$$P_{D(reverse\ recovery)} = (Irr)(V_{SUPPLY\text{-}TOTAL})(t_4-t_5)/T) = (3.28A)(97.3V)(50nsec)/2\mu sec) = 7.98W$$

Adding this power loss due to body-diode reverse recovery to the peak power loss calculated above (13.17W) yields 21.15W, and averaged over the 1msec for the 1KHz full-power case yields 14.17W as calculated below:

$$P_{Daverage} = (0.67)(P_{Dpeak}) = (0.67)(21.15W) = 14.17W$$

One of the major benefits of GaN and SiC power devices over silicon mosfets concerns the body diode. GaN devices do not have a body diode, and SiC devices have a Schottky body diode, which theoretically exhibits no reverse recovery time or reverse recovery charge. Thus, these wide bandgap (WBG) devices eliminate the problems associated with body-diode reverse recovery and the issues it causes.

Figure 11.9 provides more detail about rectifier reverse recovery in general, and provides specific application to the mosfet body diode (*). The upper waveform of Figure 11.9 is the voltage across the body diode, and the lower waveform is its current. Assume that the waveforms describe the voltage and current conditions

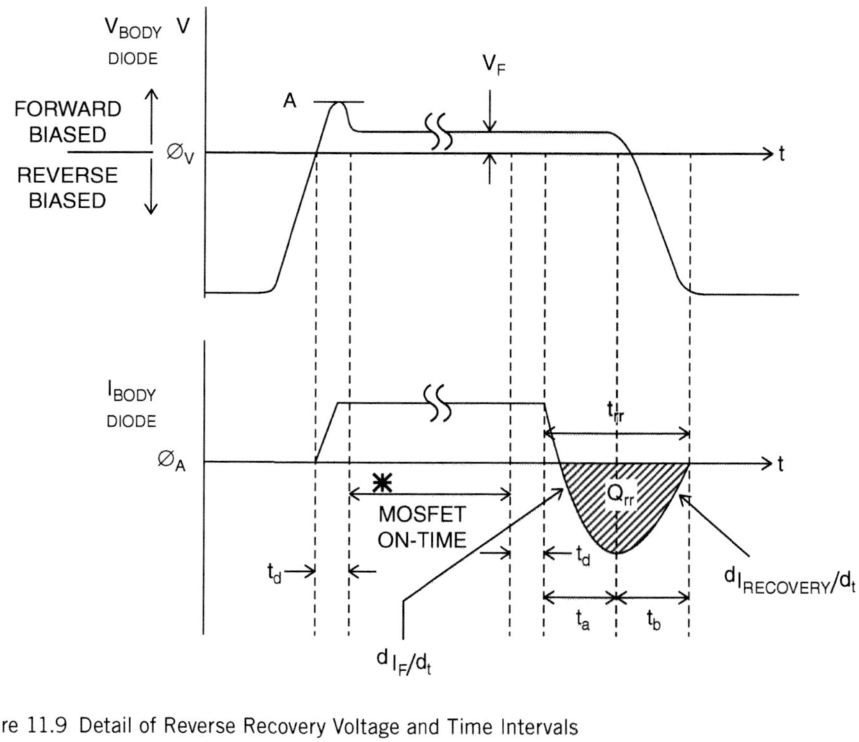

Figure 11.9 Detail of Reverse Recovery Voltage and Time Intervals

for the body diode of the lower mosfet of the half-bridge stage. Further assume that it depicts the condition where the upper mosfet is sourcing high current *to* the inductor of the L-C demodulation filter. The waveforms start with the upper mosfet on. The lower mosfet is off and its body diode is reverse biased. As the upper mosfet switches off, the reverse voltage across the body diode ramps towards 0V, and during the first deadtime (t_d) the body diode of the lower mosfet becomes forward biased. At this point, the forward voltage of the body diode will increase, and will overshoot its typical forward voltage, due to the high dI_F/dt of the forward current. The peak forward voltage reached is termed the forward-recovery voltage and is typically in the range of 5V. Next, the forward voltage settles down to the typical forward rectifier voltage. However, at the end of the deadtime, the lower mosfet will switch on. This occurs during the time interval labelled with the asterisk (*). During this interval, the V_F value of the upper waveform would assume a value of $I_d \times R_{DS}ON$. At the start of the second deadtime (t_d), current will transfer from the mosfet channel back to the body diode, which will again conduct in the forward-biased direction at a forward voltage of V_F. *Now* the reverse recovery interval begins when the upper mosfet switches on. Initially its V_{DS} voltage remains high, and the voltage across the lower mosfet body diode still remains positive (forward biased). During the reverse recovery time interval (t_{rr}), which is made up of sub-intervals t_a and t_b, the upper mosfet must supply the reverse recovery current (I_{rr}) required by the mosfet body diode, in order to sweep out the minority carriers it accumulated during the body diode's forward conduction time. During interval t_b, the voltage across the upper mosfet begins to decrease, and the voltage across the body diode begins to become reverse biased, as it is gradually able to support reverse voltage. At the end of t_b the body diode is able to support its full reverse voltage again.

Mosfet datasheets typically include at least the following specifications relative to the reverse recovery of the body diode:

- Reverse recovery time (I_{RR})
- Reverse recovery charge (Q_{RR}).

Additionally, peak reverse recovery current (I_{RM} or I_{RR}) is sometimes specified.

Another parameter that is usually specified for power rectifiers, but less often for mosfet body diodes, is "softness". This is determined by the ratio of t_b to t_a (see Figure 11.9). A softer reverse recovery characteristic results from the lengthening of the t_b interval and usually slower di/dt and dv/dt, resulting in less ringing and voltage overshoot, but longer overall switching intervals.

It is extremely important to realize that all these reverse recovery parameters are specified under a fixed set of conditions disclosed on the datasheet. These are:

- Temperature
- Forward current (before commutation)
- Supply voltage (end-state blocking voltage)
- di/dt.

However, Irr, Qrr, and I_{RM} (or I_{RR}) will all *vary* as these parameters change. For example, they will all *increase* as these above parameters increase (Texas Instruments, 2003).

A further complication, which is not obvious from the datasheet, but that is mentioned above, is that the accumulation of reverse recovery charge occurs during the forward conduction time of the body diode (Huang, 2014) (i.e. the *deadtime* intervals). If this time is kept short, less reverse recovery charge will build up. The adjustment of deadtime in a Class-D amplifier is a critical aspect of design, affecting many performance parameters. In this case, shorter deadtime can increase efficiency by decreasing reverse recovery losses and forward-biased body diode conduction losses, but this must be balanced against increased switch conduction overlap (cross-conduction), which will increase losses. Additionally, reducing deadtime will decrease (improve) open-loop and closed-loop THD.

Figure 11.10 depicts various overlap conditions for switched voltage and current, and the resultant *energy* (E). Note that energy = power × time, so each interval consists of a voltage × current product multiplied by the time interval (t). To convert this to *power* dissipation, the *energy* product must be divided by the interval over which the power is to be calculated. For example, referring to Figure 11.5 (the upper mosfet switch-off interval), switch current remains constant at 7A, while switch voltage (V_{DS}) ramps from 0V to 96V. Thus, the leftmost condition (A) in Figure 11.10 is applicable. Since the switching time interval for that specific condition was 25nsec, the *energy* product is (0.5)(96V)(7A)(25nsec) = 8.4uWatt-seconds = 8.4uJ (microjoules). If we then divide this *energy* product by the time interval over which the *average power dissipation* is to be calculated (i.e. the 2μsec switching period), the resultant average power dissipation attributable to

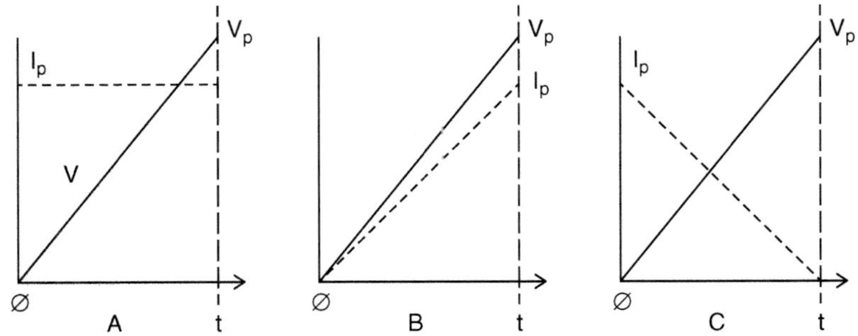

ENERGY= $\frac{1}{2} V_p I_p t$ ENERGY= $\frac{1}{3} V_p I_p t$ ENERGY= $\frac{1}{6} V_p I_p t$

Figure 11.10 Three Different Types of Switching Waveform Relationships of Voltage and Current

this specific switching time interval is 4.20W, which agrees with the calculations above for Figure 11.5.

Addition of External Schottky Diodes (Drain-to-Source)

The calculations and discussions above illustrate the significance and impact of the mosfet's body diode on many aspects of the Class-D design, including efficiency (power loss), noise and ringing, and EMI. One common design enhancement is to employ Schottky diodes in parallel across each mosfet, to shunt the internal mosfet body diode. Schottky diodes (especially lower-voltage types) have significantly lower forward voltage as compared to the mosfet body diode, and are majority carriers, exhibiting no reverse recovery. Theoretically, if the free-wheeling current flowed through the external Schottky diodes instead of the mosfet body diodes, the issues associated with the body diode could be avoided. This is essentially true, but in practice is not a perfect solution. The reasons are as follows:

1) Due to the higher supply rails necessary for Class-D design, a higher-voltage Schottky diode must be selected. These higher-voltage Schottkies have higher values for forward voltage, and thus the *differential* voltage between them and the mosfet body diode is such that not *all* the free-wheeling current is shunted to the external Schottky. Thus there is still some reverse recovery effect from the mosfet body diode.

2) Due to trace connections, internal bond wires, etc., there are inductive elements between the internal body diode of the mosfet and the external Schottky. These inductances impede the transfer of current from the mosfet channel to the external Schottky, especially considering the short switching times involved. Again, this means less current transferring to the external Schottky diode during the free-wheeling interval.
3) Schottky diodes, though having no reverse recovery, *do* have significant capacitance. This capacitance must be charged and discharged each switching cycle, exactly as C_{OSS} is (see discussion below). This contributes to capacitive switching loss.

Mosfet Capacitive Switching Losses

In addition to the conduction and switching losses discussed above, there is also a loss component due to the charging and discharging of the output capacitances (C_{OSS}) of the mosfets. For this calculation, the datasheet of the mosfet must be consulted. Some mosfet and GaN fet datasheets list two different values for effective output capacitance. These are *energy* related (ER) and *time* related (TR).

The datasheet in EPC (2022) is a good example of a device where both ER and TR values of C_{OSS} are specified.

For power loss calculations, it is C_{OSS} ER that is more relevant.

When these parameters are not specified (such as for the IRF6775 mosfet), then it is best to use the Effective Output Capacitance specified (175pF). It is important to note the conditions under which it is specified. On the IRF6775 datasheet (Infineon (International Rectifier), 2014), the $C_{OSS\text{-}EFF}$ of 175pF is essentially the TR C_{OSS}, specified over a V_{DS} range of 0–120V.

Figure 11.11 depicts the upper and lower mosfets in the Class-D half-bridge power stage as mechanical switches for simplicity. The +/−48VDC power supply rails are shown as a single 96VDC rail, and the C_{OSS} capacitance of each mosfet is shown as a discrete capacitance shunting each mosfet drain-to-source. On the *left* side of the figure, the upper mosfet is ON (switch closed). On the *right* side of the figure, the lower mosfet is ON.

When the upper mosfet is on, and the lower mosfet is off, the upper capacitance is discharged to 0V by a discharge current i_2, flowing through the upper mosfet. An additional current i_1 flows through the upper mosfet and the lower capacitance, charging it to 96V.

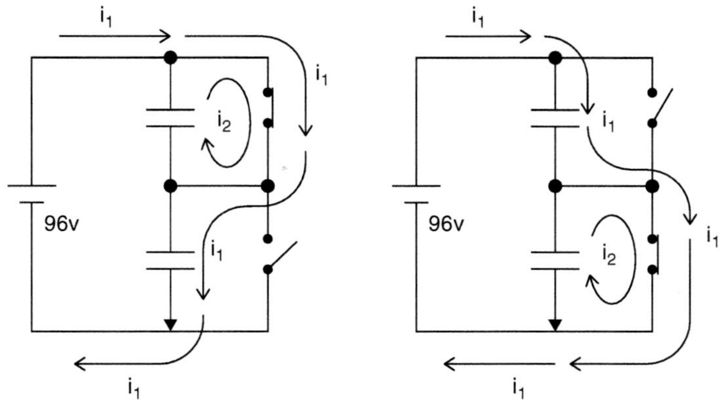

Figure 11.11 Capacitive Switching Currents in the Half-Bridge Power Stage of the Class-D Amplifier

Next, the upper mosfet is off and the lower mosfet is on. The lower capacitance is discharged from 96V to 0V by a current i_2 flowing through the lower mosfet. An additional current i_1 flows through the lower mosfet and the upper capacitance, charging it from 0V to 96V.

Thus, *each* capacitance (C_{OSS}) is charged from 0V to 96V and then discharged from 96V to 0V during *each* switching cycle at the Class-D switching frequency.

The *total power loss* due to this capacitive charge/discharge is:

$$P_{D\text{-}CAPACITIVE} = (C_{TOTAL})(V_{TOTAL})^2\,(f) = (2 \times 175\text{pF})(96)^2\,(500\text{KHz}) = 1.61\text{W}$$

Note: An ideal capacitance is a lossless component, therefore this power is dissipated by the resistive components of the circuit, which in this example are the mosfets. Each mosfet dissipates 1.61W/2 = 0.805W due to C_{OSS} capacitive switching loss. This loss is particularly significant in the IDLE condition, where very low quiescent power loss in the Class-D power stage might otherwise be expected. It is for this reason that the designer must be extremely wary and cognizant of C values when selecting mosfets for Class-D application. A trend in the semiconductor industry is to decrease a particular *figure-of-merit* for power mosfets, which is defined as the R_{DS}ON × Qg product. Manufacturers strive to produce mosfets with the lowest possible R_{DS}ON and gate charge. Unfortunately, lower R_{DS}ON usually comes at the expense of higher capacitances. This tradeoff may be fine for other applications, especially when application voltages and switching frequencies are lower; however, this is *not* the best tradeoff for Class-D design. As can be seen from the analysis above, *conduction* losses attributable to $R_{DS}ON$ are relatively low, and are lower than the calculated capacitive switching loss. This Class-D design could benefit (from an efficiency standpoint) if a mosfet could be selected with slightly higher $R_{DS}ON$, but lower C_{OSS}.

Gate Charge (Qg) Capacitive Power Loss

Another capacitive switching loss is that due to the charging and discharging of each mosfet's gate capacitance. The graph in Figure 6 of the IRF6775 datasheet (Infineon (International Rectifier), 2014) shows typical gate charge (Qg) versus gate-source voltage. As can be seen, the graph shows three distinct regions. The initial region, where the gate charge slopes from 0 to 6nC, is due to the charging of the gate capacitance C_{iss}. The next "flat" region is called the Miller plateau, and is due to current flowing in the gate-drain capacitance (the Miller capacitance), as the drain voltage decreases. This presents a local negative feedback to the gate terminal, keeping it relatively constant until the drain voltage has reached zero. During this Miller plateau region, the current flowing through the gate-drain capacitance continues to be supplied to the gate (by the external gate-drive circuitry) and thus Qg continues to increase, even though V_{GS} is not increasing during this interval. Once the drain voltage has reached zero, the Miller capacitive current goes to zero, and the gate voltage can continue to increase, and additional gate charge is required for this.

Note: The mosfet gate is fully enhanced at a V_{GS} voltage between the threshold voltage and 10V. $R_{DS}ON$ is specified at a V_{GS} voltage of 10V, so there is little $R_{DS}ON$ benefit from driving the gate voltage above 10V, and this will increase Qg and gate charge power losses. The datasheet for the IRF6775 (Infineon (International Rectifier), 2014) does an excellent job of specifying the individual gate charge components (described above) that contribute to the total gate charge.

Total gate charge versus V_{GS} for IRF6775 (from Infineon (International Rectifier) (2014), see Figure 6.1):

$$P_{D\text{-GATE CHARGE}} = (Qg)(V_{GS})(f) = (36nC)(10V)(500KHz) = 180mW$$

Qg = total gate charge from the IRF6775 datasheet (Infineon (International Rectifier), 2014). Max. value at V_{GS} = 10V is used.

f = Class-D switching frequency = 500KHz

Note: This capacitive switching loss due to gate charge is *per mosfet* in this design, and is dissipated by the resistive components of the gate-drive circuitry. If a gate resistor is used (as in the reference design), this usually has the highest resistance in the gate-drive loop. However, other "resistive" components in the gate-drive loop are: the internal gate resistance of the mosfet (specified as $R_{G(INT)}$ = 3Ω max. on the IRF6775 datasheet – Infineon (International Rectifier), 2014), and the R_{DS}ON of the internal switches in the gate driver IC (usually negligible compared to external resistances). It is important to note that if no external gate resistor is used, then the gate charge power calculated above is *unchanged*, and *will* be dissipated by the gate driver IC and the mosfet's internal R_G.

Note

1 This statement is a simplification. The inductor current consists of the audio frequency output current plus the switching frequency ripple current. This ripple current component flows through the filter capacitor(s) of the L-C filter, leaving the remaining audio frequency inductor current to flow to the output and connected speaker. For this conduction loss calculation, the inductor current (and power switch current) will be considered to be the audio frequency output current, and the switching frequency ripple current will be ignored.

Bibliography

Cerezo, J. (n.d.). Class D audio amplifier performance relationship to MOSFET parameters. International Rectifier (Infineon) Application Note AN-1070. El Segundo, CA: International Rectifier.

Cirrus Logic. (2018, March). RC snubber for Class-D audio amplifiers. Cirrus Logic Application Note, AN_0454. Rev. 1 . Austin, TX: Cirrus Logic.

Efficient Power Conversion (EPC). (2022, July). Datasheet, EPC2207 – Enhancement mode power transistor. El Segundo, CA: Efficient Power Conversion.

Haaf, P., & Harper, J. (2007). Understanding diode reverse recovery and its effect on switching losses. Fairchild Power Seminar 2007. Fürstenfeldbruck, Germany: Fairchild Semiconductor Europe.

Huang, A. (2014, March 12). Hard commutation of power MOSFET. Optimos™ FD 200V/250V. Infineon Application Note AN 2014-03 V1.0. Villach, Austria: Infineon Technologies Austria.

Infineon (International Rectifier). (2014, February 26). Datasheet, IRF6775MTRPbF Digital audio mosfet. El Segundo, CA: International Rectifier.

NXP Semiconductors. (2012, April 25). Designing RC snubbers. NXP Application Note AN11160. Rev. 1. Eindhoven, Netherlands: NXP Semiconductors.

Texas Instruments. (2003, December). Datasheet, UCC27223, High-efficiency predictive synchronous buck driver with enable, SLUS558. Dallas, TX: Texas Instruments.

Todd, P. C. (1993). Snubber circuits: Theory, design and application. Dallas, TX: Texas Instruments.

Chapter Twelve

Feedback and Compensation

Overview

Class-D (switching) amplifiers have different error sources and distortion-generating mechanisms, as compared to analog (linear/non-switching) audio amplifier topologies.

The implementation of feedback and compensation in the design of Class-D amplifiers has been a topic of much discussion and interest. Class-D amplifier designs have ranged from having no feedback at all (open-loop), through many intermediate applications of feedback, all the way to full feedback from the final speaker output, after the L-C demodulation filter. The principal compensation difficulty facing designers seeking to close the loop *after* the L-C filter, is the 180 degrees of phase shift caused by the (at least 2-pole) output L-C filter, and its load-dependent transfer function. Including the L-C filter in the feedback path has the benefit of allowing the feedback loop to correct for non-linearities in the filter response, as well as its load dependence. However, this comes at the expense of design complexity, in trying to then achieve a stable feedback control loop with high closed-loop bandwidth (BW). The conflicting goals are loop stability, defined as adequate gain and phase margins over all operating conditions, and high feedback gain over the full 20Hz–20KHz audio frequency range. The latter is necessary to provide maximum error correction of all non-linearities and error sources in the amplifier and associated power supplies, as well as load variations. It is helpful to keep in mind that the goal is to make the Class-D amplifier behave as an ideal voltage source.

DOI: 10.4324/9781003327943-12

A General Discussion of Negative Feedback

At this point, a general discussion of negative feedback, as applied to Class-D amplifiers, is justified. Traditionally this has been a sensitive topic. Some audiophiles are of the opinion that there are negative consequences to using feedback in audio amplifier designs. However, the author is of the opinion that feedback should be used, and as long as adequate stability margins can be maintained, as much feedback as possible should be used. Texas Instruments (2008) lists ample justification for this philosophy, as well as a historical perspective concerning attitudes about using feedback in audio amplifier design.

Open-Loop Designs

Note: Before diving into a detailed discussion of the benefits of *feedback*, it is crucial to understand the sources of error in the Class-D audio amplifier. These are covered extensively in Chapter 10.

The simplest approach, no feedback at all, is an amplifier consisting of just a pulse-width modulator and a power stage, followed by the output L-C filter. This approach, which lacks feedback, is termed open loop. The main drawback of this approach is that disturbances in the power supply rail(s) as well as imperfections in power stage switch timing (resulting from finite switching edge times and deadtimes) and output L-C filter characteristics are not corrected for. These will then lead to noise and distortion (THD) in the output.

Nonetheless, open-loop amplifiers can have good audio performance if they are designed with very precisely timed power stages with fast edge rates, low propagation delays and deadtime, as well as precise and tightly regulated DC power supply rails. The power supplies must have excellent line and load regulation, in order to maintain a precise voltage output. It is also important not to ignore the effects of crosstalk in multi-channel open-loop amplifiers, where the load demands of one channel can cause power supply rail disturbances which show up as THD and noise in another amplifier channel because there is no power supply rejection ratio (PSRR) benefit provided by feedback.

It is always beneficial to assess the *open-loop* performance of a Class-D amplifier before feedback is applied. A good general design approach is to minimize open-loop noise and distortion. An example is helpful to illustrate this benefit. Suppose an open loop THD+NOISE plot is run (at a specific power output level), over the audible frequency range of 20Hz–20KHz, and it is found that THD at 20KHz is 1%. Now, if negative feedback is applied around this loop, and the loop gain at 20KHz is 20dB (×10) for example, this will result in a 10× reduction in THD+NOISE at 20KHz, or a reading of 0.1%.

To run an open-loop sweep on the reference amplifier for example, the audio input signal from the audio analyzer would be applied to the PWM comparator input (R43A = E/A OUT A LOCATION in reference schematic). Since the triangle ramp amplitude applied to the other PWM comparator input is +/−4Vpeak, the audio signal applied at the other PWM comparator input should not exceed +/−4V peak, or the output voltage signal will be clipped.

The general method for applying feedback is to use an integrating difference amplifier or error amplifier. This amplifier amplifies the difference voltage between the input signal and a divided-down version of the output voltage signal (feedback signal voltage). An integrator has very high gain at DC, with decreasing gain as signal frequency increases. Gain will fall at a rate of 6dB/octave (20dB/decade) for a single-pole (first-order) integrator and 12dB/octave (40dB/decade) for a double-pole (second-order) integrator. Even higher orders are possible. It is critical to understand the phase relationship of first-order and second-order integrators, in order to be able to design stable negative feedback loops. First-order integrators will reach a maximum phase shift of 90 degrees. Second-order integrators will reach a maximum phase shift of 180 degrees. An important concept to be aware of is that phase shift increases with the frequency range over which the gain decrease applies. For example, if a first-order integrator's gain only falls over a single octave, the maximum phase shift it will have is less than if its gain is allowed to fall over a wider BW (two octaves, for example). This concept, combined with the selection of pole and zero frequencies, are used to tailor the response of the negative feedback loop to achieve the desired response, with stability.

Stability Criteria

To understand negative feedback loop stability, the concepts of open-loop gain, feedback divider gain, and closed-loop gain must be explained. First, observe that the amplifier has a forward gain path. This consists of the error amplifier (or integrator), the PWM comparator and power stage, and the L-C filter. The feedback divider is the resistor divider connecting the amplifier output voltage (after the L-C filter) to the error amplifier summing junction. It is important to keep in mind that these circuit blocks have frequency-dependent response. Using the reference schematic will be illustrative.

The PWM comparator and power stage gain can be explained as follows.

The triangular ramp applied to the "+" comparator input has an amplitude of +/−4Vpeak. This is compared against the error amplifier's output voltage (after the lead network), which is applied to the – input of the PWM comparator. So, a signal swing of +/−4Vpp out of the error amplifier-plus-lead-network will "map" into a duty cycle ranging from 0% to 100% (ignoring deadtimes for simplicity), where "0%" means the lower mosfet is on 100% of the time, and "100%" means the upper mosfet is on for 100% of the time. This means a +/−4V output from the error amplifier-plus-lead-network will "map" to a +/−48V output from the amplifier. This is a *modulator gain* of 96V/8V or 48V/4V = ×12 = 21.58dB.

Simplifying Simulation

For simplicity in simulation and analysis, the assumption can be made that the modulator gain is flat with frequency. This is indeed a simplification, because in reality, the finite deadtimes used in the power stage do appear as a phase lag in the response. But for a first-pass simulation and analysis, this simplification makes it much easier to simulate and speeds up simulation time, because the simulation (in LTspice) can be continuous time instead of including a switching stage. This is accomplished by substituting a fixed gain block of ×12 in place of the PWM comparator and power switching stage. Figure 12.1 shows the LTspice schematic representation. The circled components represent the *modulator gain*.

In this simulation method, a sine wave is applied to the L-C filter, rather than the actual rail-to-rail switching waveform of the Class-D power stage. This does not present a problem, because the principal interest is the circuit's response up to the unity gain crossover frequency (approximately 100KHz), where the L-C filter's response to the fundamental frequency will be very similar for sine-wave versus square-wave excitation. The biggest difference between the sine- and square-wave response of the L-C filter will be its response to the third harmonic of the square wave. Considering that the switching frequency is 500KHz, the third harmonic is 1.5MHz, which is beyond the frequency of interest for the simulation. Therefore, this simplification will yield accurate results in Bode plots over the audio frequency range, extending to beyond the unity gain crossover

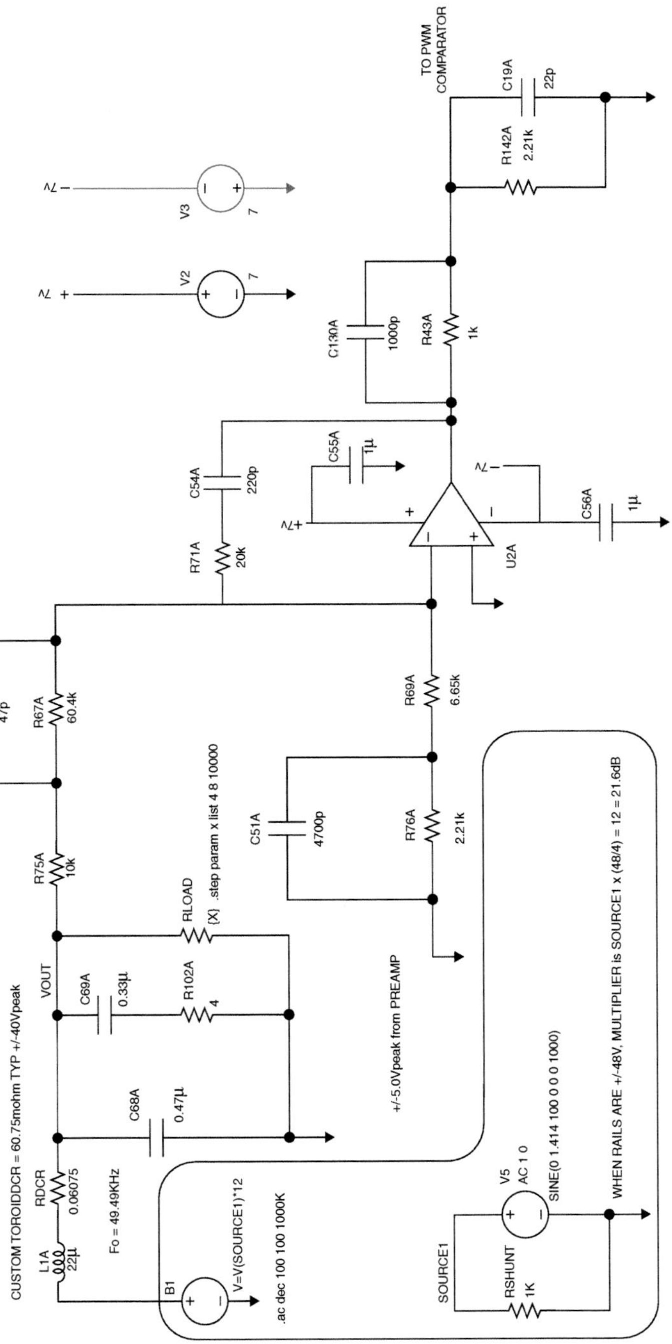

Figure 12.1 LTspice Schematic Used to Model the Entire Feedback Circuit for the Reference Amplifier

frequency, and will be valid for assessing loop stability, while greatly reducing simulation time.

The other circuit blocks in the LTspice simulation are the *error amplifier* (integrator), the L-C filter, the Zobel network, and the feedback resistor divider network.

The resistor divider network, in conjunction with the input network, set the *closed-loop gain* of the Class-D amplifier. A typical gain configuration for residential audio (as opposed to pro-level or commercial amplifiers) is to configure the amplifier to produce its full rated output power with a 2VRMS input. This is the input level provided by CD players, for example. *Note:* Audio from mobile phones and other sources will have lower output voltage and will not be capable of driving the amplifier to its full output voltage/power with this gain structure. These lower output voltage sources will need some pre-amplification before driving the Class-D amplifier, if full output voltage/power is desired.

100W into an 8-ohm load (typical speaker load impedance) = 28.28VRMS

28.28VRMS × SQRT 2 = 40Vpeak

This is the maximum peak output level achievable with +/−48VDC rails, with the deadtimes we have imposed by design.

So desired total amplifier *gain* is 28.28VRMSout/2VRMSin = ×14.14 = 23dB

However, the *preamp* circuit has been designed to boost the 2VRMS input signal (+/−2.828Vpeak) to +/−5Vpeak, which is the input level to the error amplifier.

Gain of preamp = 5/2.828 = ×1.768 = 4.95dB

This means the gain of the amplifier, from E/A in, to amplifier output, will be:

23dB − 4.95dB = 18.05dB

Checking:

AMP output voltage peak / E/A input voltage
peak = 40Vpeak/5Vpeak = ×8 = 18.06dB

The input/output voltage divider is then designed as follows (assuming the *op-amp* summing junction is maintained at 0VDC by the op-amp):

$R_{TOP}/R_{BOTTOM} = 8$ (for ×8 gain)

Select an input resistor (R_{BOTTOM}) of 8.86 Kohm (2.21 Kohm + 6.65 Kohm resistors in series).

The feedback resistor (R_{TOP}) will be 8.86 Kohm × 8 = 70.88 Kohm

Standard 1% resistor values of 10K plus 60.4K will be used (70.4K total).

An interesting observation is that this closed-loop gain, and the selected input resistance value (8.86 Kohm), will have no effect on the open loop (forward gain). This is because of the summing node of the op-amp (inverting input) being held at 0V by the op-amp's local feedback. In the forward gain path, we have the (frequency-dependent) gain of the error amplifier itself, the fixed gain of the modulator stage, and the (frequency-dependent) gain of the L-C filter and Zobel network, as well as the effect of the attached load.

For the error amp, the gain is $Z_{FEEDBACK} / Z_{INPUT}$. Both $Z_{FEEDBACK}$ and Z_{INPUT} use a combination of R's and C's to create *poles* and *zeroes* in the open-loop response.

The L-C filter has a classic double-pole response, which is discussed in more detail in Chapter 9. What is important to know regarding the L-C filter and closing the feedback loop *after* it, is that an undamped L-C filter can have very high Q. This means that it can have gain peaking at its resonant frequency, and its phase can change very quickly after resonance, approaching 180 degrees. Damping the L-C filter decreases peaking and also makes the phase shift occur more gradually beyond resonance. The ultimate phase shift still asymptotically approaches 180 degrees, but the approach is more gradual, and the asymptote occurs at higher frequency, allowing the "loop to be closed" at a higher frequency.

The phrase "loop to be closed" in the previous paragraph refers to the point where the unity gain crossover frequency of the loop is achieved. This is the frequency at which the open-loop gain has dropped to unity (0dB). *Stability criteria* dictate that a loop will oscillate (due to *positive feedback*) if the phase shift around the loop exceeds 180 degrees at the frequency where the open-loop gain has decreased to 0dB. *Note:* This definition allows that a loop will be stable (even with phase shift greater than 180 degrees), as long as the open-loop gain is greater than unity. This, however, is called *conditional stability*, because it is conditional upon the loop gain being greater than unity. If, for any reason, loop gain should drop below unity, the loop will oscillate. It is for this reason that conditional stability is not an advisable design goal. Diving a little deeper into this concept, it is important to understand that an unstable loop does not oscillate at some arbitrary frequency. It in fact oscillates at the exact frequency at which the open-loop gain reaches 0dB, if the phase shift around the loop exceeds 180 degrees. It does so at this frequency because at this frequency, positive feedback occurs, since phase in excess of 180 degrees means that the feedback signal *adds* to the input signal at the error amplifier, instead of subtracting from it (as it should with *negative* feedback).

Before leaving the topic of conditional stability, a little extra discussion is warranted. As shown in previous chapters, the open-loop gain of the modulator stage is directly proportional to the supply voltage rails. If these rails decrease for any reason, as will occur during amplifier turn-on or -off, or during high output current demand (in combination with "soft" or unregulated power supply rails), during these rail-sag events, oscillation could occur in a conditionally stable design, because of the decreased forward path gain.

Looking at the frequency response of the elements making up the forward gain path: the error amplifier has the following response shown in Figure 12.2. (Note that the input path's 8.86 Kohm resistance is irrelevant here.)

The error amp is followed by a passive "lead" network, which serves the purpose of providing positive phase compensation. It provides a critical 10 degrees of phase lead in the frequency range of 100KHz–200KHz, where the loop will have its unity gain crossover. A critical detail in the design of this lead network is that the output of the lead network (the junction of the 1K and 2.21K resistors) must be able to produce an output voltage *at least equal to* the +/−4Vpeak voltage of the triangular ramp, in order to take advantage of the full duty cycle (modulation index) of the Class-D amplifier. This then requires the op-amp's output swing to be at least +/− ((4V) (1K + 2.21K) / 2.21K) = +/−5.81Vpeak. The op-amp chosen for the reference design meets that output voltage specification while running from +/−7VDC rails.

Figure 12.4 shows the response of the lead network and Figure 12.5 is the LTspice schematic for the lead network.

The L-C filter has the responses below. The first simulation (Figure 12.6) shows the response without the Zobel network. The LTspice schematic without the Zobel network is shown in Figure 12.7. The second simulation (Figure 12.8)

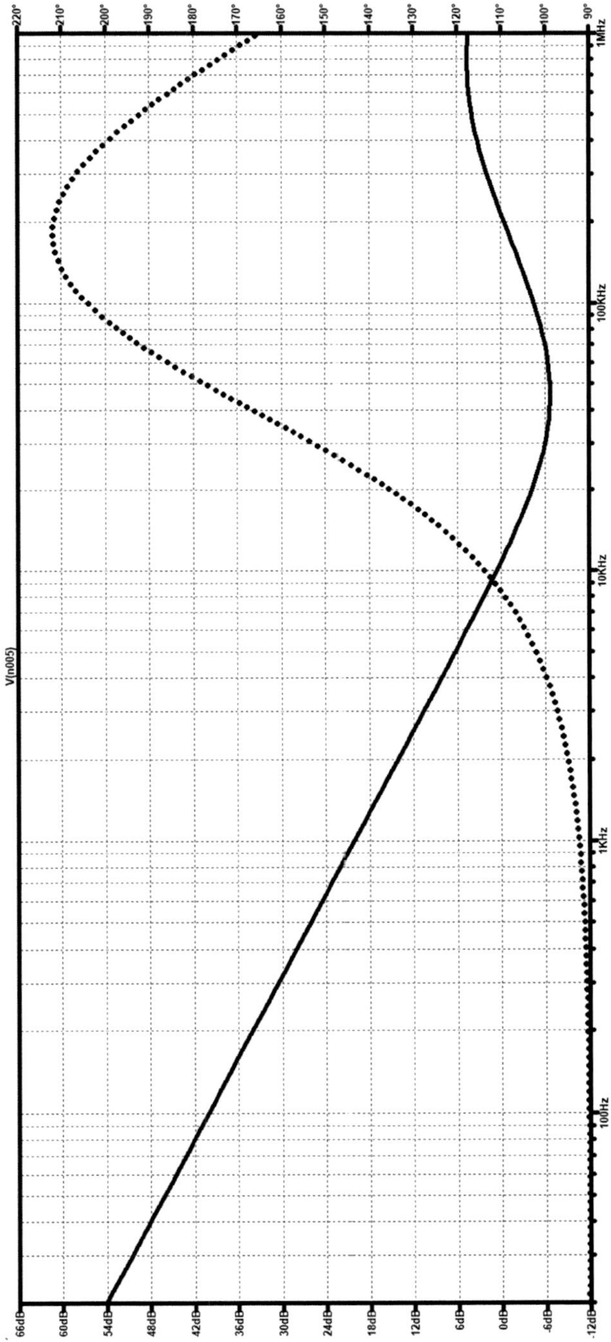

Figure 12.2 Response of Error Amplifier Alone (Circuit of Figure 12.3)

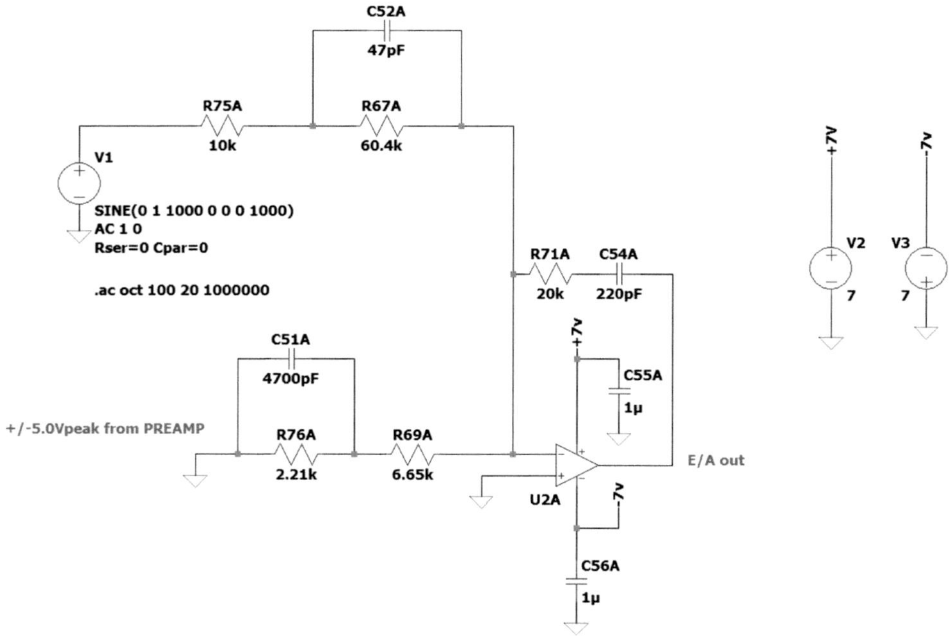

Figure 12.3 LTspice Schematic for the Error Amplifier

shows the response with the Zobel network. The LTspice schematic with the Zobel network is shown in Figure 12.9.

For both simulations, the LOAD is stepped through the following values: 4 ohm, 8 ohm, and 10 Kohm. These are intended to simulate expected speaker loads (4 and 8 ohm) as well as the “NO-LOAD” condition (10-Kohm load).

(*Note:* For these examples, the speaker is modeled as a simple resistive load; however, in reality, speakers have very complex impedances, that are frequency dependent as well.)

Observations concerning L-C filter responses: The addition of the Zobel network adds a critical 18 degrees of phase lead in the region of unity gain crossover (between 100KHz and 200KHz) compared to the no-Zobel case, for all 3 values of load resistance.

Figure 12.10 shows the complete loop response, with loading stepped between 4 ohms, 8 ohms, and 10 Kohm (NO LOAD) and Figure 12.11 shows the complete circuit used for simulation. The unity gain (0dB) crossover occurs very close to 100KHz, and *gain margin* is:

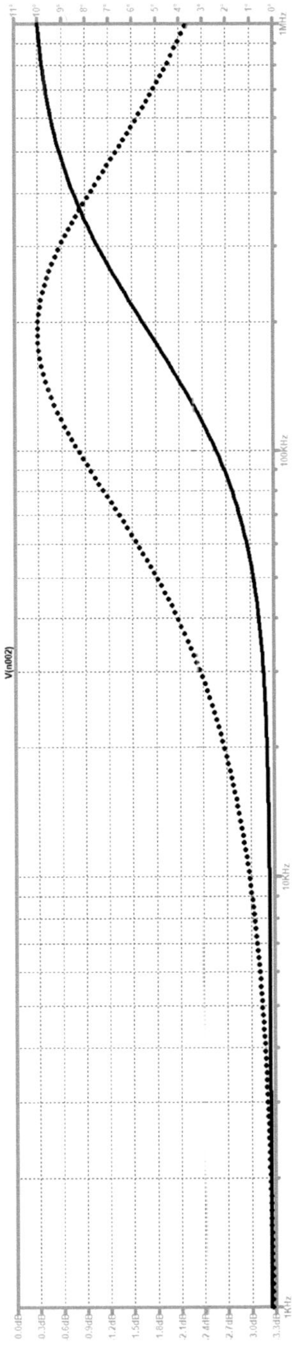

Figure 12.4 Response of the Lead Network

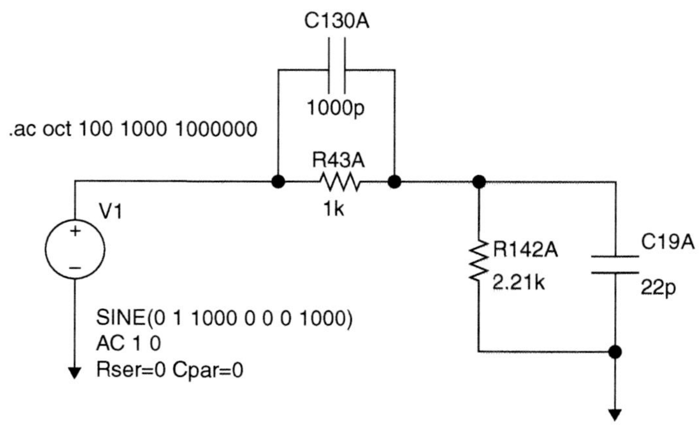

Figure 12.5 LTspice Schematic for the Lead Network

- 70 degrees with 4-ohm load
- 60 degrees with 8-ohm load
- 50 degrees with 10-Kohm load.

The LTspice simulation in Figure 12.10 includes the NO-LOAD simulation (i.e. no speaker attached). The LOAD resistor has been set to 10 Kohm. With this high-value load resistance, the L-C filter does not benefit from the damping that an 8- or 4-ohm speaker load provides. The peaking effect of the L-C resonance at 40KHz is evident. The Zobel network consisting of C69A and R102A in the LTspice schematic helps control L-C damping and make the response manageable. (*Note:* The tradeoff between Zobel damping and Zobel power dissipation is discussed in Chapter 9, L-C Demodulation Filters.) Even in this NO-LOAD simulation, a phase margin of 50 degrees minimum is attained. With 4- and 8-ohm loads, the peaking contribution of the L-C filter has been damped much more effectively, and the *phase margin* at the unity gain crossover frequency has increased.

The simulation shows the characteristic high loop gain at low frequency (close to 58dB @ 100Hz), a loop gain of 16dB @ 20KHz, and a unity gain crossover frequency of 100KHz (approx. $1/5^{th}$ of Fswx = 500KHz).

The loop gain at 20KHz is very modest: 16dB. As noted before, loop gain will decrease as frequency increases, so, generally, loop gain at 20KHz will be an indication of the error-correcting ability of the feedback loop in the audio pass-band. A design goal of 30dB loop gain at 20KHz is desirable, and would yield

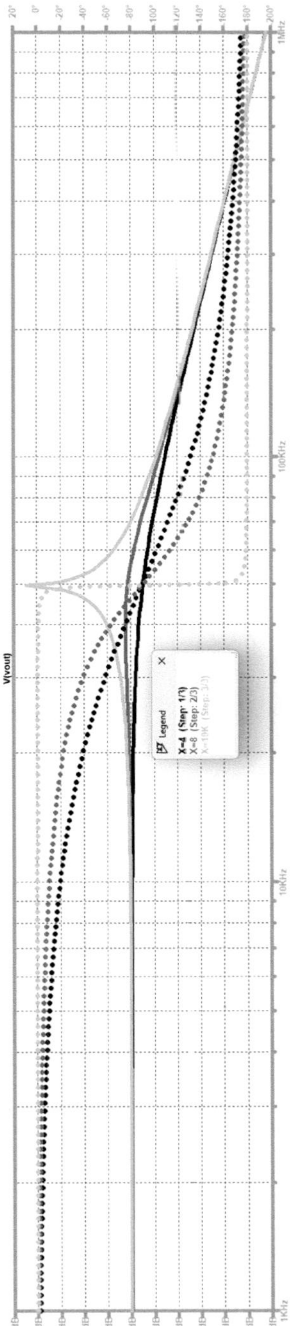

Figure 12.6 L-C Filter Response without Zobel Network

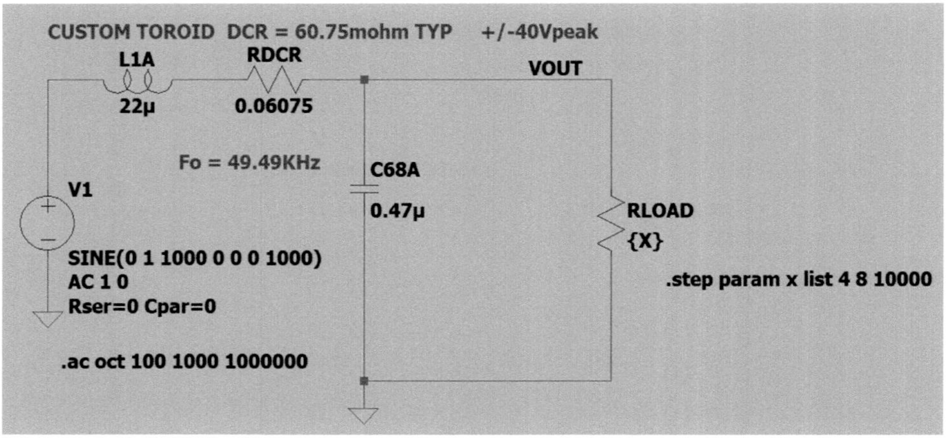

Figure 12.7 LTspice Schematic for L-C Filter without Zobel Network

better audio performance in terms of lower THD+N at 20KHz, and lower noise overall. This is difficult to achieve in fixed-frequency designs, but is doable, with greater circuit complexity and (generally) higher switching frequency. However, the goal of the reference schematic is to provide a practical design as a vehicle for explaining Class-D amplifier design, and therefore simplicity and adequate performance are favored over complexity and premium performance for this task.

Feedback before the L-C Filter

One approach to feedback compensation of Class-D amplifiers has been to take feedback from the switching node (before the L-C filter). This approach eliminates the phase-shift problems of the L-C filter, and still allows the feedback loop to correct for power supply rail deviations (thus providing power supply rejection, for good PSRR performance). This approach allows the feedback loop to have high gain (and therefore correction capability) over the entire audio range, but has the drawback that the non-linear effects of the L-C filter and its load dependence are not corrected by feedback. This results in varying frequency response with different speaker loads, and a decreased damping factor, due to the fact that the output impedance of the amplifier will never be less than the impedance of the L-C filter. This topology is very popular among integrated-circuit (IC) Class-D amplifier controller solutions from various IC

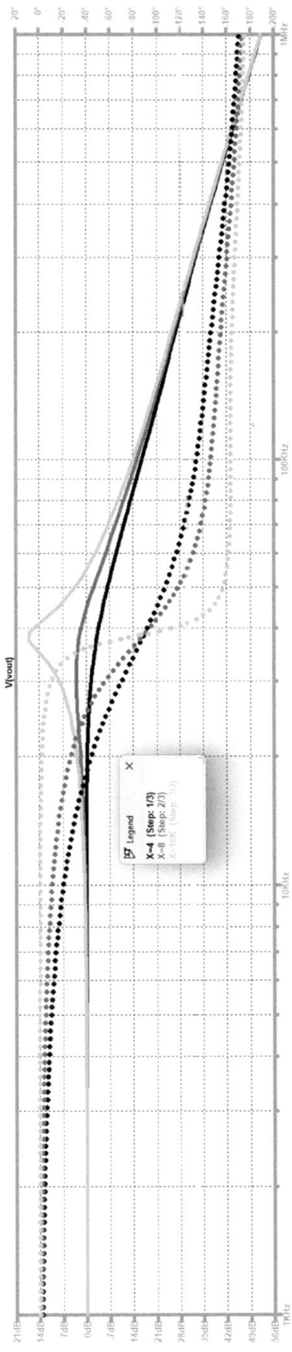

Figure 12.8 L-C Filter Response with Zobel Network

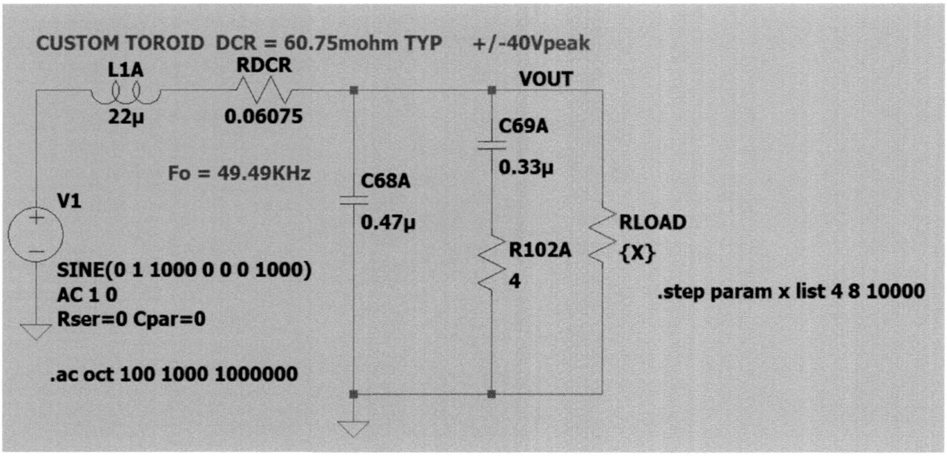

Figure 12.9 LTspice Schematic for L-C Filter with Zobel Network

manufacturers. A reason for this is that the switching node is already a pin on the IC package and is therefore conveniently sensed. To sense the voltage *after* the L-C filter would require an additional package pin. Another intermediate feedback approach is to take relatively heavy feedback from the switching node, as described above, and additional, but less feedback from the amplifier output, after the L-C filter. Having less feedback gain from the output means it is easier to achieve adequate gain and phase margins, but this is at the expense of error-correcting gain available in the feedback loop. Yet another alternative scheme is to take feedback *after* the L-C filter for frequencies below the L-C corner frequency (when excess phase shift will begin to rise), and to take feedback from the switching node *above* that frequency. This compromise allows post-filter feedback over the entire audible frequency range, with its concomitant benefits, and then shifts to the more easily compensated pre-L-C filter feedback above the L-C filter resonance frequency. It is possible, however, in fixed-frequency Class-D designs, to take *all* feedback directly from the amplifier output (post-filter), and still achieve loop gain approaching 30dB at 20KHz, while still maintaining greater than 45 degrees of phase margin under worst-case NO-LOAD conditions. As stated earlier, however, this is beyond the scope and purpose of the reference design.

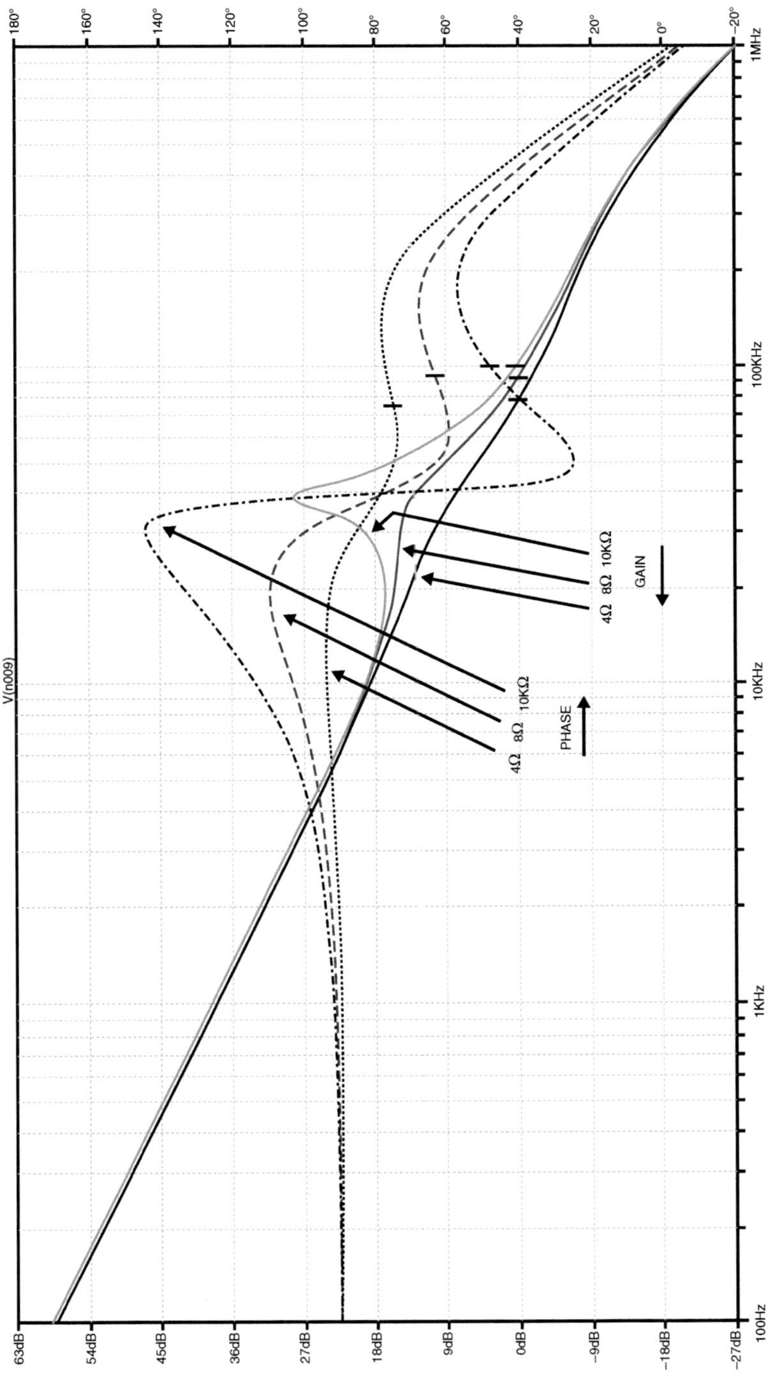

Figure 12.10 LTspice Bode Plot Simulation for the Complete Feedback Loop

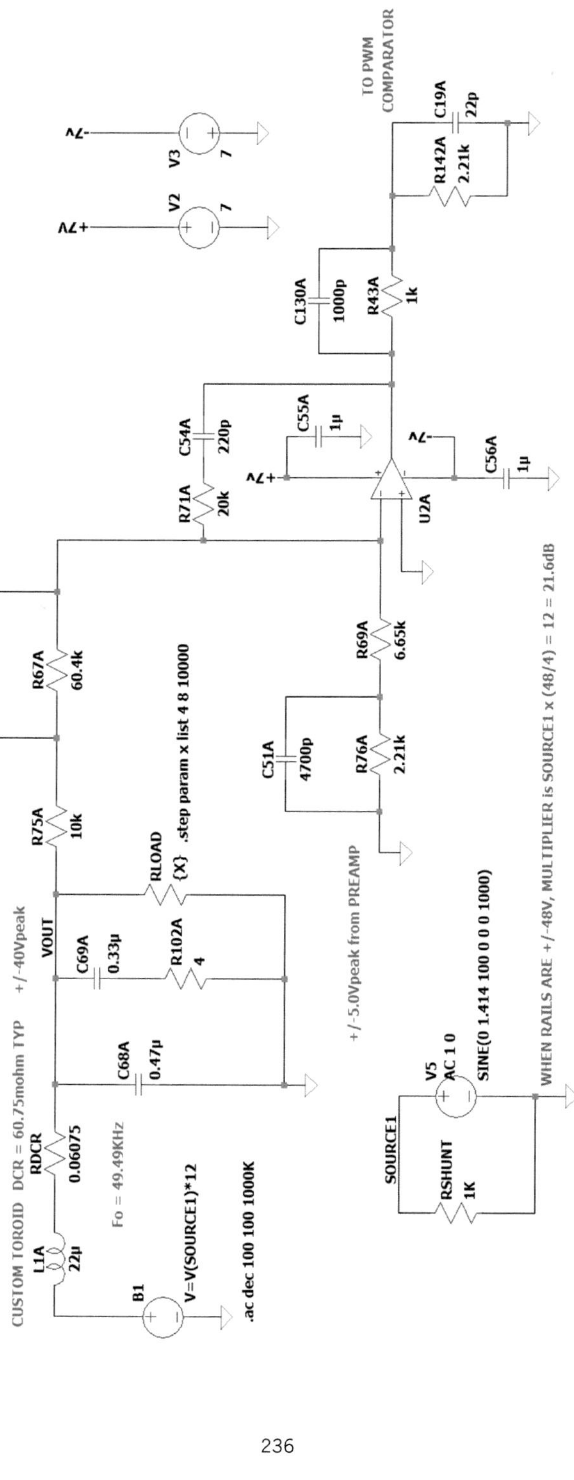

Figure 12.11 LTspice Schematic for the Full-Loop Bode Plot in Figure 12.10

Historically, Attwood (1983) describes full feedback from the output as early as 1982. These designs are fixed-frequency designs. Variable-frequency (self-oscillating) designs require feedback in order to oscillate. Some of these self-oscillating topologies have the advantage (from a feedback and BW standpoint) that they *intentionally* are designed to oscillate at their feedback loop's unity gain crossover frequency. Since this frequency is typically in excess of 400KHz, it is much easier to achieve a loop gain of 30dB at 20KHz than it is for fixed-frequency designs, whose unity gain crossover frequency is much lower than their switching frequency (usually by a factor of 4 or more).

There are also digital feedback methods being developed, whereby the analog amplifier output is digitized by high-speed (low-latency) analog-to-digital converters (ADCs), and the digital output of these ADCs is then processed by a purely digital control core which calculates error correction, and is then applied to the PWM drive stage. This is discussed further in Chapter 22.

Shu and Chang (2008) states that THD in closed-loop class-D amplifiers is comprised of two different contributors. The first is open-loop distortion effects that *can* be reduced by Loop Gain +1 (feedback). The second is a byproduct of feedback itself, and that is phase and duty cycle error caused by feedback. This second effect dominates as the source frequency rises (>3KHz). This reference makes the useful claim that this latter distortion effect can be decreased by the following actions:

1) Increasing the switching frequency.
2) Increasing feedback loop gain at the third harmonic frequency of the input.
    - *Note:* An input signal of 6.66KHz is the highest source frequency for which the third harmonic falls within the 20KHz audiophile BW. It is for this reason that typical THD plots feature this frequency as one of the test frequencies (with 1KHz being the most typical).
3) Decreasing the feedback loop gain at the switching frequency.

It is important to note that this characteristic of phase and duty cycle error actually *increasing* THD in Class-D amplifiers employing feedback, is a characteristic that linear audio amplifiers do not share. This is one way in which Class-D amplifiers present unique challenges to designers, which do not have to be conquered in Class-AB designs.

Feedback Loop Measurement Techniques

It is extremely helpful to actually *measure* loop stability, characterized principally by gain and phase margins, as figures of merit. There is a well-documented measurement technique that will be described here (Middlebrook, 1975; Texas Instruments, 2008; LeCroy, n.d.; Tektronix, n.d.). See Figure 12.12.

In this technique, a low-value resistor (R_{INS}) is inserted at a judiciously selected point in the feedback loop. An isolation transformer (T_{INS}) is connected across this added resistor, and a sinusoidal voltage source (DRIVE) drives this transformer to create a differential sinusoidal perturbation voltage across this added resistor. Suppose for example that the DRIVE signal is adjusted such that 100mVRMS is impressed across R_{INS} from point A to point B. Oscilloscope PROBES A and B are then connected to points A and B so that the voltage on each of these points can be measured *independently* at these points with respect to circuit GND. As the *loop gain* of the feedback loop changes (with frequency), the voltages appearing at

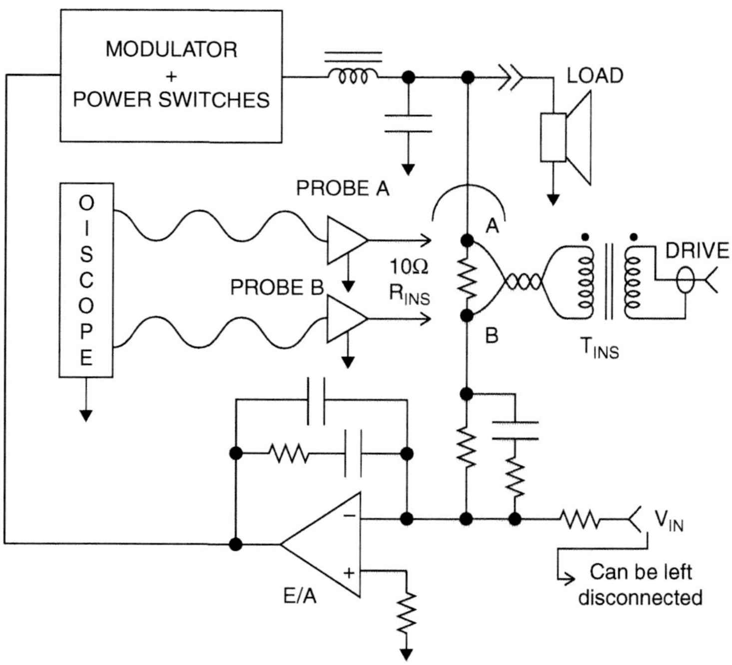

Figure 12.12 Loop-Injector Technique Used to Measure Feedback Loop Stability

points A and B with respect to circuit GND will vary, although the differential voltage measured from A to B will always remain constant at 100mVRMS. This can be better understood as follows: At low frequencies, the loop gain (measured from point B, and going around the loop to point A) will be extremely high. Therefore, a small voltage (measured from point B to GND) will produce a large voltage measured at point A to GND. Conversely, as frequency increases, and loop gain decreases, the portion of the 100mVRMS appearing at point B with respect to GND will increase, and the portion of the 100mVRMS appearing from point A with respect to GND will decrease. Most importantly, at the frequency where the voltage appearing from point A with respect to GND is equal to the voltage appearing at point B with respect to GND (50mVRMS at each point with respect to GND), this is the unity gain crossover frequency for the feedback loop. The *phase difference* between these two equal-amplitude signals at A and B is the *phase margin*. Increasing the DRIVE signal frequency further allows *gain margin* to be measured. At this frequency, the *phase* of the sinusoidal signal at A will match the phase of the sinusoidal signal at B. At this frequency, the amplitude of the signal from A to GND will be *less* than the amplitude of the signal from B to GND, by an amount (measured in dB) equal to the gain margin.

Where to "break" the feedback loop to insert the injection resistor and transformer?

Figure 12.12 shows the injection resistor and transformer positioned in a location at the amplifier output, in the circuit leg feeding the error amplifier. Another location that works well is at the output of the error amplifier (feeding the modulator block) in Figure 12.12. The criteria for selection of the location are that one side of the injection location be a low-impedance driving point, and the other side a much higher impedance. Also, it helps to select a location where noise is minimized, as this will help in measuring the loop signal voltages in the presence of (switching) noise.

Practical Feedback Loop Measurement Tips

It can be appreciated that these measurements described above are being made in a high-noise environment. The switching mosfets are creating large transients, contributing to noise in the oscilloscope readings for PROBES A and B. Furthermore, the driving signal levels are kept low, to be sure that the amplifier is being measured in its linear range. If the signal levels are too large, you will see non-linearities in the sinusoidal signals. So this is a classic case of trying to measure small signal levels in the presence of high noise. To facilitate these measurements, an oscilloscope with an *averaging* function is very helpful. To use this function, a solid trigger signal is required. Typically, the DRIVE signal itself can be used as the oscilloscope trigger. It may also be helpful to use an isolation transformer with a large *step-down* ratio, such that the DRIVE signal has an even greater amplitude, as the trigger to the oscilloscope. When the averaging function is used, the switching noise of the Class-D amplifier will be averaged out, because the switching of the Class-D amplifier's power stage is asynchronous to the sinusoidal DRIVE signal frequency (oscilloscope trigger). Please note that there are a few commercially available feedback loop analyzer solutions (see the References) that use this technique and provide the necessary hardware, as well as software, to drive the loop and measure the results, with the necessary noise suppression. However, it is possible to make these measurements manually, as long as the noise issues are handled, and the waveforms are constantly monitored for signs of non-linearity. It is helpful in terms of signal-to-noise in these measurements to increase the DRIVE signal level as much as possible, provided linearity is maintained. Although these manual measurements are tedious, it is helpful even if only the two measurements of unity gain crossover frequency (with phase margin) and gain margin are made.

Bibliography

Attwood, B. E. (1983). Design parameters important for optimization of very high-fidelity PWM (Class-D) audio amplifiers. *Journal of the Audio Engineering Society, 31*(11), 842–853.

Basso, C. (2008). The link between the phase margin and the converter transient response. Toulouse, France: ON Semiconductor.

Dixon, L. H., Jr. (n.d.). Closing the feedback loop. Texas Instruments Application Note slup068. Dallas, TX: Texas Instruments.

Feldman, R. (2024). How to calculate a DC-DC compensation network. *Electronic Design* (May/June).

Freet, J. (2021, December 1). Two methods for SPICE-based amplifier stability analysis. *planetanalog.*

Honda, J., & Cerezo, J. (2003, October 8). Class D audio amplifier design. El Segundo, CA: International Rectifier.

LeCroy. (n.d.). Measuring phase margin. LeCroy Application Brief, No. LAB 748. Chestnut Ridge, NY: LeCroy.

Lehmann, C. (2009). A simple post-filter feedback topology for Class-D amplifiers. Paper presented at the AES 37th International Conference, Hillerød, Denmark, August 28–30, 2009.

Middlebrook, R. D. (1975). Measurement of loop gain in feedback systems. *International Journal of Electronics*, *38*(4), 485–512.

Putzeys, B. (n.d.). The F-word, or, why there is no such thing as too much feedback. In: *Linear Audio*, vol. 1 – *Circuit Design*.

Shu, W., & Chang, J. S. (2008). THD of closed-loop analog PWM Class-D amplifiers. *IEEE Transactions on Circuits and Systems*, *55*(6), 1769–1777.

Tektronix. (n.d.). Measuring the control loop response of a power supply using an oscilloscope. Tektronix Application Note. Beaverton, OR: Tektronix.

Texas Instruments. (2008, October). AN-1889, How to measure the loop transfer function of power supplies. Application Report SNVA364. Dallas, TX: Texas Instruments.

van der Hulst, P., Veltman, A., & Groenenberg, R. (2002). An asynchronous switching high-end power amplifier. Paper presented at the 112th AES Convention, Munich, Germany, May 10–13, 2002. New York: Audio Engineering Society.

Chapter Thirteen

Housekeeping and UVLO Circuits

The Roles of Housekeeping Circuitry and UVLO

Most amplifier types with output power greater than a few watts will require housekeeping and **U**nder**V**oltage **L**ock**O**ut (UVLO) circuitry. The role of this circuitry is monitoring and control. In general terms, operating conditions of the amplifier are monitored, and control is exerted by various circuit blocks and components in response to operating conditions.

More specifically, the typical operating conditions that are monitored are voltage and current conditions, and thermal conditions. The various types of control that is exerted can be:

- Enable/Disable control and sequencing
- Relay contact closure
- PWM duty-cycle control
- Gain control.

Note: There can be some overlap in the above functions, between housekeeping control and FAULT response. This chapter is concerned with housekeeping circuits, so an attempt will be made to differentiate housekeeping circuit functionality from FAULT detection and response circuits' functionality.

This discussion will refer to the "Housekeeping Circuits" page of the reference schematic. The role of the circuitry on this page is to monitor the low-voltage and high-voltage power supply rails used in the Class-D amplifier, and

DOI: 10.4324/9781003327943-13

prevent the power circuitry (Class-D power stage) from switching, unless and until all voltages are above minimum operational voltage levels. The reason and necessity for this monitoring function should be made clear. The power mosfets used in the power switching stages can switch high levels of voltage and current, and produce high levels of output power. The *control* of the high-power switching stages is provided by the modulation, feedback, and gate drive circuitry. If those circuits are not properly biased by adequate supply voltages, they may produce erroneous output signals, which can cause errors in the control of the power switching stage, and uncontrolled outputs, which can at the very least cause audible pops, clicks, thumps, and other spurious audible artifacts in attached speakers, and at worst can cause component failures and speaker damage. It is therefore critical for reliable operation of the amplifier that power processing and switching be inhibited until all critical rail voltages have been qualified.

UVLO is a term used quite often in electronic device datasheets, especially for devices that have a power control function. One ubiquitous example of a device type that almost always features a UVLO specification and datasheet notes, is mosfet gate driver ICs. Mosfets can be distinguished by gate-drive voltage range. Typical mosfets intended for high power, and moderate to high voltages, usually have a nominal 10V gate-drive voltage specification. This is the gate voltage required in order to fully enhance the gate and turn on the device (for an enhancement mode mosfet). This voltage is usually listed as V_{GS} (voltage from gate to source, accompanied by values of on-state drain-to-source resistance ($R_{DS}ON$)). Usually a minimum $R_{DS}ON$ specification is provided for a V_{GS} voltage of 10V. As the V_{GS} voltage decreases from 10V, $R_{DS}ON$ will increase. A reliable design will ensure that the mosfets can never be driven with a V_{GS} voltage less than some minimum value. For this type of mosfet, specified with 10V V_{GS}, a safe minimum V_{GS} is usually around 8V. Therefore, it is prudent to have circuitry which does not allow the mosfet's gates to be driven unless the gate-drive voltage is at least 8V. Furthermore, since the outputs of most gate drivers saturate to the positive rail, the gate-drive voltage from the driver is usually equal to the power supply rail (Vcc). Therefore, it is sufficient to monitor the IC's rail voltage for the 8V UVLO level.

Most modern gate driver ICs perform this function internally, and call it UVLO. The datasheets must be read to understand the details and nuances of the specific UVLO features of the device, but typically, the UVLO function exhibits hysteresis. That is, there is a voltage threshold for rising Vcc, and a different, lower voltage threshold for falling Vcc. This hysteresis is typically on the order of 1V for a nominal 10V gate driver. For example, the rising threshold might be 9V, and the falling threshold 8V. It is helpful to understand the reasons for hysteresis in this UVLO function. As with any comparator block, if only a single voltage threshold is used (i.e. without hysteresis), any voltage noise superimposed on the monitored voltage, at the threshold, will cause the comparator's output to switch multiple times. This is very undesirable for a power control circuit. It is important for the monitoring comparator's output to provide a single transition from off-to-on and on-to-off, so that the power circuitry ultimately gets a clean and stable enable/disable signal. Another, equally important reason to have hysteresis in the monitoring circuits is because the monitored voltage may actually be affected by the power circuit's turn-on. This is best explained with an example. Everyone is familiar with incandescent lights dimming when there is a large power draw on an AC distribution network. For example, the lights in the kitchen might dim when the refrigerator's compressor switches on. This is because the large surge current drawn by the compressor causes the AC line voltage to drop, due to voltage drops in the house's wiring. Now suppose the refrigerator employed circuitry which monitored the AC line voltage, and only allowed the compressor to switch on when the AC voltage exceeded 105VAC. If there were no hysteresis used, the comparator might switch the compressor on when the AC voltage were 105.001VAC, and off when it dropped to 104.999VAC (just a 2mV drop). Problems will arise because when the compressor switches on, the AC voltage will drop much more than 2mV. We know this, because we can see the lights dim due to the drop in the AC voltage. So, what will happen is that the compressor will switch on. The AC voltage will drop as a result, and the comparator will immediately switch off, switching the compressor off. This then will cause the AC voltage to recover and the comparator and compressor will switch on again, causing the cycle to repeat indefinitely. This is informally termed "motorboating" because the rapid switching on/off of the power circuitry can cause an unpleasant and audible buzz or hum, or relay "chatter" if an electromechanical relay is used. This is definitely *not* a desirable operating mode. Hysteresis avoids this problem. If the designer knows that when the compressor switches on, the line voltage in an adequately wired home should not drop more than say 5V, he can safely use a hysteresis level of 10V and prevent motorboating. For example, the ON voltage threshold for the comparator could be set to 105VAC, and the OFF voltage threshold would be 95V, and motorboating would be prevented.

The previous example of the use of hysteresis involved AC voltage sensing; however, the voltages sensed in the housekeeping and UVLO circuits in the reference design are all DC voltages, which simplifies the design of the voltage comparator circuits.

The best housekeeping and UVLO circuits use a coordinated design approach that takes into account the characteristics of all the relevant circuitry in the design. In the reference design, the half-bridge gate driver IC selected (U4A) has its own internal UVLO function, which prevents its outputs from driving the half-bridge mosfets until the +5VD rail (VDDI supply voltage at pin 3 of U4A) has reached a threshold level. For the Si8244BB, that level is 3.3V minimum. This feature is used to coordinate the design of the housekeeping circuitry on the Housekeeping Circuits page. When designing housekeeping and UVLO circuits, it is critical to know the minimum valid operational supply voltage(s) for each part. For example, the LM339 comparators used are guaranteed to have a valid output state when the supply voltage to the comparators is at least 2.0V. The 2.5V reference, which is generated by Q1, U8, and associated passives, will produce a precise 2.5V reference voltage, capable of delivering up to 1.8mA, when the +5VD supply has reached 3.2V. All the LVC logic gates are guaranteed to produce valid logic states when Vdd (+5VD) is at least 1.65V. What this means is that, as the +5VD supply rail starts to ramp up from 0V (during an amp power-up event), the 3.3V UVLO function of the gate driver will prevent any spurious turn-on of the power mosfets until +5VD has reached at least 3.3V. Below this voltage level on +5VD, it does not matter what logic states are produced by the (underbiased) (Skyworks Solutions Inc., 2021) comparators and logic gates in the housekeeping circuitry, as their outputs will be ignored by the gate driver. Then, once the +5VD supply rail has reached the +3.3V UVLO threshold of the gate driver, it is guaranteed that the output states of the comparators and logic gates are valid, and therefore turn-on control is now transferred to these circuits. This is what is meant by a "coordinated" approach to housekeeping and UVLO design. Once UVLO control has shifted to the comparators, then each comparator sets a UVLO threshold for *each* of the critical supply rails that are monitored, and *all* monitored supply rails need to be above their UVLO rising threshold voltages before any power circuits are enabled.

Detailed Circuit Description of the Reference Design

Referring to the Housekeeping Circuits page of the reference schematic, the housekeeping circuitry is comprised of comparators and LVC "single-gate" logic.

The comparators chosen are low-power, bipolar analog types. The main features of these comparators are wide supply voltage range (2V–36V), open-collector outputs, and single or dual supply (+/−) operation. They draw less than 200uA supply current per comparator, and have relatively short propagation delay of 200nsec typical. Input current is low at 500nA max, so little error is introduced in input resistor divider circuits using typical (10 Kohm–100 Kohm) resistor values. Another convenient feature is that the inputs can be driven above the supply rails without damaging the part. A main feature is that the output structure is open-collector, which allows a wired-or connection of multiple individual comparator outputs. The open-collector output pulls down to the negative supply rail. The LVC single-gate logic family is chosen because of its flexibility. The individual logic gates allow great placement flexibility in terms of PCB layout. Using a more integrated part (with more gates in a single package) means each individual gate is non-optimally placed, requiring longer PCB traces. The other important features of the LVC logic family are 1.65V–5V supply operation, and no input loading when the gates are unpowered. The LVC1G57 multifunction gate is particularly interesting, because this single gate can be configured to perform the function of many different types of logic gates.

Quad comparator U6 qualifies the low-voltage rails: +5VA, −5VA, +7VA, −7VA. Each individual comparator has hysteresis to provide clean switching and prevent motorboating, as described above. Since the comparators have an open-collector output structure, the outputs can be wire-or'ed, such that any single comparator output can keep the wire-or'ed output node low, and all comparator outputs must be high (i.e. all the low-voltage DC rails must be qualified) for the combined output to be high. In this sense, "qualified" means that the voltage rail is above the rising UVLO threshold voltage for each low-voltage rail.

When selecting the rising and falling thresholds for these qualification comparators, the designer must be mindful of the minimum operational voltage for each component biased by each voltage rail being monitored. In the case of the high-voltage rails, both the minimum operational voltage and the maximum operational voltage must be known, and the rising and falling comparator thresholds must be set with these ranges in mind. LTspice is an excellent simulation tool that can be used to easily set these thresholds by selecting the appropriate resistor values for the comparator circuits. The lower end of a resistor value range is dictated by component size and power dissipation limits, whereas the upper end is limited by noise pickup concerns as well as leakage current effects. Typically, the lower resistor in each input voltage divider can be set to 10 Kohm, while the upper resistor divider value and hysteresis (positive feedback resistor) are set by the circuit requirements for minimum operational voltage and hysteresis requirements mentioned above. A single +2.5V reference voltage (+2.5VREF) is used for all comparators. It is desirable to bias the voltage-sensing comparators from a single low-voltage rail (in this case +5VD). This means that some circuit modifications are required in order to sense the negative voltage rails (−5VA and −7VA), because the input voltage range for this comparator is from the negative rail (0V) up to within 2V of the positive rail (5V − 2V) = 3V. In order for this input voltage range to be accommodated, while sensing negative voltages, the circuit in Figure 13.1 is employed.

In Figure 13.1, the +2.5V reference voltage is divided down and applied to the positive input of the comparator. When the comparator output is *low*, the threshold voltage set at the positive input pin will be +994mV (which is within the input voltage range of the comparator). In order for the comparator output to switch *high*, the negative input pin must be brought more negative than the positive input pin, or in this case, more negative than +994mV. The negative input pin is connected to a voltage divider that is biased from the +2.5V reference at the

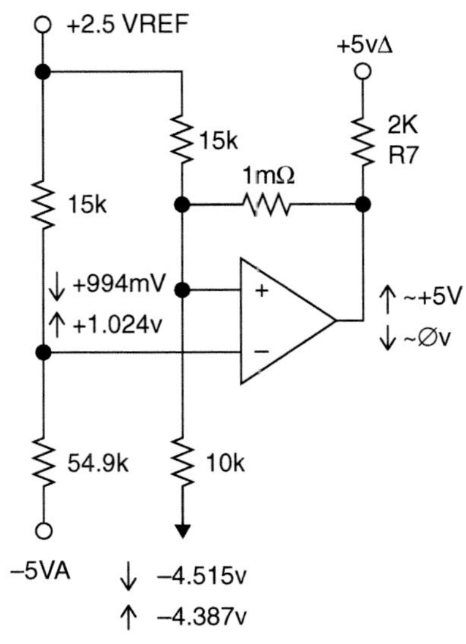

Figure 13.1 Circuit Technique Allowing a Comparator, Biased from a Single Positive Voltage Rail, to Sense Negative Voltages

top end, and the sensed negative low-voltage rail at the bottom end. In the case of the −5VA output, for example, the sensed negative voltage rail needs to be more negative than −4.515V, in order to pull the comparator's negative input pin more negative than +994mV. When this occurs, the "rising" UVLO threshold for the −5VA rail will have been achieved, and the comparator output will swing positive (open collector).

With the comparator's output high (open collector), the threshold voltage set at the positive input pin of the comparator will now be +1.024V (ignoring the small loading effect of the other comparators connected to the wire-or'ed output and R7). In order to detect an undervoltage (UV) event on the −5VA rail, the voltage on this rail must drop (towards 0V) becoming more positive, sufficiently to make the comparator's negative input pin more positive than +1.024V. This occurs when the −5VA rail has drooped towards 0V and is more positive than −4.387V. LTspice simulations will confirm these voltages.The component values in the reference schematic result in the switching thresholds for the low-voltage rail-monitoring comparators (U6) that are listed in Table 13.1. U9 alone sets the UVLO thresholds for the +5VD rail, and these are also listed in Table 13.1.

The combined output of the low-voltage rail comparators connects to the MR/ pin of U9. U9 is a voltage detector IC with a reset pin (MR/). It qualifies the +5VD rail, with its own internal voltage detection comparator selected for a 4.697V (rising) UVLO threshold. The output pin of U9 (RESET/) will be *low* whenever +5VD is below the UVLO threshold, *or* the MR/ pin is held low. When +5VD is above the UVLO threshold, *and* the MR/ pin is released (i.e. wire or'ed comparator output from U6 is high = open collector), U9 output will still remain low for a one-second delay interval, before switching high. This delay interval is set by the selection of C_T = C129. At the end of this delay, the RESET/ output of U9 will go high, driving ON the gate of Q15, which in turn will drive

Table 13.1 Switching Voltage Thresholds for the Low-Voltage Rail-Monitoring Comparators

Rail Voltage	*Rising Threshold Voltage*	*Falling Threshold Voltage*
+5VA	+4.285V	+4.215V
+7VA	+6.033V	+5.963V
−5VA	−4.515V	−4.387V
−7VA	−6.012V	−5.845V
+5VD	+4.697V[a]	+4.65V

a Rising threshold for U9 is typically 1% higher than the falling threshold voltage.

Q14 ON. The one-second delay in U9 is used to allow all low-voltage supplies to settle to their final regulated output voltages. The signal at the drain of Q14 indicates all **L**ow **V**oltage rails are qualified = "OK" and is applied to one input of logic gate U11 as "LV_OK". Multi-function logic gate U11 is configured as an AND gate. Both the LV_OK and LATCH/ signals must be *high* in order for U11 output = HV_ENABLE to be high. The HV_ENABLE signal is used to switch on the HV rails, (+/−48VDC) for the amplifier power switching stage.

U7 is another quad comparator which monitors the high-voltage DC rails from the ACDC power supply. These DC rails have an enable/disable function, which is a critical part of the protection circuitry in this design. Two of the comparators in U7 (wire-or'ed to pullup resistor R134) provide UV detection, and the other two comparators of U7 (wire-or'ed to provide the HV_OV/ signal) provide overvoltage (OV) detection for the high-voltage DC rails. *Note:* The terms "low-voltage" and "high-voltage" as used in this design are relative. In this case, "high-voltage" refers to the power rails for the Class-D switching stage which are +/−48V. "Low-voltage" refers to all the remaining DC rail voltages. These are used for biasing all the signal processing and housekeeping circuits. LTspice is again used to select the component values for the UV and OV thresholds for the HV rail comparators.

Note that although hysteresis is used for the UV thresholds, no hysteresis is used for the OV thresholds, because OV is a FAULT condition that requires the fastest possible response. The OV comparator circuits have been designed such that the first falling edge at wire-or'ed OV comparator output pins U7-13/14 will trigger the fault LATCH circuitry (to be discussed along with the fault and protection circuits in the next chapter).

See Table 13.2 for a listing of the UV and OV thresholds for the HV rail comparators.

Once the HV rails have exceeded their rising UV threshold voltages, the wire-or'ed output of U7-1/2 will go high (open collector). This will begin an R-C timing delay at the input to multi-function logic gate U10, which is wired as an AND gate. This R-C has a charging time constant of (R134 + R135)(C130) = (10,221 ohms)(10μF) = 102.21msec, and a discharging time constant of (R135) (C130) = (221 ohms)(10μF) = 2.21msec. The LVC1G57 gate has Schmitt trigger inputs. The hysteresis present in the input logic switching thresholds allows for clean switching, even with slow-moving input signals, as are created by the R-C input circuit. With this R-C input, U10 provides a delayed enable for the Class-D

Table 13.2 UV and OV Switching Voltage Thresholds for the HV Rail-Monitoring Comparators

HV Rail Voltage	*Rising UV Threshold Voltage*	*Falling UV Threshold Voltage*	*Rising OV Threshold Voltage*
+48V	+35.77V	+34.47V	+57.75V
−48V	−34.84V	−34.11V	−58.13V

amplifier gate driver IC (U4A). Once again, this delay interval allows some time for the HV supplies to reach their nominal regulated output voltage levels and settle, before the Class-D amplifier power switching stage is enabled. The delay is intentionally asymmetrical, because while a delayed enable for the amplifier gate drivers is beneficial, there is no reason to delay switch-off. In fact, since switch-off occurs due to a UV condition on the HV rails, it is beneficial to disable the amp gate drivers as soon as possible, before the HV rail voltages droop even further below the falling UV threshold voltage. Multi-function logic gate U12 is configured as an AND gate, requiring the AMP_EN signal from U10-4 to be high, AND the OCA/ signal to be high. OCA/ is an OverCurrent signal, generated by the *protection* circuitry, discussed in the next chapter.

+2.5V Reference

U8 and R137 form the 2.5V reference that is used by the comparators on this page. The design of the reference circuitry includes some important details for proper operation. Since we know that all the comparator circuitry relies on this reference for UV detection, *and* that the U4 gate driver keeps the power circuits disabled until its internal UVLO (based on the +5VD rail) is satisfied, the +2.5V reference voltage must be up and regulated at a rail voltage *lower* than U4's minimum falling UV threshold for the +5VD rail (+3.30V). U8 is a SHUNT-style voltage reference which requires a bias current to regulate. There are many varieties of this type of reference, differing not only in reference voltage, but also in parameters such as minimum bias current for regulation, and maximum bias current. The simplest way to bias this type of reference is with a resistor to a voltage rail with voltage greater than the reference voltage (as is done in the reference design). A potential problem with this (if used in other designs) is that the resistor must be selected such that enough current is provided from the positive rail to supply the circuits loading +2.5VREF (all the comparator resistor dividers) as well as the minimum bias current required by U8. In this design, the current which must be provided by the 2.5V reference to the housekeeping circuits is 1mA. The minimum bias current required by U8 for regulation is 70uA. If the biasing resistor value is selected such that 1.07mA is provided when the +5VD rail is 3.30V, it will need to be (3.30V − 2.50V)/1.07mA = 746 ohms. 750 ohms is selected as a standard 1% value. This means that when the +5VD rail is regulated at +5.00VDC for example, this 750-ohm bias resistor will be providing (5.00V − 2.5V)/750 ohm = 3.33mA. Of this 3.33mA, 1mA will go to the comparator resistor dividers and 2.33mA will be shunted by U8. The power dissipation of U8 will be (2.33mA × 2.50V) = 5.83mW, and the power dissipated by the biasing resistor (R2) will be ($3.33mA^2$ × 750 ohm) = 8.32mW. These dissipations are sufficiently low to be acceptable for this high-efficiency design.

Reference Biasing Option for Designs with Higher Rail Voltage

The biasing of the U8 voltage reference is discussed in detail above, because *power dissipation* is a critical part of Class-D design. All circuitry, including the low-power monitoring and housekeeping circuits, must be designed for the lowest possible power dissipation, if maximum overall power efficiency is to be achieved. A biasing option for U8 is presented here. This option provides power dissipation benefits for cases where U8 needs to be biased from a higher-voltage rail, as may be the case in other amplifier designs.Figure 13.2 illustrates the case where simple resistive biasing of U8 is attempted from a rail voltage of +15V instead of +5V. In this case, if there is still a requirement for the +2.5V reference to be fully regulated by the time the biasing rail has reached +3.3V, then a bias resistor value of 750 ohms is still required, to supply the necessary 1.07mA of current to fully bias U8 *and* supply the reference load current of 1mA at a biasing rail voltage of +3.3V. However, when the rail voltage reaches its nominal value of +15V, there will now be 12.5V across the 750-ohm resistor, causing it to dissipate 208.3mA (necessitating a relatively large 1/2W resistor package). U8 will need to shunt 15.67mA of excess current with a power dissipation of 39.13mW, for a total circuit dissipation of 247.48mW.Figure 13.3 shows the power dissipation advantages afforded by biasing U8 with a constant current source, instead of a simple resistor. In this example, the first condition that will be analyzed is when the +15V rail is at +3.3V. In that case, there will be 0.6V across the 332-ohm resistor, establishing

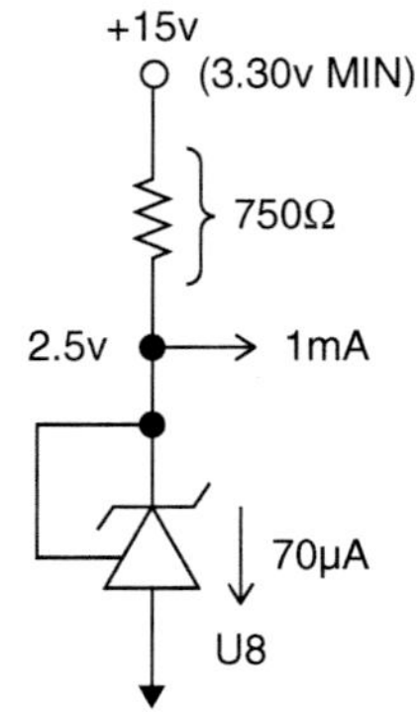

Figure 13.2 Schematic for a Shunt Voltage Reference Utilizing Simple Resistor Biasing

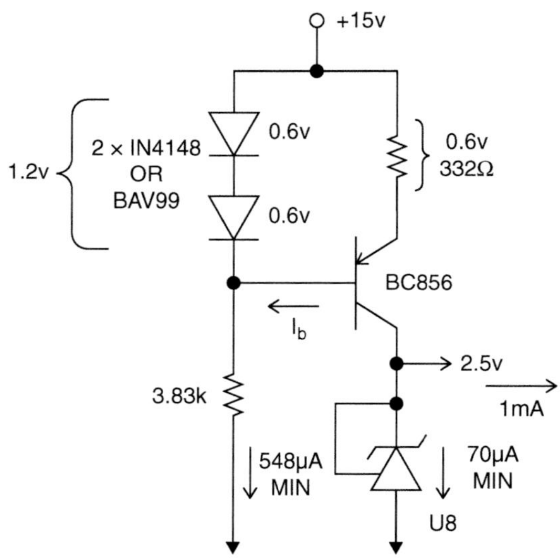

Figure 13.3 Schematic of a Current-Source Biasing Technique for the Shunt Voltage Reference

an emitter current (I_E) through the PNP transistor of 1.80mA. The base current required will be approximately 1/20th of that value (90uA), because the PNP will be operated in saturation, with a Collector-Emitter voltage (V_{CE}) of only 0.2V. The collector current will be 1.80mA − 90uA = 1.71mA. The resistor dividers loading the +2.5V reference will sink 1.0mA, leaving 710uA to be shunted by U8 itself. The base terminal of the PNP transistor will be at +2.1V, and a current of (2.1V/3.83 Kohm) = 548uA will flow through the 3.83K base resistor to ground. Of this 548uA, 90uA will be the base current from the PNP, and the remaining (548uA − 90uA) = 458uA will flow through the two series-connected diodes, which will each have a forward voltage drop of 0.6V (approximately).

Now, when the rail voltage has risen to its full nominal voltage of 15V, the voltage across the 3.83K base resistor will have increased to 13.5V. A current of (13.5V/3.83 Kohm) = 3.52mA will flow through the series-connected diodes. Their forward voltage will increase slightly, to approximately 0.75V each, due to this increased current. The voltage across the 332-ohm emitter resistor will be approximately (1.5V − 0.6V) = 0.9V, establishing an emitter current (I_E) of (0.9V/332 ohm) = 2.71mA. The voltage across the PNP transistor (V_{CE}) will be (15V − 0.9V − 2.5V) = 11.6V. At this V_{CE} voltage, the current gain (beta) of the PNP will be well over 100, so base current will be negligible, and collector current (I_C) will essentially equal I_E = 2.71mA. U8 will shunt 1.71mA, leaving 1.0mA to flow to the +2.5V reference load circuits. Power dissipation of this circuit is then:

332 ohm = (0.9V^2)(332 ohm) = 2.44mW

PNP = (11.6V)(2.71mA) = 31.44mW

U8 = (2.5V)(1.71mA) = 4.28mW

3.83 Kohm = (13.5V)^2(3.83 Kohm) = 47.6mW (an 0603/100mW resistor is used at 50% derating)

2 series diodes = (2)(0.75V)(3.52mA) = 5.28mW

Total = 91.04mW, for *power savings* of (247.48mW − 91.04mW) = 156.44mW.

Bibliography

Skyworks Solutions Inc. (2021, July 31). Datasheet, Si824x Class D audio driver with precision dead-time generator. Rev. 1.0. Irvine, CA: Skyworks Solutions Inc.

Texas Instruments. (2008). Datasheet, LM4040, Precision micropower shunt voltage reference, SLOS456K. Originally published January 2005, revised March 2008. Dallas, TX: Texas Instruments.

Texas Instruments. (2016). Datasheet, SN74LVC1G57 Configurable

multiple-function gate, SCES414P. Originally published November 2002, revised November 2016. Dallas, TX: Texas Instruments.
Texas Instruments. (2018). Datasheet, LM339, LM239, LM139, LM2901 Quad differential comparators, SLCS006U. Originally published October 1979, revised November 2018. Dallas, TX: Texas Instruments.

Chapter Fourteen

Protection Circuits (Overcurrent, Overtemperature, DC-Offset)

Class-D Amplifier Fault Conditions

A Class-D audio amplifier can function without the protection circuits described in this chapter. It would be able to produce output signal within its design limits, and, if never driven beyond those limits, might never fail. But this is a totally unrealistic operating scenario for most audio amplifiers. In actual real-life usage conditions, most audio amplifiers are attached to separate, external speaker loads, which can vary greatly in impedance and power requirement. The amplifier outputs can be mis-wired, or inadvertently shorted to earth ground or to each other. They can be overdriven and overheated, either due to driving and load conditions, or due to inadequate cooling and ventilation. All these conditions are very likely to occur sometime over the life of an amplifier. Even amplifiers designed with adequate cooling, either passively via conduction, through ventilation holes, or actively by convection using fans, can have these ventilation holes blocked, or fans stalled, due to mean time to failure (MTTF) life limitation (wear out) of the fan, or even by dust build-up on heatsinks and/or filters. Even though the efficiency of Class-D amplifiers is extremely high, and still increasing with newer devices and innovations, there is still a thermal challenge, because designers take advantage of this great power efficiency to design smaller, denser amplifiers, with less or no cooling in smaller packages. So it is still possible to overheat these efficient designs. The bottom line is… protection circuits are still

DOI: 10.4324/9781003327943-14

required for reliable, long-life Class-D amplifier designs. Protection circuits are mainly designed to handle at least the following faults:

- DC output condition
- Overcurrent (OC) on output
- Overtemperature (OT).

Additional protection circuitry is sometimes added to handle:

- Excessive high-frequency output protection
- Over/under voltage (supply pumping)
- Over power protection (power supply overload).

These protections will first be discussed within the context of the reference design, and then an expanded discussion will follow.

DC Output Condition (DC Fault)

An audio amplifier intended for reproduction of music or voice is usually designed to handle audio signals over the full range of human hearing (20Hz–20KHz). High-pass filters, usually incorporating AC coupling capacitors in the signal path, set the low-frequency −3dB corner frequency, and "block" DC signals. This high-pass corner frequency typically falls in the range of 1Hz–7Hz, in order for the attenuation at 20Hz to be minimal (less than 0.5dB for a single-pole filter). In the reference design, the high-pass (R-C) filter components responsible for this response are CXA (C50A) with the series sum of RXA (R76A) and RYA (R69A). EQ. 14.1 shows the high-pass corner frequency to be 1.8Hz:

$$f_{HP} = 1/2\pi(Rx + Ry)Cx \qquad \text{EQ. 14.1}$$

For the component values used in the reference design, $f_{HP} = 1.8\text{Hz}$.

This high-pass network will prevent DC signal content from propagating through the amplifier. This is important for two reasons:

1) Loudspeakers are not designed to handle DC current. DC current flowing through a loudspeaker voice coil will result in a fixed cone excursion, producing

no audio output (except for the initial cone movement to the fixed voice coil position), and potentially high power dissipation from the I^2R power losses in the voice coil. Refer to Figure 14.1 for a cutaway view of a typical electrodynamic loudspeaker, illustrating its major component parts. The power dissipation problem is exacerbated further by the fact that some voice coils are designed to dissipate power by virtue of their movement. Voice coil movement results in air movement over the voice coil, which helps to dissipate power. If the voice coil is dissipating power due to DC current, it is also not moving, which makes its temperature rise even further. Another potentially damaging effect of the DC voice coil current is due to excessive voice coil excursion. This can result in damage to the speaker surround, as well as the spider which helps to center the voice coil in the magnetic "gap".

2) In the case of a Class-D amplifier employing a half-bridge output stage operating from positive and negative supply rails (as is the case for the reference design), average current drawn from the positive and negative supply rails is expected to be "balanced" over time. Any unbalance results in supply pumping (discussed in other chapters). A DC voltage/current output condition will result in severe supply pumping and overvoltage (OV) on the less loaded power supply rail.

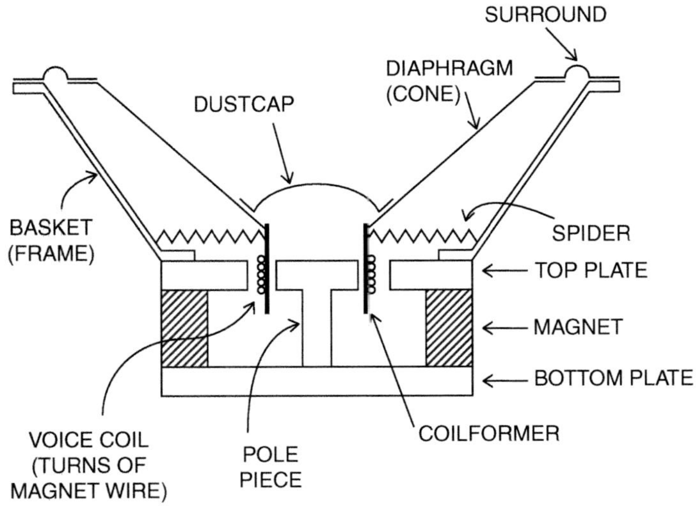

Figure 14.1 Cutaway View of a Typical Electrodynamic Loudspeaker

Notwithstanding the above explanations, the overriding justification cited for incorporating DC output fault protection in an audio amplifier is "speaker protection". The justification can be explained as follows.

Although there are components in the amplifier design, such as the high-pass filter described above, which prevent a DC signal from propagating, it is always still possible that there can be a component failure in the amplifier that can result in DC voltage and current being applied to a connected loudspeaker. Some audiophile loudspeakers are extremely expensive, and protecting them in the case of an amplifier fault or failure is prudent.

Amplifier protection response in the case of component failure requires new design considerations. For example, the first protection response (and the fastest one possible) is to simply stop the output stage mosfets from switching. If the power stage is operating nominally and is undamaged, this is a valid response and will protect the amplifier and any connected loudspeaker. However, if one or more output stage mosfets has failed shorted (which is a common failure mode for power mosfets), then there will be a direct circuit connection made between the high-voltage DC power supply rail(s) and the loudspeaker. This can result in extremely high and uncontrolled DC current flow through the loudspeaker's voice coil. For example, using the reference design for illustration, if the output power mosfet attached to the positive supply rail were to fail shorted, there would be a direct connection between the +48V rail and the loudspeaker. If the loudspeaker were 8 ohms, this would result in a DC current of 48V/8ohms = 6ADC, and a continuous power dissipation of ($6A^2$)(8ohms) = 288W in the speaker.

It is imperative, therefore, that the protection response to a DC fault work properly even when components in the amplifier have failed. For example, in the above scenario, if one or both of the amplifier's output power mosfets have failed shorted, there must be a way to disconnect the power supply rails from the loudspeaker, or to disable (switch off) the high-voltage power supply rails.

Disconnect Response

The standard disconnect response to DC fault is to incorporate a power relay in each amplifier output path. These relays can *open* the circuit path between the amplifier output and loudspeaker.

Relay Advantages

One important advantage of a disconnect relay is that it can be used to disconnect the output for reasons other than DC fault. For example, it can keep the output disconnected during amplifier power-up and power-down intervals. This results in the ultimate pop and click protection (i.e. "popless" start-up and shutdown).

The relay can be selected to be Normally Open (NO). This means that the relay contacts are *open* when the relay coil is unpowered. This is a safe default state for the relay contacts. It requires that the relay coil be actively driven in order to close the relay contacts and thereby connect the loudspeaker to the amplifier output. This active drive allows the possibility of adding separate voltage qualification circuits to the relay drive circuitry, ensuring that the coil can only be driven when certain conditions are qualified. This can be the UVLO (**U**nder**V**oltage **L**ock**O**ut) circuits discussed in Chapter 13, or a separate UVLO circuit dedicated to relay drive qualification.

Relay Disadvantages

There are some disadvantages to adding power disconnect relays to an amplifier design.

Size and Cost

The most obvious disadvantage is the added cost of the components. In addition, these are typically large through-hole parts which take up significant PCB space. They must be sized to handle the maximum RMS output current of the amplifier, but in addition, the DC current rating for the contacts must be reviewed (see next section).

Ability for Contacts to Open and Close with DC Current Flowing

Electromechanical relay contacts open a defined physical distance to stop current flow. When DC current is flowing, this current will continue to flow in the form of an arc (or plasma), as the relay contacts initially separate (due to the initially small distance between contacts). As the contact distance increases, the arc stretches and eventually extinguishes, but this extinguish distance depends on the final DC voltage across the contacts. The higher the voltage difference, and the higher the current being interrupted, the farther the contacts need to separate before the arc extinguishes. Contact arcing results in surface damage to the contacts, due to the extremely high temperatures of the arc plasma, and can cause contact pitting, wear out, and eventual failure. The failure mode can be either contacts welded together permanently (i.e. the contacts will no longer open) or high contact resistance or open circuit.

When selecting a relay for output disconnect duty, it is important to check the DC contact ratings of the relay. These are usually much lower in current and voltage than the AC contact ratings. Relays have a much easier time disconnecting an AC current as opposed to a DC current, because the AC current passes through 0A, as it changes polarity. Any arcing that occurs at initial contact separation is extinguished when the current amplitude passes through 0A. Typically, it will not be possible to find relays of suitable size and cost that have contacts with a DC current interrupt rating suitable for the amplifier application. In this case it is best to contact the relay manufacturer and explain the usage condition to them. In some cases, they may already have relays that have been used in this type of application, or they may be willing to test particular relays to qualify them for the intended usage condition.

Further Information

Higher-power relays (much higher power than those used in audio amplifier design) feature "magnetic blow-outs" to help in rapidly extinguishing the arc as the contacts open. These blow-outs consist of a permanent magnet adjacent to the contacts. The arc plasma is repulsed by the magnetic field and pushed and stretched out of the contact gap. The stretching of the arc causes it to break sooner than it would if the blow-out magnets were not present. Special relays purposely designed to interrupt high-current DC circuits are called "contactors", but these are meant for very high current circuits, and are not appropriate for Class-D amplifier design.

Relay Power

Another disadvantage to using electromagnetic relays is the power that the coil requires to close the relay contacts, and hold them closed. Typically, for relays of the size required for most amplifier applications, a relay coil power of from 0.3W to 1W is typical. Relays are available with coil voltages ranging from 5V to 48V. For the reference design, a relay with a 12V coil would be appropriate, because it could be powered from the 12V housekeeping supply rail. If a lower-voltage rail were used, it would incur greater overall power dissipation losses due to the extra power dissipation of the voltage stepdown from the 12V housekeeping rail to any of the lower-voltage DC rails used in the reference design.

Note: The 48V rails (positive and negative) could possibly be used to power the relay coils, because the relays will only need to be closed (ON) when these rails are up and regulated. For any condition where these rails were in an undervoltage condition, the relay coils should be in the open (undriven/default) state for an NO (Normally Open) relay type.

Wear out

As mentioned above, electromechanical relays have wear out mechanisms, and therefore have a finite lifetime. Besides the "abnormal" contact damage considerations mentioned above, there is "normal" operational wear out, evidenced by the endurance specifications listed on relay datasheets. Typically, this is specified in terms of numbers of operational cycles under various contact load conditions.

Latching Relays

Electromechanical relays are also available in LATCHING varieties. The advantage of these is that the relay coils only require power in the dynamic state, i.e. when they are actively moving the contacts into the closed or open states. They require no power in either static state. When the relay contacts are open or closed, the relay coil requires no power. The main disadvantage to this type of relay is availability. There are much fewer latching relays produced, compared to the non-latching type. However, if a suitable latching relay is found and available, some important considerations (in addition to all the same considerations listed above for non-latching types) must be addressed.

Latching Relay Coil Drive

Latching relays are driven into each of the latched contact states by a pulsed (short-duration) signal, applied to the coil(s). It is important to consult the datasheet to determine the coil driving conditions, including minimum and maximum coil drive voltage, as well as minimum and maximum pulse width. The minimum pulse width is limited by the mechanical time constants involved in the construction of the part, i.e. how long it takes to energize the coil and for the mechanical movement of the contacts into the new latched state. Typical minimum pulse widths are in the range of 50msec. The maximum pulse duration of the relay drive pulse is limited by coil power dissipation and temperature rise. Latching relays come in different varieties of coil drive. For example, there are types that use a single coil that is driven in two different current directions. A pulse in one direction will close the contacts, and a pulse in the opposite direction will open the contacts. Another type uses two separate coils that are driven independently. Driving one of these coils will close the relay contacts. Driving the other coil will open the contacts. Another important consideration in selecting a latching relay is its vibration and shock tolerance. It is possible to move the relay contacts to the other latched state by shock or vibration above the specified level.

Latching Relay Control

Particular care is needed to always control the contact state for latching relays. The contact position (state) of a latching relay is determined by its most recent coil drive pulse (whenever that happened to occur). Some latching relay manufacturers will state that the relays are delivered with the contacts in a specific state (either open or closed). But, it is prudent to positively control latching relays in any amplifier design where they are used for output disconnect duty. One way to do this is to actively drive (pulse) the relay contacts to the open position during the UVLO interval. If the contacts were already in the open condition, there is no problem, the pulse is redundant, and it causes no contact movement. However, if the contacts had inadvertently moved to the closed position, due to a physical shock or vibration experienced by the amplifier since its last power-up, this pulse will reset the relay contacts to the open position, before the high-voltage rails are enabled. Essentially this resets the relays to a known starting condition during each amplifier power-up sequence. In addition to this power-up control, the relay contacts should again be actively pulsed into the OPEN state during amplifier shutdown or power-down. This can be triggered by *any* monitored power supply rail dropping below its UVLO falling threshold. Again, this provides a level of housekeeping redundancy, in that it will ensure that the relay contacts are left in the OPEN state upon each amplifier shutdown/power-down event.

Detecting DC Fault

Now that the DC output condition has been explained, along with the reasons it should be prevented, the circuitry involved in the detection of DC fault will be explained. Referring to the reference design, dual comparator U26A detects DC fault. The voltage divider ladder consisting of R83A–R86A sets the threshold voltage for the two comparators of U26A (configured as a window comparator). In selecting the threshold voltage, it is important to consider the power dissipation by connected loudspeaker loads during an output DC fault. Table 14.1 lists the power dissipation in 8-ohm and 4-ohm loudspeaker voice coils at different voltage levels of output DC fault.

Typically, a DC fault trip level at the output of approximately +/−3VDC is used, because the power dissipated in the speaker voice coil is low enough at this voltage level to prevent overheating.

The output voltage of the amplifier can be as high as +/−48V, not including any effects from supply rail pumping or voltage regulation by the power supply itself. This voltage is too high to be applied directly to the comparator inputs. The comparator (U26A) is biased from +/−5VDC rails. A conservative design would prevent the voltage at any comparator input pin from exceeding its supply rails. In the case of some comparator types there can be output phase reversal when an input is driven below the Vss (negative) rail. The datasheet should be consulted for this behavior (see TI datasheets for LM393 (Texas Instruments, 2018) and LM339 (Texas Instruments, 2023)). The design approach taken here will be to ensure that the output state is correct when the rail voltages are above the UVLO levels. For the LM393 and LM339 comparators, the output state will be valid when at least *one* of the inputs is within the common-mode (CM) input voltage range (Vss to Vcc −2V), and the other input does not violate the input voltage limits (Vss −0.3V to +30V). The input voltage at pins 3 and 5 is +273mV and −273mV, respectively (calculated below). These are within the input CM voltage range. This means that the input voltages at pins 2 and 6 should not exceed the input voltage limits (Vss −0.3V to +30V). The worst-case falling UVLO rail thresholds for the +/−5VA rails are +/−4.215VDC minimum. Therefore, when the output is at +/−48VDC the comparator input pins 2 and 6 should not be more negative than

Table 14.1 Power Dissipation in 8-Ohm and 4-Ohm Voice Coils at Different DC Voltage Levels

DC Output Fault Voltage	*Power Dissipation in 8-Ohm Voice Coil*	*Power Dissipation in 4-Ohm Voice Coil*
2V	0.5W	1W
3V	1.125W	2.25W
4V	2W	4W

−4.215VDC − 0.3V = −4.515V. The positive input voltage limit of 30V will never be approached. The values of the resistor divider R100A-R101A are selected to comply with these input voltage range limits. *Note:* Generally it is not good design practice to generate comparator thresholds off of supply rails, as is done in the reference design. The preferred method is to use voltage references to generate more accurate and stable DC reference voltages. However, in this application, the actual trip levels do not require very tight accuracy. The approximately +/−7% accuracy provided by the +/−5% voltage rails and 1% resistor divider values provides more than adequate accuracy for this application. The connection to the amplifier output voltage is made through R100A. This resistor is sized at 0805 to handle the maximum voltage rather than power dissipation (*Note:* an 0805 resistor is rated for a continuous working voltage of 150V). R100 forms a divider with R101A, producing approximately +/−4.36V maximum, when the output is +/−48V respectively. With this constraint met, the comparator threshold voltages are set by calculating the voltage on R101, when the output is +/−3VDC (or the designated DC fault level). That threshold voltage is +/−273mV. Voltage dividers R83A and R84A, and R85A and R86A are then selected to set these threshold voltages for the dual comparators of U26A.

DC Fault Low-Pass Filter

The network consisting of R100A, R101A, and C63A forms a voltage divider and low-pass filter. This network attenuates the output voltage in order to meet the input voltage limits of the comparator (U26) as discussed above, but also provides a single-pole rolloff in order to allow sensing of the DC component of the output signal, while rejecting higher-frequency signal components. A design constraint to keep in mind when selecting component values for this low-pass filter, is that the amplifier should be able to produce a 20Hz signal at full amplitude without tripping the DC fault response. This means that when producing a 20Hz signal at full amplitude, the peak voltage across R101 should not exceed +/−273mV. It is useful to simulate this simple filter circuit in LTspice. An important finding in this simulation is that the initial half-cycle at 20Hz will produce an output of approximately *twice* the steady-state peak-to-peak voltage at R101. So in order to keep comparator U26 from triggering, the low-pass rolloff must be more severe than initially thought.

The first simulation (Figure 14.2) shows the initial response to a 12Hz, full-amplitude output signal.

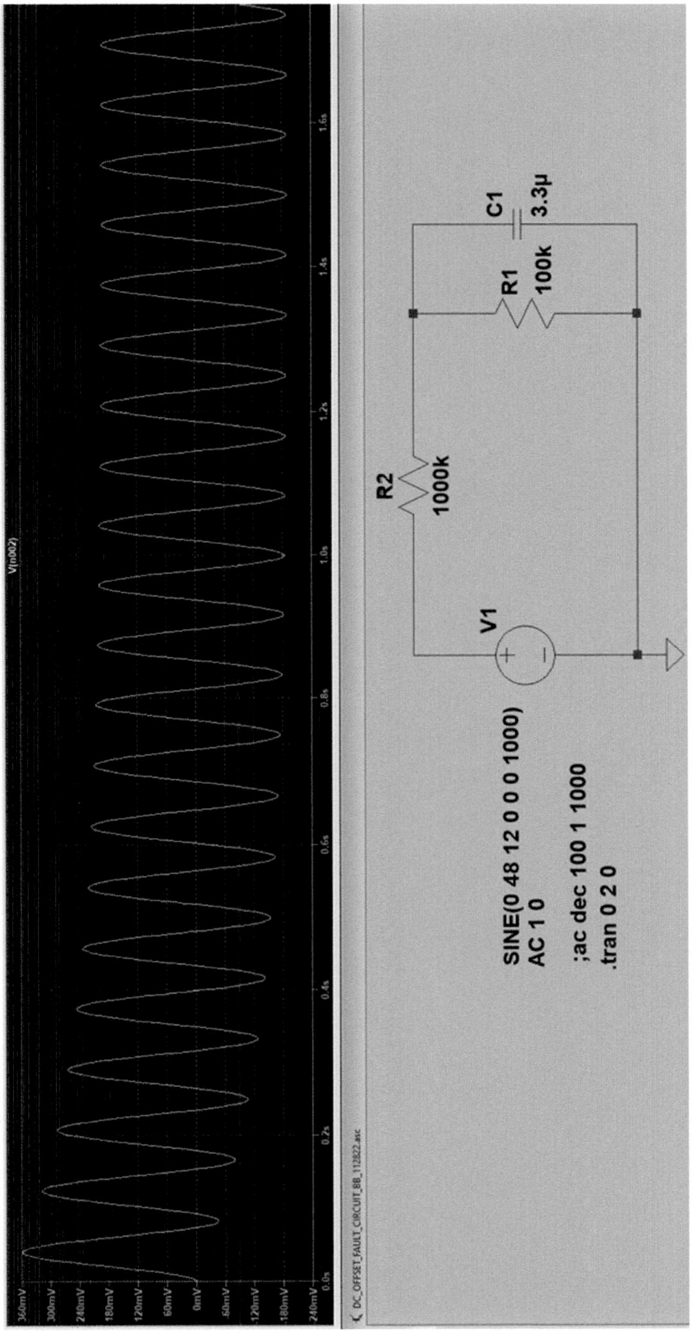

Figure 14.2 Time-Domain LTspice Simulation of the Response of the DC Fault Low-Pass Circuit, to a 12Hz Full-Amplitude Output Signal from the Amplifier

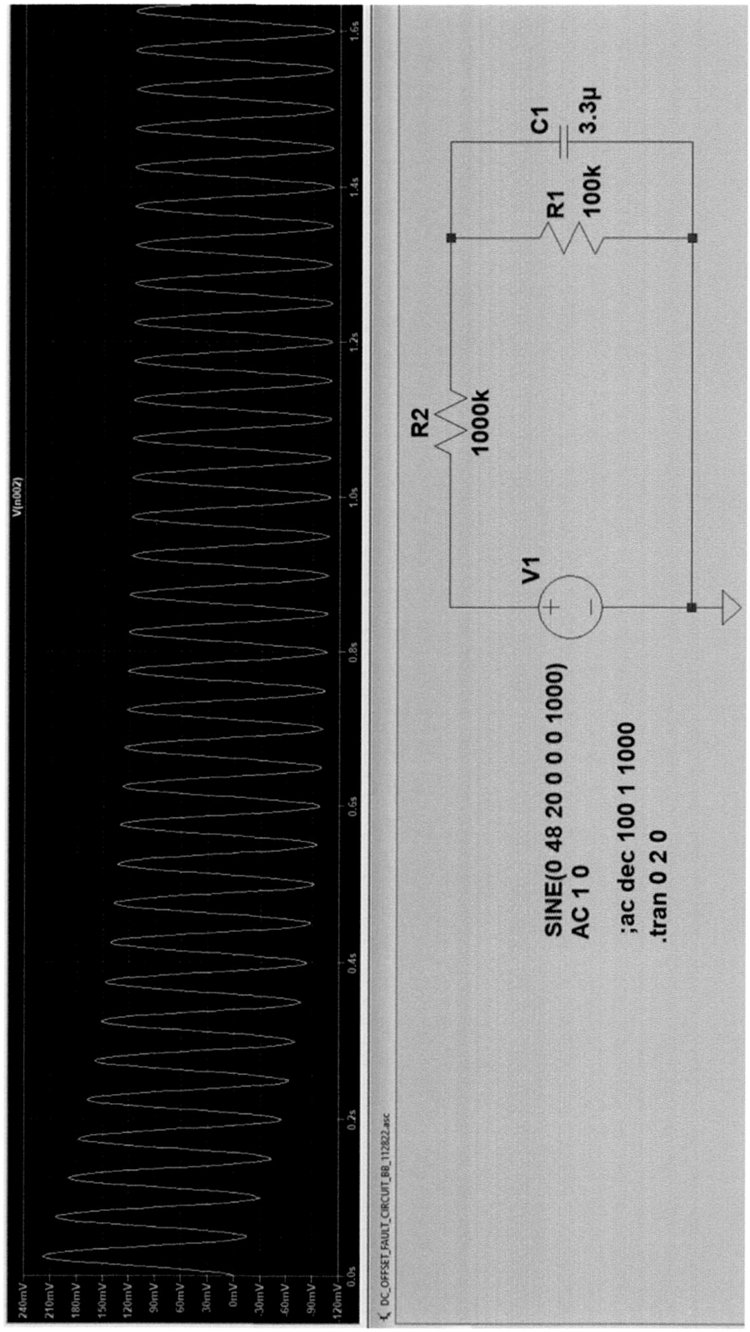

Figure 14.3 Time-Domain LTspice Simulation of the Response of the DC Fault Low-Pass Circuit, to a 20Hz Full-Amplitude Output Signal from the Amplifier

With the component values shown, the level seen by the comparator input will be +360mVpeak. This is just slightly above the +/−273mV trip thresholds of the comparator. A 20Hz signal will yield a lower voltage at the comparator input (+/−210mVpeak), thus providing margin against tripping DC fault (see Figure 14.3).

If it is desired to adjust the component values (and filter corner frequency) with more precision, such that a 20Hz signal at full amplitude (+/−48Vpeak) just reaches the +/−273mV comparator trip thresholds, then it will be necessary to size C63A with greater precision. This would require placing an additional capacitor in parallel with C63A, so that the sum of the two capacitances could be adjusted to a value between 2.2µF and 3.3µF. However, for this application, such precision is not required, and a capacitor value of 3.3µF was chosen for C63A.

It is important to check how this filter will respond in the event of a shorted mosfet in the power stage (since that is the worst-case scenario for which we desire the DC fault protection response). Figure 14.4 shows the response at the comparator input to a *step change* in the output voltage, from 0V to +48V (simulating a short of the upper mosfet connected to the +48VDC rail). The step change occurs at the 100msec graticule on the X-axis. The simulation shows that the DC fault comparator trip threshold of +273mV will be crossed at the 120msec graticule. Thus, it will take the DC fault circuit 20msec to respond to this worst-case fault. This will result in a loud "thump" in any connected loudspeaker, but should not cause damage to the loudspeaker if it is rated appropriately for the 100W output power capability of the reference amplifier.

The Use of Hysteresis with Voltage Comparators

As a general rule of good design practice, hysteresis should be used around voltage comparators to provide for clean monotonic switching, and the elimination of "runt" output pulses. In some cases however, hysteresis can be eliminated when there is good justification. In the case of the DC fault detection circuitry, hysteresis around the U26A comparators has been eliminated, because the wired-or outputs trigger a LATCH, and the latch trigger is level sensitive, not edge sensitive. So it does not matter if the comparator's output transition is non-monotonic, or is a runt pulse with too small or too short a duration to trigger the latch. As soon as there is an output DC fault level of sufficient magnitude to force the comparator output(s) to fully switch low, the latch will trigger. So in this case it does not matter how cleanly or monotonically the DC fault comparators switch.

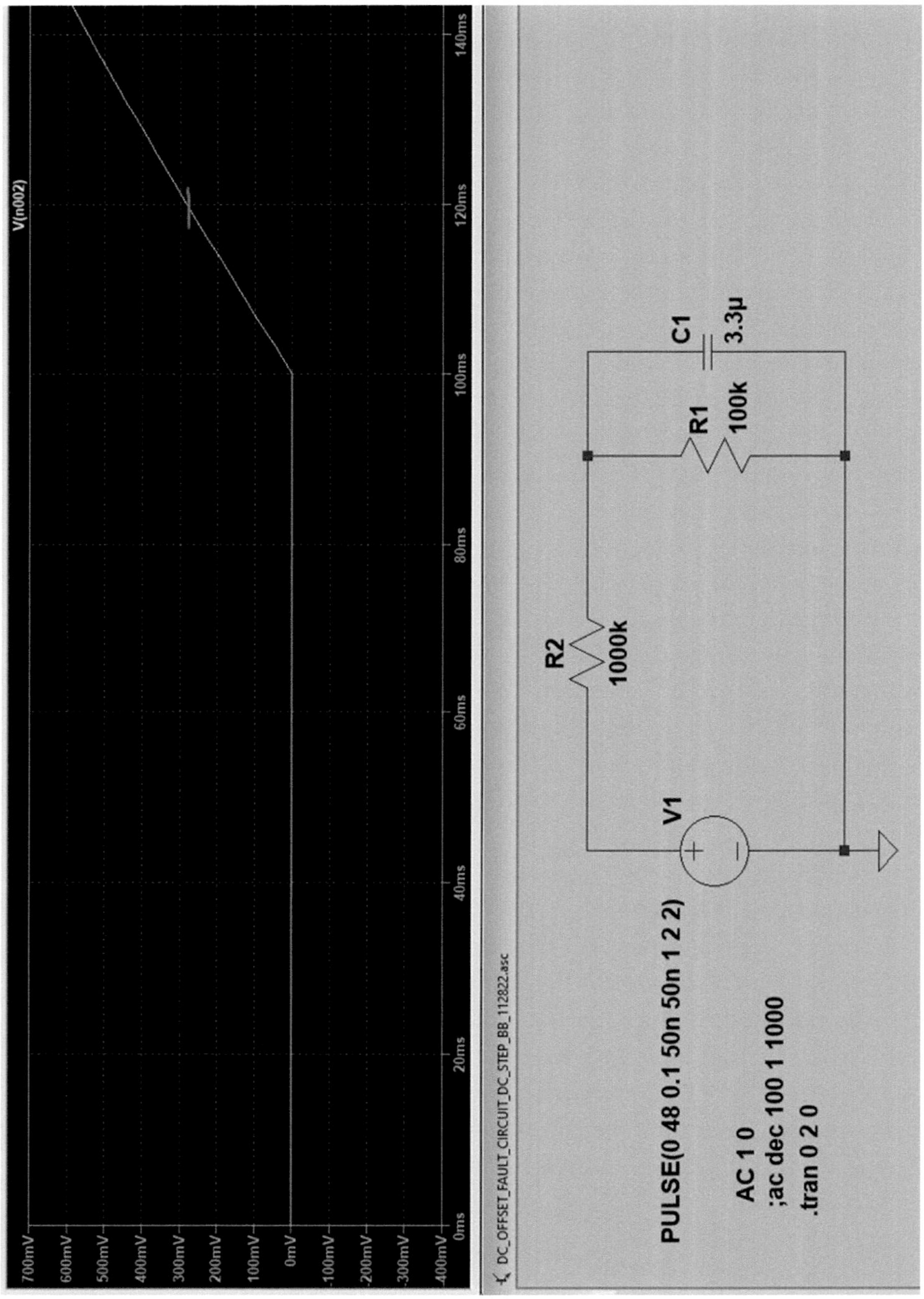

Figure 14.4 Time Domain LTspice Simulation of the Response of the DC Fault Low-Pass Circuit, to a Step Change in Amplifier Output Voltage from 0V to +48VDC

DC Fault LATCH Operation

A latched OFF response of the amplifier was chosen as the desired reaction to an output DC fault condition, because this type of fault is indicative of a serious internal amplifier fault, that can damage connected loudspeakers if the amplifier output is not disabled quickly. This is in contrast to other faults which are transient in nature and *not* indicative of failure (such as an OC or OT condition). For this reason, the reference design has the DC fault comparators (U26A) triggering a LATCH. The latch circuit is comprised of transistors Q11 (PNP) and Q16 (NPN). These are connected in an "SCR" arrangement, except for the addition of resistors R92 and R93. These two resistors allow the "SCR" arrangement to trigger and latch ON without "crowbarring" the +5VD rail. A typical SCR does not include these two resistors. Instead, when triggered, a typical SCR is meant to act as a crowbar, shorting the rail voltage to 0V (or as close as possible to 0V). That is why SCRs are used for crowbar applications, as overvoltage protection devices. In this application, however, the latching feature of the SCR is desired *without* the crowbar feature. Resistors R89 and R94 keep transistors Q11 and Q16 OFF in the absence of trigger current from R90 or R91.

LATCH Triggering

When either comparator of U26A detects a DC fault, their respective output will switch low. This will cause Q11 to be driven on through R90. Once ON, Q11 will drive Q16 ON through R93. Once ON, Q16 will provide regenerative drive to Q11 through R92, keeping it on.

LATCH Reset

PNP transistor Q12 keeps the LATCH in reset (unlatched) when Q12 is ON. It can also reset the latch once it has been triggered. The LV_OK signal is used to control Q12. As has been explained in the housekeeping logic section in Chapter 13, no high-power circuits (Class-D amplifier or high-voltage (HV) rails) can be enabled unless the low-voltage (LV) rails have been qualified, so this signal is also used to keep the LATCH in reset, to prevent false triggering of the latch during start-up and shutdown sequences.

LATCH Triggering from HV Rail Overvoltage (HV_OV)

There is another trigger input to the LATCH circuit. That is HV_OV. If there is an overvoltage (OV) detected on the HV rails by the two OV comparators of U7 on the Housekeeping page of the reference schematic (discussed in Chapter 13), the LATCH will be triggered. The justification for this is that OV on the +/−48VDC rails (+/−HV) above the expected and allowed supply pumping margins is indicative of a voltage regulation loop failure in the HV power supply, and for that reason, the amplifier and HV power supply should be shut down. The OV comparators trigger the LATCH circuit through R91.

Once triggered, the latch will remain in the ON (latched) state until the LV_OK signal (from U9) goes low. This occurs at power-down, when AC power is switched OFF, and also during the power-up sequence, when the power rails are ramping up, and initial undervoltage is detected by UVLO comparators U6 and U7 (2 of the 4 comparators). Therefore, the latch will be reset by cycling the AC power to the unit OFF then ON again.

Overcurrent (OC) Detection and Response

Over**c**urrent **P**rotection (OCP) is essential for protecting the power stage components of a Class-D amplifier from damage. The power switches (mosfet, GaN, or SiC devices) have power dissipation and drain current limits which must not be exceeded. In addition, the Class-D inductor will have a saturation current specification which must not be exceeded. In Chapter 9 on the output L-C filter, this is covered in more detail, but here it is sufficient to say that different core materials used for the Class-D inductor offer different saturation characteristics. The two commonly used core types of gapped ferrite and powdered iron differ in that gapped ferrite has a sharp, rather unforgiving saturation characteristic, whereas powdered iron has a softer, more gradual approach towards saturation, which provides further protection against OC events. It is obvious though that detection speed and response speed are critical requirements for the OCP circuits. The example below will illustrate why this is so.

This example uses component values from the reference design, specifically a Class-D inductance of 22μH, rails of +/−48VDC, and a Class-D switching frequency of 500KHz (T = 2μsec). It depicts the rise in output current (inductor current) under a shorted output condition.

Class-D Amplifier Output Current versus Inductor Current

In this discussion on OC protection, the terms "output current" and "inductor current" are sometimes used interchangeably. These are discussed in greater detail in Chapter 9 on the L-C demodulation filter, but it is worthwhile to discuss this in some detail here as well. The current flowing through the Class-D filter inductor consists of the output current *and* the AC ripple current. The ripple current is the linear ramping current that flows through the inductor at the Class-D switching frequency, as well as through the filter capacitor in the L-C filter. This current ramps in an upward direction when the upper mosfet is on (causing the positive inductor current to exceed the (average) output current by a small amount), and a downward direction when the lower mosfet is on (causing the negative inductor current to exceed (be more negative than) the (average) output current by a small amount). When inductor current is sensed, it is important to calculate OC trip levels, keeping in mind the actual peak inductor current, which will include both the average output current plus the ripple current. When the Class-D amplifier audio output current and voltage are 0A and 0V, respectively, the inductor current consists *only* of the AC ripple current, which will be a triangle wave at the Class-D switching frequency, with linear rising and falling slopes.

In the case of a hard-short on the output, the output *voltage* will remain at 0V, while the current increases on each switching cycle. Since the amplifier design (typically) uses a high-gain negative feedback loop, even a very small discrepancy between the desired output voltage and the actual output voltage will cause the feedback loop's error amplifier to swing to its local output extreme and call for minimum or maximum duty cycle. In this design, a minimum pulse width has been programmed for 100nsec. The example below (Figure 14.5) shows what would happen if the desired output were a signal even slightly more positive than 0V, with the actual output shorted (= 0V). In this case, the error amplifier will call for the upper power switch (mosfet) to remain on for its maximum duty cycle (1.90μsec ON/100nsec OFF). During the first ON cycle, assuming a starting output current of 0A for convenience, the inductor current will ramp from 0A to 4.145A. During the following minimum ON time for the lower mosfet, the current will only ramp down by 218mA, due to the short 100nsec time interval. On the next ON cycle of the upper mosfet, the inductor current will again ramp up by 4.145A, but from a starting current of 3.927A = (4.145A − 218mA). By the third switching period, the inductor current has reached 12.00Apeak. This illustrates how quickly the inductor current will climb. Of course (as illustrated by the equations in Figure 9.10 and Figure 14.5), the situation gets worse as the rail voltages increase or as the output inductance decreases.

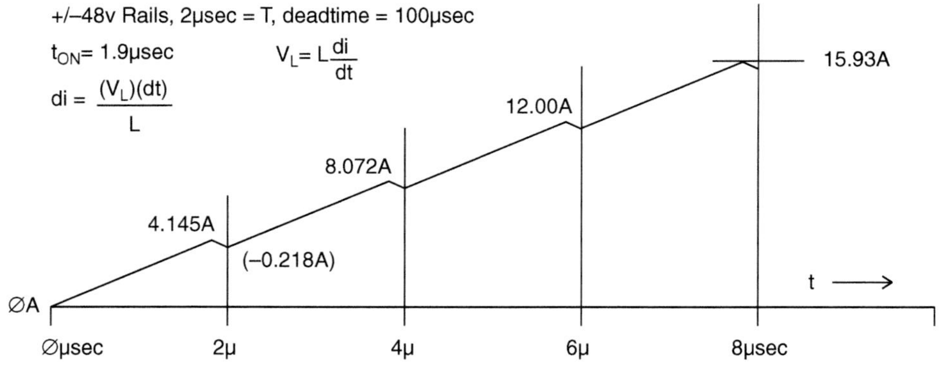

Figure 14.5 Time-Domain Response of Inductor Current, Ramping up Due to a Short Circuit at the Output of the Amplifier

This further illustrates how interactive and interrelated the various aspects of a Class-D amplifier design are. In this example, it can be seen how inductor selection affects OCP design. In other chapters we will see how inductor selection also affects THD, efficiency (power dissipation), EMI, and feedback loop design.

In the reference design, a 100W/4-ohm output will demand the highest operational output current.

100W into 4 ohms = 20VRMS = 5ARMS = 7.07Apeak

In selecting a Class-D inductor, a conservative design would use an inductor capable of handling 5ARMS for power dissipation and temperature rise, and (at first pass) handling 7.07Apeak as a saturation limit. But as can be seen, some margin must be added to handle the additional current overshoot resulting from the OCP detection and response delay. One approach for the reference design would be to set an OCP detection threshold of 8Apeak. As will be seen, the detection circuit delay is 4µsec, so the peak inductor current will climb to 16A worst-case. Therefore, the Class-D inductor must be selected to handle 16A peak without excessive saturation. Some saturation (up to 30%) is tolerable here, since any resultant THD increase caused by this 30% drop in inductance can be ignored during an OC event. The justification here is that the amplifier should not be expected to meet its THD specification when in severe OC, since this is a fault condition, with an out-of-spec load condition.

Current Sensing

A critical design consideration is the *location* where current should be sensed in the Class-D design. Some designs sense drain current in the mosfets of the full- or half-bridge power stage. This can be done either by inserting low-value current-sense resistors in series with either the drain or source terminals of the upper and lower mosfets, or by "R_{DS}ON *voltage* sensing" across the drain-source terminals of these mosfets. Both types of sensing will be discussed.

Current-Sense Resistor Location

When current-sense resistors are used to sense mosfet drain current, they are usually inserted in the source lead of the lower mosfet (see Figure 14.6). This has the advantage that one terminal of the lower mosfet's sensing resistor will be referenced to a fixed (relatively static) DC voltage (the negative HV rail). The upper mosfet's sensing resistor, however, is referenced to the very dynamic switching node.

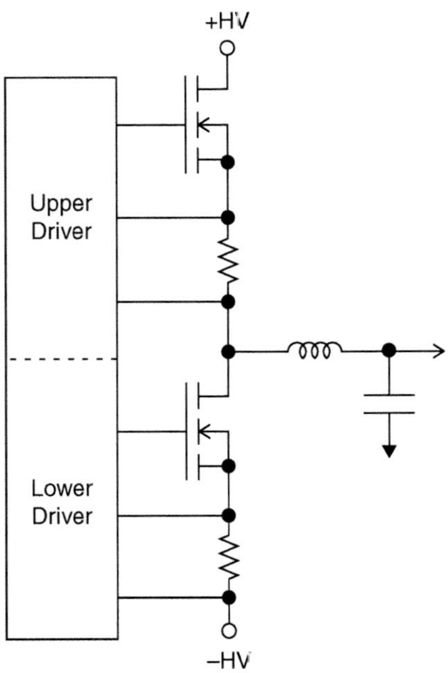

Figure 14.6 Current-Sensing Resistors Inserted in the Source Lead of Both the Upper and Lower Mosfets in a Half-Bridge Switching Circuit

Since the power dissipated by the current-sensing resistors must be minimized, this necessitates using as low a value as possible, while still generating a current-sense *voltage* signal across this sense resistor that is large enough to be used by the detection circuits monitoring this signal. As an example, suppose it was desired to trip at 7Apeak, and a 10-milliohm current-sense resistor were used; the peak power dissipated in this resistor would be $7^2 \times 0.01$ ohms = 0.49W, and the current-sense signal generated would be 70mVpeak. The current-sensing fault comparators would need to trip at +/−70mV. It is important that this signal level be detectable by the comparators in the presence of extraneous noise. A larger current-sense signal is more easily detectable above the noise level, but also dissipates more power.

If this sensing position were used for the current-sensing resistor for the lower mosfet, the detecting circuits could be conveniently referenced to the negative rail. Also, the signal could be easily processed within the gate drive circuitry for the lower mosfet as well. The upper mosfet, however, presents a more difficult current-sensing situation. Since the source terminal of the upper mosfet is the switching node, this is a very dynamic node. Referencing a current-sensing resistor to this node means the sensing circuitry also needs to be referenced to this node, unless more level-shifting circuitry is used. This circuitry would require excellent high-frequency CM rejection to reject the huge high-frequency CM switching voltage swing at this node. Alternatively, the sensing circuitry could reside within the gate driver circuitry of the upper mosfet, which is already referenced to the switching node. If a scheme is adopted whereby the current-sensing circuitry is integral to each mosfet's gate drive circuitry, then the current-sensing response will be local to each mosfet. This has the advantage of allowing a very fast response, such as terminating the ON pulse to the overcurrent mosfet. However, if a coordinated response of *both* mosfets is desired, in response to an OC event in just *one* of the mosfets, then communication is required between the upper and lower gate driver fault detection circuits, or communicating the OC fault to a centralized controller, which can then initiate a more global response, such as inhibiting both mosfets and driving a fault LED.

Another possibility for the upper mosfet is to locate the current-sensing resistor in the drain lead connection (see Figure 14.7).

In this location, the resistor can no longer easily interface to the gate driver circuitry of the upper mosfet (without a level-shifter), but it gains the advantage that one terminal of the current-sense resistor is again referenced to a relatively

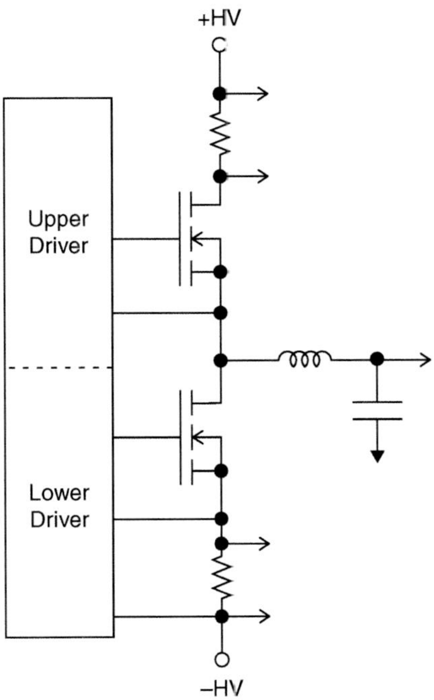

Figure 14.7 Current-Sensing Resistors Inserted in the Drain Lead of the Upper Mosfet and the Source Lead of the Lower Mosfet in a Half-Bridge Switching Circuit

static DC rail (the positive HV rail). So, as was the case for the lower mosfet, the signal developed across this resistor will be "cleaner", since it no longer has a high DC CM dynamic voltage on it. It only has a static high DC CM voltage. A level-shifter will be required to translate the small current-sense voltage generated across the resistor to a different reference potential such as ground or the negative HV rail, but this level-shifting circuitry no longer needs to have excellent high-frequency CM rejection. If, however, the OC processing is being handled locally, by the upper gate driver circuitry (referenced to the switching node), then any level-shifting circuitry *would* still require excellent dynamic CM rejection specifications.

Locating the current-sensing resistors in the positions above allows sensing drain currents in the individual mosfets. This means they can detect output current, as well as cross-conduction currents. The next current-sensing location can only monitor output current, and cannot monitor cross-conduction current.

Inductor Current Sensing

Placing a current-sensing element (such as a low-value current-sensing resistor) in series with the output inductor allows sensing of the output current, in addition to Class-D ripple current, but does not sense cross-conduction currents (see Figure 14.8).

One advantage to this current-sensing location is that the current is much less dynamic than in the mosfet locations described above. The currents are audio frequency currents (20Hz–20KHz) with the Class-D switching frequency ripple currents superimposed (see Figure 14.9).

This oscilloscope capture shows the voltage across the current-sensing resistor located in the location shown in Figure 14.8 (triangle waveform). With a resistance value of 0.1 ohms, and an oscilloscope setting of 100mV/div, this equates to 1A/div. The larger sinusoidal waveform is the amplifier output voltage at 500mV/div. With the amplifier at idle, this is the residual output voltage ripple of the amplifier. The smaller sinusoidal waveform is from a Hall-effect current probe monitoring output current to the speaker at 1A/div. This capture helps to show the difference between output current to the speaker and inductor current. Inductor current includes the triangle-wave ripple current, the great majority of

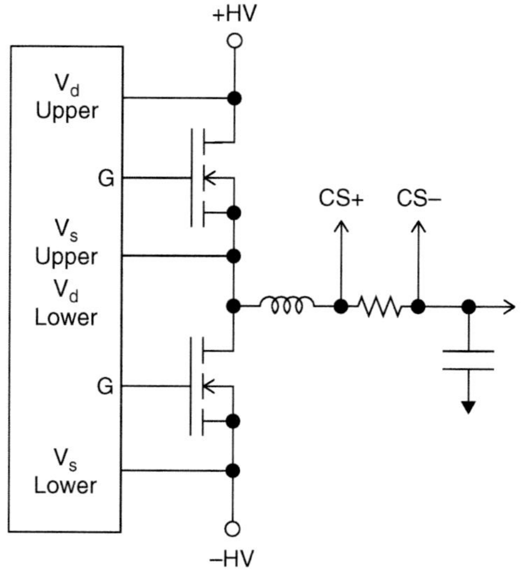

Figure 14.8 Current-Sensing Resistor Inserted in Series with the L-C Filter Inductor in a Class-D Amplifier Output Circuit

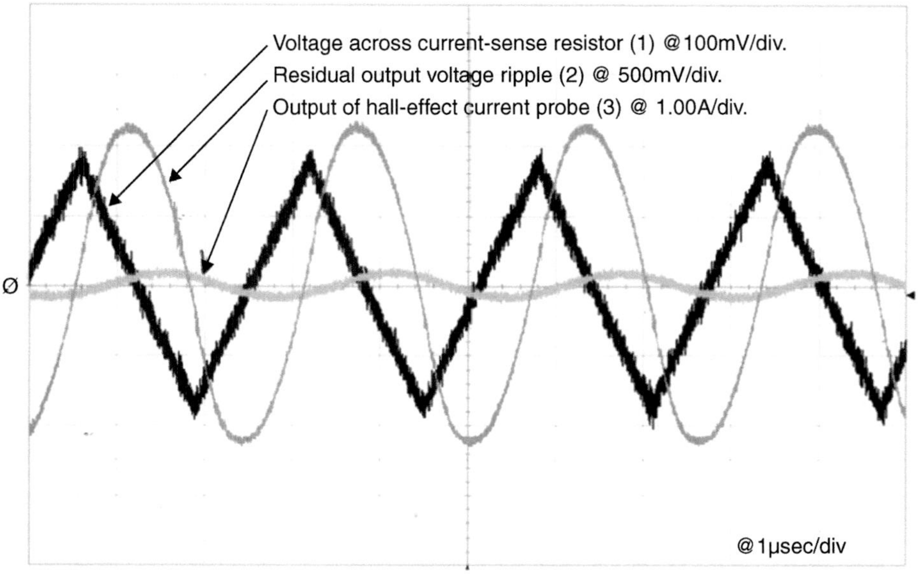

Figure 14.9 Oscilloscope Capture Showing Voltage across Current-Sensing Resistor, Hall-Effect-Sensed Amplifier Output Current, and Amplifier Output Voltage at Idle

which flows through the Class-D filter capacitor(s), while much less ripple current flows out to the speakers.

The oscilloscope capture in Figure 14.10 shows the same three voltages, but with the amplifier producing a sine-wave (speaker) output voltage (@ 10V/div) of 23.6Vpeak (16.7VRMS) across a 4-ohm speaker load. The current probe on the speaker wire (@ 2A/div) shows an output current of 5.9Apeak. The differential voltage across the 0.1-ohm current-sensing resistor (@ 500mV/div, which equates to 5A/div) shows the same 5.9Apeak current level, but again has the residual switching ripple current superimposed on it. The lower set of traces is an expanded view of the upper traces in the highlighted region. This shows how the change in duty cycle of the power switches has changed the appearance of the switching ripple across the current-sense resistor. When the amplifier is producing a positive output voltage, the duty cycle increases above 50%. This is shown by the increased ramp-up time compared to ramp-down time in the expanded trace.

The oscilloscope capture in Figure 14.11 shows the response of the amplifier to an output *short-circuit*. For this capture, the amplifier was producing a *square-wave* output signal into a 4-ohm resistive load when the short was applied at the end of

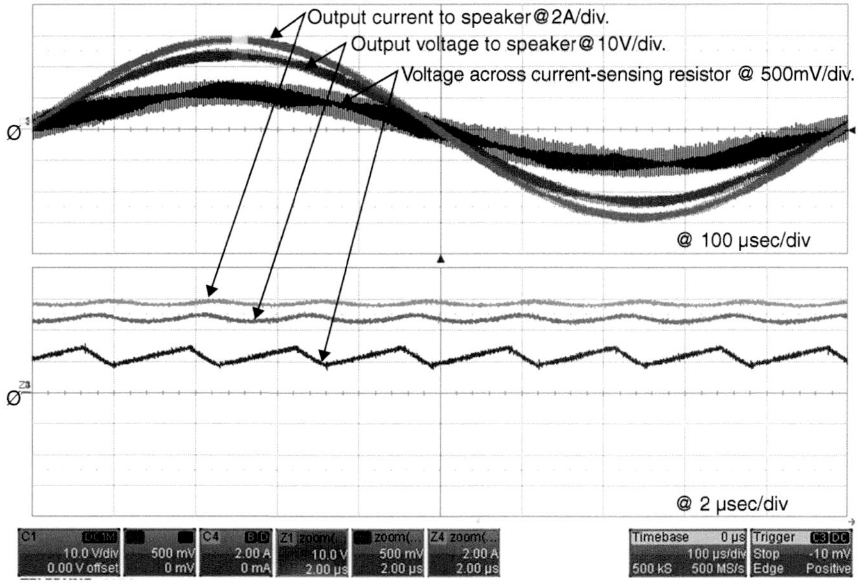

Figure 14.10 The Same Three Voltages as in Figure 14.9, with the Amplifier Producing 16.7VRMS across a 4-Ohm Speaker at 1KHz

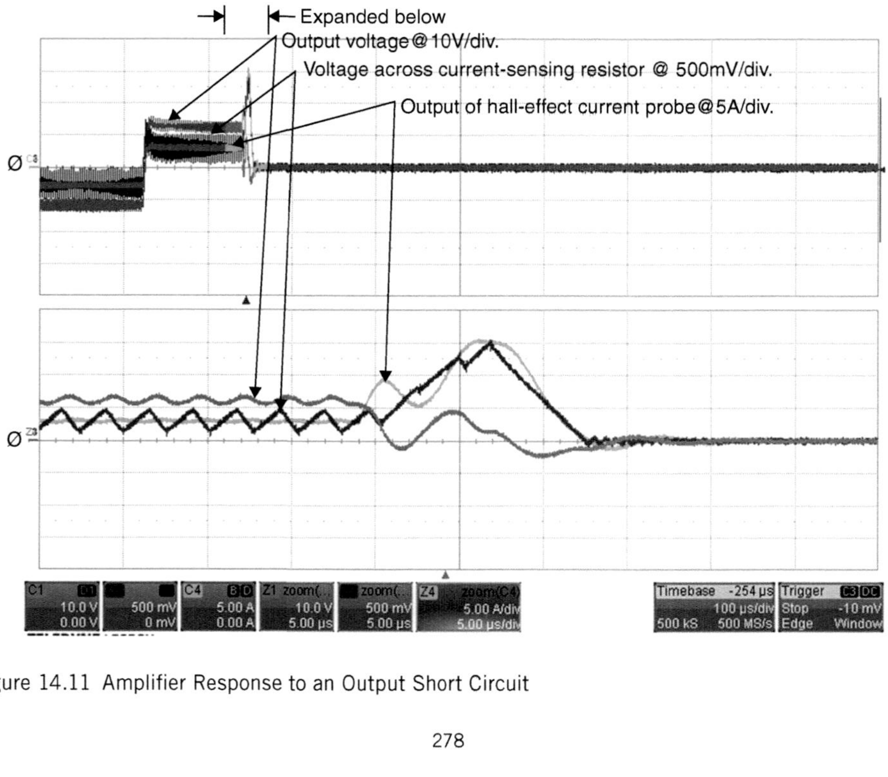

Figure 14.11 Amplifier Response to an Output Short Circuit

3 feet of speaker wire. The lower traces are an expansion of the highlighted area of the upper traces. The output voltage trace shows the output voltage decaying quickly to 0V (due to the short), with a little ringing due to the inductance of the intervening speaker wire. The *bold* trace shows the differential voltage across the 0.1-ohm current-sensing resistor in series with the Class-D inductor at an equivalent scale of 5A/div. The *Hall-Effect* trace shows the current in the speaker wire as sensed by the Hall-effect current probe. It can be seen that the current in the sense resistor quickly ramps up to the OCP trip point of 15Apeak. The duty cycle increases as well, to maximum. At the 15A OCP trip level, the amplifier is disabled by the OCP circuitry, and switching stops. The linear downslope of the *bold* trace is due to the flyback of the switching node to the negative rail, after both mosfets switch off. Since there is 15A flowing through the inductor in the direction *towards* the output, that current direction will continue after both mosfets are disabled. The switching node will commutate to the negative rail (in this case −57VDC) and the inductor current will ramp down as:

$$di/dt = (V_L)/L \qquad \text{EQ. 14.2}$$

Substituting 57V for V_L, with an inductance value of 22μH, yields 2.59A/μsec.

Estimating from the capture, the current ramps down from 15A to 0A in approximately 6μsec, which is 15A/6μsec = 2.5A/μsec.

This capture also illustrates how effective inductor current sensing is in protecting the amplifier from an OC fault. The peak current sensed by the sense resistor is equal to the peak output current sensed by the current probe.

Note: It is important for an output short-circuit test that the amplifier be producing an output voltage before the short is applied. The reason for this is that without an output voltage, the negative feedback loop will not saturate during the short, in an attempt to maintain that output voltage during the short, resulting in high output current. Another way to explain this is that if the amplifier is at idle, the output voltage is at 0V already, so shorting the output produces *no change* in the output voltage or output current. In actuality though, there will always be a small offset voltage at the amplifier output (typically below +/−50mV). Suppose a +10mV output state were shorted at the speaker output terminals, and the output voltage was sensed by the negative feedback circuitry closer to the Class-D output capacitor, as illustrated in Figure 14.12. In this output circuit, there is a CM inductor L_{CM} located between the negative feedback voltage pickoff points and

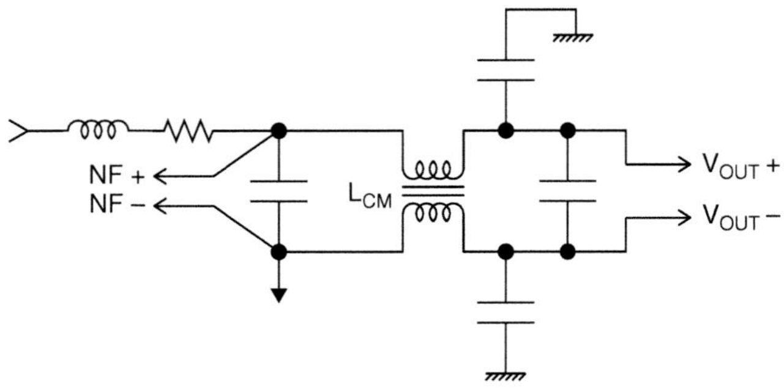

Figure 14.12 Possible Output Configuration for a Class-D Amplifier, Employing a CM Choke at the Output

the speaker outputs. If the DCR (DC resistance) of L_{CM} were 10 milliohms in each winding, then shorting the speaker terminals would only cause a current of 0.5A to flow, and OCP would not be triggered. The reason is that 0.5A through the two 10-milliohm windings of L_{CM} will create a voltage of 10mV, which will satisfy the negative feedback loop. Therefore, when testing OCP, it is necessary to have sufficient output voltage demand to drive the OCP loop above its trip threshold current.

It is important to note (as will be explained later in discussions of supply pumping) that this particular condition of shorting a low-voltage output state, resulting in output current below the OCP trip level, can result in supply pumping when a half-bridge power stage topology is used. This is another reason why including **O**ver**V**oltage **P**rotection (OVP) for the +/−HV rails is important.

Another advantage of locating current-sensing components in the Class-D inductor path (which will be discussed further in Chapter 16 on PCB Layout) is that it simplifies the layout of the critical half- or full-bridge power stages, which need to be as clean and tightly placed and routed as possible,[1] for best performance. Using the inductor location allows for using a *single* sensing element instead of two.

A disadvantage to this output inductor path location for current sensing is that the current-sensing element is not referenced to a static DC level. Instead, it is a differential signal (the signal across the current-sensing resistor), with a large CM voltage component. The CM voltage component is the audio output voltage signal of the amplifier. Any current-monitoring circuitry will have to reject this large CM voltage in order to detect the small differential voltage generated across the current-sensing resistor by the output load current. It is important to note that when the current-sensing resistor is located *after* the Class-D inductor, the CM signal (amplifier output voltage signal) is much less dynamic than the CM signal that exists at the switching node (20KHz sine wave, as opposed to 500KHz square wave). Nevertheless, this still presents a CM rejection problem for the level-shifting and detection circuits processing the current-sense signal from this circuit location.

Current-Sense Resistor Development

There has been a great proliferation of current-sensing resistor types in recent years, with a great deal of technical progress. These are resistors purpose-designed for sensing current. They feature low self-inductance to minimize di/dt-induced ringing in the voltage waveforms across these resistors. They also feature small size and high power density, allowing high power dissipation in small surface-mounted device (SMD) packages. Another feature is reverse geometry packages, where the terminations are on the long ends of the rectangular SMD package, instead of the short ends, as is the case for conventional SMD resistors. This geometry also helps to reduce self-inductance, as well as enhance power dissipation capability, since the power dissipated by the current-sense resistor is conducted to the PCB through the wider solder terminations. The range of resistance values has increased, as well as the precision and tolerance. Extremely low values of resistance are available. There are also four-terminal packages available, which include two power terminations for the high-current path, and two sensing terminations for (low-current) kelvin sensing.

R_{DS}ON Current Sensing

Another popular method for current sensing is mosfet R_{DS}ON sensing. In this scheme, the internal ON-resistance (R_{DS}ON) of the power mosfet is used as the sensing element, instead of a separate discrete current-sensing resistor. This has the advantage of not requiring a separate discrete current-sensing resistor, which would otherwise add cost, power loss, and layout complexity to the power switching circuitry. The main disadvantage is that R_{DS}ON has a strong positive temperature coefficient. That is, it increases with junction temperature. This increase is typically a factor of 2.5:1 over a junction temperature range of 25 degrees C to 150 degrees C. This makes the method less accurate than other current-sensing methods. Another disadvantage is that "intelligence" is required in the OC detection circuitry, such that R_{DS}ON is only sensed during the ON state of the mosfet. R_{DS}ON sensing is an indirect current-sensing method, whereby the drain-source *voltage* (V_{DS}) of a mosfet is sensed during its ON state and the drain current is inferred by V_{DS}/R_{DS}ON, where R_{DS}ON is known based on the datasheet parameters of the mosfet. During the OFF state, the mosfet V_{DS} voltage is very high (at least the *sum* of the voltages of the supply rails, in the case of a half-bridge power stage, as used in the reference design). The intelligent R_{DS}ON sensing circuitry in the OC detection block must reject this high off-state V_{DS}, and measure the very low voltage $V_{DS}ON = (R_{DS}ON \times I_{drain})$ during the mosfet's ON state. Still, if adequate design margins can be maintained, knowing the parameter variation over the expected junction temperature operating range, then this can be an efficient and low-cost OC protection method (see Infineon (International Rectifier), 2016).

Hall-Effect Current Sensing

Another solution to the current-sensing problems mentioned above (using current-sense resistors, or R_{DS}ON sensing) is to use Hall-effect current sensing in the inductor circuit leg. This approach has many advantages. Devices of this type feature a Hall-effect current sensor integrated with drive and detection circuitry, which linearizes the Hall-effect sensor's output and stabilizes its DC operating point. Devices of this type are available for sensing unidirectional and bidirectional currents in a variety of current ranges. The wider the current range chosen, the lower the sensitivity and resolution. The measured current travels through a low-resistance path, in close proximity to, but isolated from, the Hall-effect sensor. Devices can be selected with various working isolation voltages. This scheme eliminates the need for a level-shifter, since the input circuitry is the output current path, and the output circuitry of the device can be referenced as desired, which is usually circuit ground (the same ground that the OCP comparators are referenced to). Another very important feature of these devices is their very wide bandwidth (BW): typically DC to 100KHz (4μsec typical delay time) or 1MHz (1μsec typical delay time). The delay time must be checked in the datasheet when these parts are selected, as that will determine how much the output current will ramp up before OTP is tripped and the mosfet is switched off. This delay will account for the majority of the OTP circuitry response delay.

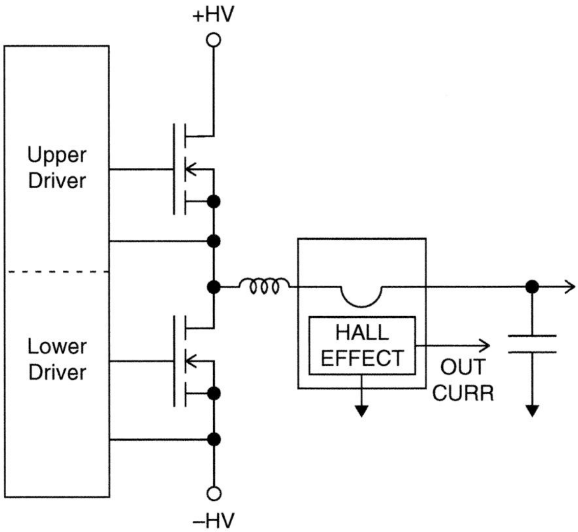

Figure 14.13 Class-D Amplifier Output Employing a Hall-Effect Current Sensor in Series with the Output Inductor of the L-C Filter

The Hall-effect current sensor is usually positioned in the output leg of the Class-D inductor, i.e. between the Class-D inductor and the Class-D filter capacitor (see Figure 14.13).

In this position, it accurately senses inductor current. The difference between the inductor current, and the actual output current to a loudspeaker, is the high-frequency ripple current that flows into the Class-D filter cap and the Zobel network. It is possible to place the Hall-effect current sensor *after* the Class-D output capacitor, such that it only senses output current to the loudspeaker. That is the position used for Hall-effect sensor U28A in the reference amplifier. When located in series with the L-C filter inductor, the Hall-effect sensor can detect inductor saturation, as well as output OC. In the reference amplifier design, the toroidal inductor is designed for such a high-saturation current that inductor current sensing is not needed. Instead, output current is sensed, and this provides adequate protection of the amplifier components (principally the mosfets). When the current sensor is placed after the L-C filter, then output current to the loudspeaker is monitored. This arrangement finds utility in load monitoring and sensing applications. With this added monitoring functionality, and associated circuitry, load impedance can be determined, based on monitoring both output voltage and current. It is important to understand the different applications and intentions of OC protection, versus output current monitoring. In protection mode, only a single, or at most two, OC "trip" levels are utilized, whereas in a current monitoring application, the full output current range of the amplifier needs to be monitored, so it is desirable for the current sensor to be linear and accurate over the intended current range.

The Hall-effect current sensor selected for the reference amplifier design (U28A) has a quiescent output voltage centered at 2.5VDC (one-half of the 5V supply voltage). It features bidirectional current sensing, with a sensitivity of 110mV/A, and a BW of 100KHz, with 4.6μsec propagation delay. Note that the sensitivity listed in the datasheet is with a 3.3V supply rail. The sensitivity is ratiometric with the supply rail, so the sensitivity with a 5V rail is 110mV/A × 5V/3.3V = 167mV/A. The optimized sensitivity range is +/−12.5A, yielding an output voltage of (12.5A × 167mV/A) = +/−2.09V centered around 2.5V. Since the reference amplifier is designed to deliver 100W, and the minimum specified loudspeaker impedance is 4 ohms, the maximum output current into 4 ohms is 5ARMS = +/−7.07Apeak. In order to allow a comfortable margin above +/−7.07Apeak for audio transients, without exceeding the current-sensing range and output voltage range of the ACS711 (Allegro Microsemiconductor, 2022) (U28A), the OC trip thresholds are set for +/−10A. This results in an output voltage trip level of +/−1.67V = (10A × 167mV/A) centered around 2.5VDC. So, the comparator trip levels are set to 4.17V and 0.83V. These thresholds are set by voltage divider R78A, R79A, R80A, operating from the +5VA rail. Again, it is usually best to derive comparator trip levels from a precision voltage reference, but in this application, that level of precision in setting the OT thresholds is not required. It is sufficient that the +5VA rail is sensed and qualified by the housekeeping/monitoring circuits, which are discussed in Chapter 13 on housekeeping. The voltage output of the Hall-effect sensor (U28A) goes to window comparator U24A so that an OC in either output current polarity can be detected. The wired-or open-collector outputs of the U24A comparators connect to multifunction gate U25A, which is configured for fast falling edge (OC FAULT = OCA/ = LOW) and delayed rising edge. The R-C consisting of R82A is used to adjust the rising edge delay. When OCA/ is low, mosfet switching is disabled (both mosfets are OFF). This will result in audio dropouts, and therefore this time constant can be adjusted for best overall sonic performance during current limiting (refer to "Types of OCP Response" below).

Current Transformers

Current transformers should be mentioned as a method of output current sensing. Though current transformers are proliferating for switch-mode power supply applications, are available in a great variety of current ranges, and feature excellent isolation between the sensed current circuitry (primary side path) and the output (secondary winding), they have the disadvantage that they can sense AC current only. They have no ability to sense DC current. In a Class-D amplifier, the operational current-sensing frequency must extend to 20Hz, but in the case of a shorted mosfet, DC output current must be detected as well. Current transformers, however, *can* be used quite successfully in the output current *monitoring* function mentioned above. In a monitoring application, only "normal" audio output current is monitored, so an AC current-sensing response down to 20Hz is sufficient. It is, however, a challenge to design a current-sensing transformer that has a flat frequency response over the full audio BW of 20Hz–20KHz. Flat frequency response down to 20Hz requires a large high-permeability core, and/or many winding turns. Wide high-frequency response without rolloff or low Self-Resonant Frequency (SRF) requires minimizing intra-winding capacitance. So, the multiple turns must be managed carefully to control capacitance and SRF.

Types of OCP Response

Now that the various modes of current sensing have been discussed, it is worthwhile to discuss what the desired OC *response* should be.

Typical OCP responses range from latching off the channel, to smoothly attenuating the signal and "playing through" the OC event. Intermediate responses are switching off the mosfets momentarily and then resuming normal operation. Depending on the OFF time of the mosfets, this can result in "audio holes" or noticeable (audible) dropouts.

Obviously, adding digital signal processing (DSP) capabilities to a Class-D amplifier opens up many possibilities for how the OCP response can be managed.

It is important to note and understand that the output *voltage* of the Class-D amplifier is typically being tightly controlled by a very high gain negative feedback loop (in closed-loop designs), and any (even very small) deviations in output voltage will cause the voltage error amplifier to saturate to one of its local rail limits or the other. Therefore, when the OCP circuitry activates, and switches off a mosfet, this will cause an output voltage error. The voltage error amplifier will rail in an attempt to quickly try to bring the output voltage back to nominal (closing the feedback loop again) once normal mosfet switching control is restored. The recovery characteristics of the error amplifier after being driven to its rails must be considered. Some error amplifiers have a "sticky" characteristic which appears as a delay in coming out of saturation. These characteristics must be taken into account when designing the OC response for the Class-D amplifier (see Cordell, 2019). Some Class-D designs, having more intelligent control over mosfet switching, will switch off the upper mosfet (for example) in the case of an OC event detected in the upper mosfet, and immediately switch ON the lower mosfet for the remainder of the switching cycle (or for a fixed time), allowing the current to ramp down rapidly. This is done independently of the duty cycle commanded by the error amplifier and PWM comparator. After a time, this intelligent OC mosfet drive will relinquish gate drive control back to the error amplifier. If this intelligent control is combined with controlled attenuation of the audio signal in response to the detected OC event, the amplifiers' mosfets can continue switching in this manner, without any audio dropouts. The audible response can be much more pleasant, sounding more like conventional voltage clipping.

Overtemperature (OT) Detection and Response

Although Class-D amplifiers run more efficiently than Class-AB and other classes of audio amplifiers, it is still prudent to add thermal protection to Class-D amplifier designs. An example will help to illustrate why this is so.

Thermal Example (Class-AB vs Class-D)

Suppose a Class-AB amplifier of 100W/channel uses TO-3P plastic-package power transistors as the output devices. These large devices will be mounted to a substantial heatsink with some type of thermal interface material for heat transfer. This material may be a thermal grease, or another thermally conductive compound, or a thermally conductive, electrically insulating interface sheet material may be used. This arrangement has a great deal of thermal mass, which means the temperature of the entire assembly, including the transistor junctions, rises comparatively slowly. There may even be fan-cooling involved. It is reasonable that an assembly of this type dissipating a worst-case power of 50W would rise at a rate of 10 degrees C per minute (0.166 degrees per second). Now consider a Class-D amplifier with the same output power rating. The mosfets will dissipate only about 3W each, at full output power. This is low enough that SMDs can be used (such as DirectFets) and these can be soldered to, and therefore heatsunk directly to, PCB copper areas. Depending on how the PCB is constructed (how much copper is available for heatsinking, how many copper layers the PCB has, and how the PCB is mounted to a chassis), this alone may provide sufficient heatsinking to keep the junction temperatures of the DirectFets within a safe operational range. However, it still may be possible to exceed the maximum allowable junction temperature under conditions of high ambient temperature, or with blocked enclosure ventilation holes. The thermal mass of the components to be cooled (DirectFets, PCB copper areas, and chassis mounting areas) is much less than for the components described above for the Class-AB amplifier. For this reason, the junction temperature rise in the DirectFets will happen at a much faster rate, of perhaps 60 degrees per minute (1 degree per second). This means that in order to provide adequate OT protection, the thermal sensor must have a fast response time and have low thermal mass itself. Negative Temperature Coefficient (NTC) thermistors in small surface-mount packages are ideal for this application, because they are available in very small SMD packages with low thermal mass and fast response time.

NTC Surface-Mount Thermistors

This type of thermistor is available with a precise resistance at 25 degrees C, and exhibits an accurate and linear decrease in resistance as temperature rises from 25 degrees C to 150 degrees C. *Note:* Most NTC thermistors of this type have a maximum operation temperature of 150 degrees C. This is convenient, because that is also the maximum operational junction temperature of most silicon and GaN devices (although some SiC devices and silicon mosfets have 175 degrees C maximum operational junction temperature ratings). Of course, it would be prudent to set a maximum overtemperature (OT) limit of *less* than this maximum. A maximum OT junction temperature limit of 120–125 degrees C provides some margin for the design, and also allows for the small amount of thermal sensing delay and overshoot which may occur due to variations in the thermal design.

The reference design employs an NTC thermistor (NTC1A) (Panasonic Industry, 2020) for sensing OT. Its nominal value at 25 degrees C is 100 Kohm. At 120 degrees C, its value drops to 3 Kohm. The OT threshold voltage is set by voltage divider R96A and R97A. Hysteresis is added to the OT sensing comparator for two reasons (through hysteresis resistor R98A). The first is to provide clean switching, instead of multiple switching events, as would occur if there were just a single threshold voltage at the comparator input. The second reason is that hysteresis allows for switching thresholds at two different temperatures. There will be one temperature threshold as the unit is heating up towards the OT threshold, and then a second temperature threshold, to which the unit must cool, in order to reset the OT comparator and allow amplifier operation to resume. In the reference design, 120 degrees C is chosen as the OT trip temperature, and 100 degrees C as the restart temperature after cool-down.

Hysteresis Circuit Design

LTspice is used to determine the resistor values for comparator U27A-1.

The reference schematic design allows for two different options in the connection of the OT comparator. These allow for two different responses to an OT condition.

Types of OTP Response

If the OT comparator output (U27A-1) is connected via path "A" (NON-LATCHING), it will be wire-or'ed into the OC path, and will force the Class-D amplifier to stop switching (both mosfets OFF) until the unit cools down to the restart temperature (100 degrees C). Alternately, if U27A-1 is connected via path "B" (LATCHING), it will cause a latched shutdown of *both* the Class-D amplifier as well as the +/−HV rails. Once latched, the AC power will need to be cycled OFF then ON again to restart the amplifier. As long as the unit temperature is below the 120 degrees C trip threshold when AC power is switched on again, the amplifier will restart.

There are many other possibilities for OT fault responses in Class-D amplifiers. These require circuit modifications not present in the reference design. One common response is to reduce the volume (and thus output power) of the Class-D amplifier, to allow continued operation, while the amplifier is allowed to cool. Another possibility is to maintain, or decrease, the output power, while enabling a cooling fan, or increasing the speed of a variable-speed cooling fan. If the fan is indeed variable speed, then the control of the fan speed would be accomplished through a servo control circuit, where a temperature-sensing device (such as an NTC thermistor) is used as part of a fan-speed control circuit. However, in that application, the full linear range of the NTC is used, rather than just specific resistance values at specific trip and reset temperatures, as is the case in the reference design.

Location of the Thermal-Sensing NTC Component

The location of the NTC thermal-sensing component is critical to the success of the OT protection circuit design. The objective is to locate the NTC thermistor sensor so that it will be in optimal thermal contact with the component whose temperature is being monitored. In Class-D amplifier design, the power switching transistors are usually a prime candidate for thermal sensing, as they are typically the highest power dissipation part in the design. A second reason for sensing the temperature of the power transistors is that it may be a way to reduce heatsink size even further, because the heatsink can then be sized to limit transistor operating junction temperature to just below the NTC thermistor trip point under a specific set of load conditions.[2]

The alternative (without transistor temperature monitoring) would be to size the heatsink to prevent excessive transistor junction temperature under worst-case conditions, including fault conditions and impeded cooling conditions. This latter case would result in a much larger heatsink, in order to prevent transistor OT failure. If a heatsink is used in the Class-D design, the location of the monitoring thermistor is straightforward. It can be mounted to the heatsink in close proximity to the power transistors. However, due to the high thermal conductivity of the heatsink (usually aluminum), it does not need to be extremely close, as long as the temperature delta between the transistor's location on the heatsink and the temperature monitor device is known. It is possible to find NTC thermistors in packages that can be conveniently heatsink mounted. Typically, these are embedded in Ring Terminals, with leadwires (Thermometrics (Amphenol), 2023), or TO-220 packages (Littelfuse Inc., 2018). There is also a popular temperature sensor available in a TO-220 package that converts sensed temperature (heatsink mounting temperature) to a linear output voltage (LM35; Texas Instruments, 2017). It should also be mentioned that *thermostats* can also be employed. These are mechanical switches that open and close at selected temperatures. These can be selected to interrupt either a power path or a small-signal path, in response to temperature.

For Class-D designs that do not use a heatsink (such as the reference design), the placement of the sensing thermistor is more critical. Surface-mounted (SMD) NTC thermistors are ideal for this application, because they have extremely low

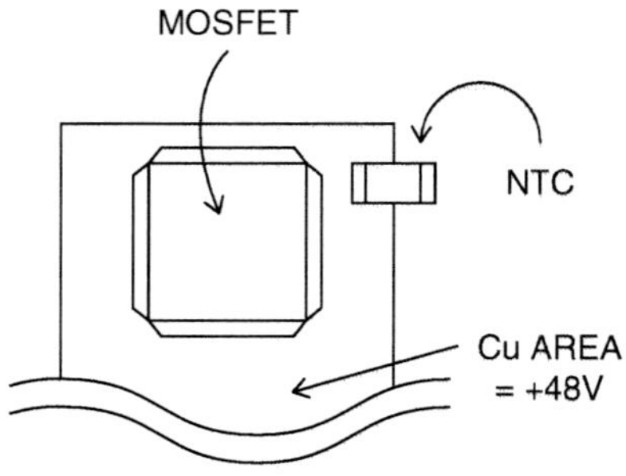

Figure 14.14 NTC Thermistor, Sensing the Temperature of a DirectFet Transistor, without Isolation

thermal mass, allowing them to heat up quickly, and they can be directly soldered to copper traces or copper areas on the PCB. Since copper is an excellent thermal conductor, the copper traces and areas on the board conduct the heat from the sensed SMD component (such as a surface-mounted mosfet or DirectFet) to the NTC thermistor.

Ideally, there will be a direct solder connection between a PCB thermal pad of the SMD power mosfet and one terminal of the SMD NTC thermistor. Figure 14.14 shows an example of this circuit arrangement.

Sometimes, however, it is inconvenient to connect one terminal of the SMD NTC thermistor to a heat-generating thermal pad of a power mosfet, because that pad is electrically connected to an "inconvenient" voltage. Using the DirectFet as an example, the optimum thermal connection for the device is the metal shell, which is electrically connected to the DRAIN of the mosfet. In a Class-D half-bridge topology, the DRAIN terminal of the lower mosfet is connected to the switching node, and the DRAIN terminal of the upper mosfet is connected to the positive HV supply (+48V in the reference design). Usually the NTC thermistor will need to interface to *grounded* circuitry, so having one terminal of the NTC connected to either the switching node or the HV rail is inconvenient, and adds complexity to the sensing circuitry. One solution to this problem is to use a thermal bridging device (International Manufacturing Services, Inc., 2020). This is an SMD component made of highly thermally conductive material such as Aluminum Nitride, and having solderable terminals. Notably, the thermally conductive material is electrically insulating. This device conducts heat from one terminal, through the body of the device, to the other terminal. The primary application for these devices is to conduct heat, and therefore provide a cooling

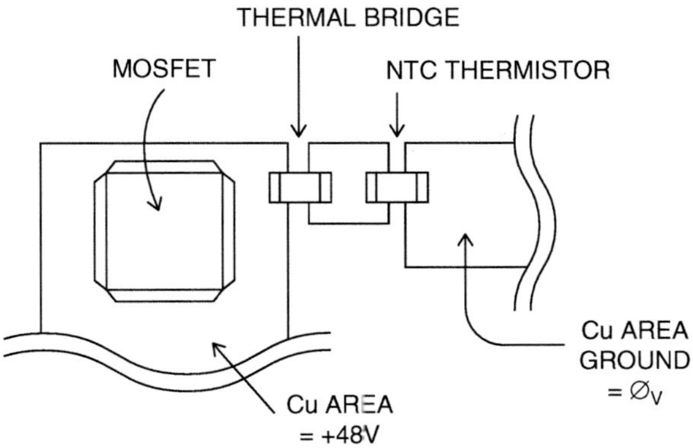

Figure 14.15 NTC Thermistor, Sensing the Temperature of a DirectFet Case in an Electrically Isolated Configuration

function. However, due to their high thermal conductivity, they will experience a very small temperature differential across the device when conducting a small amount of heat from one terminal to the other, as is the case for a thermal-*sensing* application (in contrast to a *cooling* application, where a larger amount of heat is conducted across the device, with a resultant higher temperature differential across the device, measured in degrees C per watt). Figure 14.15 illustrates how this thermal bridge device could be used to sense mosfet temperature with high thermal conductivity and excellent electrical insulating properties.

An alternate sensing solution requires level-shifting circuitry. In this case (shown in Figure 14.16), the NTC thermistor is used to directly sense the temperatures of the upper or lower mosfets in the Class-D half-bridge without a thermal bridge.

In this circuit arrangement, it is still advisable that one terminal of the sensing NTC thermistor be connected to a DC voltage (such as the positive or negative voltage rail of the half-bridge), and that the level-shifting circuitry feature good common-mode rejection (CMR) in order to reject any noise or ripple present on these rails. This allows the circuitry to translate the differential voltage across the NTC element to a ground-referenced voltage, which can then be sent to a ground-referenced comparator circuit, or fan-speed control circuit if desired.

Ux in Figure 14.16 is an adjustable shunt voltage reference. It regulates the voltage differential between its REF pin and CATHODE pin at 1.24V, by adjusting its cathode (shunt) current. If this voltage drops below 1.24V, it will decrease its shunt current. If the voltage rises above 1.24V, it will increase its shunt

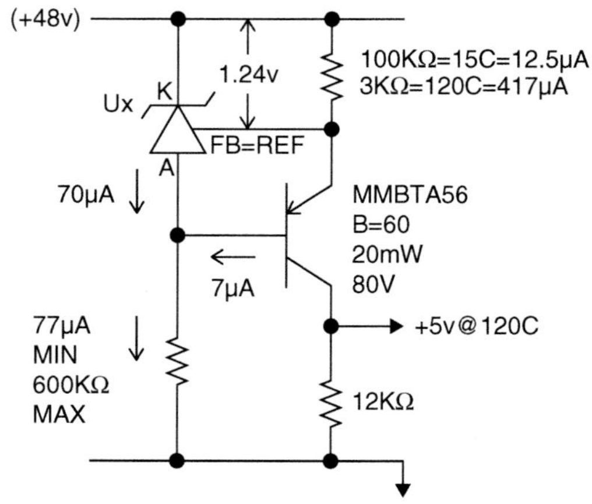

Figure 14.16 Thermal Sensing Arrangement Where the NTC Thermistor is Electrically and Thermally Connected to the Same Copper Fill Area as the Drain of a High-Side Mosfet Switch, and a Level-Shifter Circuit is Used to Translate the NTC Voltage to a Ground-Referenced Voltage

current. Therefore, if a resistor is placed between the REF and CATHODE pins, this regulated voltage will enforce a regulated current in that resistor. Thus, in the circuit arrangement shown in Figure 14.16, the circuitry will function as a high-side current source. An important consideration in this design is the minimum operational cathode current required by Ux. The LM385 (Texas Instruments, 2013) operates with as little as 10uA. The LM4041 (Texas Instruments, 2016) operates with as little as 60uA cathode (shunt) current. This consideration is important for achieving the lowest possible power dissipation. When the NTC thermistor (emitter resistor in Figure 14.16) is at its lowest value of 3 Kohm (at 120 degrees C), a voltage of 1.24V across it will produce an emitter current of 417uA. At a minimum current gain (beta) of 60, for the PNP transistor (Diodes Inc., 2023), this will require a base current of 7uA. The base resistor must be sized to allow the *sum* of the base current *plus* the minimum cathode current of Ux to flow, at the minimum operational HV rail voltage. In this case, Figure 14.16 illustrates that this base resistor could be as high as 600KΩ, with a +48V rail, and therefore dissipate only $(48V - 1.84V)_2/600K\Omega = 3.55mW$. In reality, it is desirable for this circuit to function at the *minimum* operational rail voltage (dictated by the UVLO threshold of 34.47V). Therefore, the maximum value allowed for the base resistor would be (34.47V − 1.84V)/77uA = 424K. A conservative design could use 402K, and dissipate $(48V - 1.84V)_2/402K\Omega = 5.30mW$.

This would require a maximum cathode current of (48V − 1.84V)/402KΩ = 114.8μA from Ux, for regulation under all conditions (resistance vs temperature) of the NTC thermistor. The PNP transistor will dissipate a maximum of (48V − 1.24V − 5V)(417uA) = 17.4mW. If the LM4041 is used in place of the LM385, the minimum cathode current is higher, so these resistor values have been calculated to accommodate both devices.

The emitter current, which is determined by 1.24V/NTC resistance, is essentially equal to the collector current of the PNP (minus the small base current), and flows in the ground-referenced load resistor, producing a voltage which can then be sensed by ground-referenced OT comparator circuits.

Additional OT Sensing Locations

In addition to sensing OT at the Class-D mosfets in the power stage, it can also be beneficial to sense OT in the Power Supply section of the design. In the reference design, commercial ACDC power supplies are used, which already incorporate OT protection. However, if an ACDC or DCDC power supply is custom-designed to power a Class-D amplifier, it will be prudent to include OT protection separately in that ACDC or DCDC design. In that case, the components which typically experience the most thermal stress are the power mosfet switches in the power supply and the power transformer or inductor. The power mosfet temperatures can be sensed using the same methods discussed previously for the Class-D mosfets, but the designer must be mindful of the voltages involved and any creepage or clearance requirements which must be maintained for safety. The transformer can be designed to include an NTC thermistor placed within or between the winding layers of the transformer. Again, the designer must be mindful of isolation voltages, and creepage and clearance requirements. If an ACDC power supply is used, the OT sensing can be referenced to either the primary side (LINE side) or the secondary side of the isolation barrier, depending on where the OT control and monitoring circuitry is located and referenced.

The need for OT protection of the ACDC power supply can be greater in a multi-channel Class-D amplifier. The reason is that it is possible to run multiple Class-D channels at a relatively low output power level per channel, causing the Class-D channels themselves to run cool, but the total power sourced by the power supply (being the sum of all the Class-D amplifier channels' input power) can be high, thus causing thermal stress for the power supply.

Notes

1 The PCB layout of the Class-D switching power stage is the most layout-critical area of the design. As will be seen in Chapter 16 on PCB Layout, the components in this stage need to be placed as close together as possible to minimize trace lengths and current loop areas. If current-sensing resistors are added in this area, they will increase the PCB area needed, and will increase current loop areas.

2 The trip temperature of the NTC thermistor could be set to the temperature measured at the mosfet case, when operating at the highest expected *average* long-term power level, at the highest rated ambient temperature for the design. Typical values are 1/8th rated output power, and 40 degrees C ambient. If sized in this way, the amplifier will be well protected, able to operate normally, without tripping OT fault, with most types of musical material (with normal crest factor), driven to moderate clipping, but *will* experience OT shutdown if driven above this power level, and/or at higher ambient temperatures, or with impaired cooling.

Bibliography

Allegro Microsemiconductor. (2022, January 21). Datasheet, ACS711 Hall-effect linear current sensor with overcurrent fault output for <100V isolation applications, ACS711A-DS. Rev. 6. MCO-0000224. Manchester, NH: Allegro Microsemiconductor.

Cordell, B. (2019). *Designing Audio Power Amplifiers* (2nd ed.). Abingdon: Routledge. (Clipping and sticking: Chapter 21.2.)

Diodes Inc. (2023, April). Datasheet, MMBTA55 / MMBTA56, 60V PNP medium power transistor in SOT23. Document number: DS30054. Rev. 14 – 2. Plano, TX: Diodes Incorporated.

Infineon (International Rectifier). (2016, September 19). Datasheet, IRS20957S Protected digital audio driver. Datasheet No – PD. El Segundo, CA: International Rectifier.

International Manufacturing Services, Inc. (2020, October). Thermabridge™, electrically isolated AlN thermal management device, ver. D. Portsmouth, RI: International Manufacturing Services, Inc.

Littelfuse Inc. (2018). Datasheet, TO-220 Series thermistor probe. Revised: March 2, 2018. Chicago, IL: Littelfuse Inc.

Nielsen, O. N. (2011). True current limiting. US Patent 7936216 B2.

Panasonic Industry. (2020, March 1). Datasheet, Multilayer NTC thermistors (automotive grade), ERTJ-M Series. Tokyo: Panasonic Industry.

Texas Instruments. (2013). Datasheet, LM185/LM285/LM385 Adjustable micropower voltage references, SNVS741F. Originally published February 2000, revised April 2013. Dallas, TX: Texas Instruments.

Texas Instruments. (2016). Datasheet, LM4041-N-xx Precision micropower shunt voltage reference, SNOS641G. Originally published October 1999, revised January 2016. Dallas, TX: Texas Instruments.

Texas Instruments. (2017). Datasheet, LM35 Precision centigrade temperature sensors, SNIS159H. Originally published August 1999, revised December 2017. Dallas, TX: Texas Instruments.

Texas Instruments. (2018). Datasheet, LM339, LM239, LM139, LM2901 Quad differential comparators, SLCS006U. Originally published October 1979, revised November 2018. Dallas, TX: Texas Instruments.

Texas Instruments. (2023). Datasheet, LM393B, LM2903B, LM193, LM293, LM393 and LM2903 Dual comparators, SLCS005AF. Originally published October 1979, revised August 2023. Dallas, TX: Texas Instruments.

Thermometrics (Amphenol). (2023, October). Datasheet, RTS Series, ring terminal temperature sensor, AAS-920-801E. St Marys, PA: Amphenol Thermometrics, Inc.

Chapter Fifteen

EMI Mitigation

Reasons Why Class-D Amplifiers Generate EMI

Electro**M**agnetic **I**nterference (EMI) mitigation in Class-D amplifiers is a critical topic, deserving its own chapter. Whenever current or voltage is switched with fast rise and fall times at high switching frequency, there is the potential for EMI. This is one area where traditional linear (Class-AB) amplifiers have an advantage over Class-D. Since there is no high-speed switching occurring in that topology, it does not generally emit EMI. In contrast, Class-D amplifiers feature power switching stages that switch high voltages and currents at frequencies greater than 250 kilohertz with fast rise and fall times. To make matters worse, the goals of achieving high efficiency and low EMI are in direct opposition. In order to increase efficiency, switching losses need to be minimized. To do this, switching speed is increased such that the power switches spend less time in transition, and more time (as a proportion of the period) in either the fully on or fully off state. Increasing switching speed increases the potential for EMI.

Current Loops (B-Fields)

One of the most common sources of EMI in Class-D design is due to high-amplitude current switching in current loops. Switched current in a circuit loop constitutes a transmitting antenna. Reducing EMI can be approached as reducing the area of current loops in the design. In order to do this, the designer must

 DOI: 10.4324/9781003327943-15

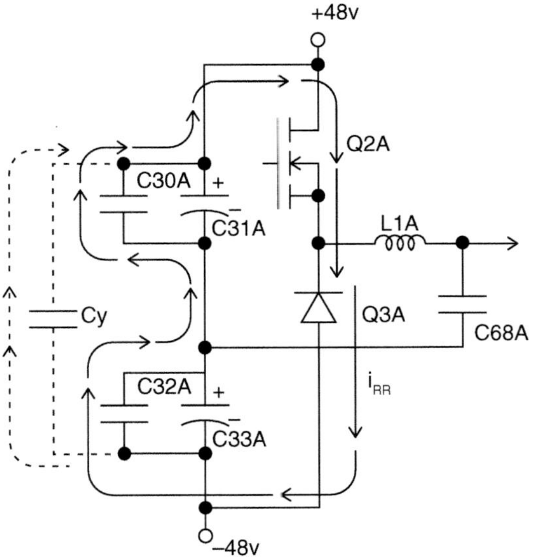

Figure 15.1 Switching Current Loops in Half-Bridge Power Circuit

understand where switched currents are flowing in the circuit and PCB layout. Figure 15.1 describes this analysis.

Figure 15.1 describes the high-frequency current paths during reverse recovery of one of the mosfet's body diodes in the half-bridge switching stage. In this example, Q3A is represented as only the body diode. Reverse recovery current (i_{RR}) is being supplied by the upper mosfet from the +48V rail. Since i_{RR} is an extremely high-frequency current pulse, trace and component inductances prevent it from being supplied by any rail capacitances at any distance from Q2A/Q3A. Instead the pulse current will be drawn from the lowest-inductance rail capacitance (which in this case is the ceramic capacitor C30A). Note how the current preferentially flows through the ceramic C30A, instead of the aluminum electrolytic capacitor (C31A), which has much higher equivalent series inductance (ESL). An important concept to understand here is that the *entire current* loop must be considered. In this case, the i_{RR} current flowing through the body diode of Q3A will also flow into the ceramic rail capacitor C32A, attached to the −48V rail, and return through the ground connection between C32A and C30A, and ultimately back to its source, the +48V rail, completing the current loop.

In some Class-D designs, a capacitance Cy is added, directly from the drain of the upper mosfet to the source of the lower mosfet, to provide an even shorter and lower-inductance path for this i_{RR} current pulse.

Current loops must be considered not just as transmitting antennae, but also as potential receiving antennae (see later in this chapter) as they can be vulnerable *input* ports for EMI and RF noise in general. Desensitizing inputs (and other vulnerable circuit nodes) to high-frequency interference will be discussed in this chapter as well.

The shortest length (of a PCB trace, or a current loop) required to be an efficient antenna is one-quarter of a wavelength or $\lambda/4$.

The general equation for *wavelength* is:

$$\lambda = \text{Velocity of Light} \; / \; \text{Frequency} \times \text{SQRT(permittivity)}$$

where: Velocity of Light is 300×10^6 m/sec and permittivity of *air* = 1, and permittivity of typical PCB material (FR4 or glass-epoxy PCB) = 4.8.

PCB material has a "wavelength shortening effect" (Dahl et al., 2016). A simplification sometimes applied is that the wavelength (and therefore the quarter-wavelength) of a signal on FR4 is approximately one-half of its value in air.

Voltage Nodes (E-Fields)

An electric field can propagate when a circuit node switches at high dv/dt. One such location where this occurs in Class-D amps is at the switching node, where the drain of the lower switch (mosfet) connects to the source of the upper switch (mosfet). Depending on the package type used for the power mosfets, the source and/or drain terminals may already have significant copper area, which cannot be reduced and serves as a minimum area size for this circuit node. In addition, from a *thermal* consideration, it is sometimes desirable to heatsink the mosfets by drawing heat from the packages, either by drain or source connection to a heatsink, or to PCB copper. In these cases, larger areas of PCB copper might be employed as a heatsink. However, this is in direct opposition to the EMI consideration of minimal area. One can see how, once again, in Class-D design there are opposing goals, requiring compromise.

General *layout* considerations to minimize EMI are as follows:

1) Route traces of current loops parallel to (or over) return paths to reduce inductance (*Note:* This also results in minimum loop area, for minimum radiated EMI)
2) Use an unbroken ground plane under potential EMI sources
3) Separate potential radiators and sensitive circuits as much as possible
4) Employ EMI filtering where possible.

Cables

Cables are usually the main RF emission source in Class-D designs. This is because cables make the most efficient radiating antennae for radiated EMI, and also because cables are the conduit for conducted EMI. Note that both radiated and conducted EMI are considered by regulatory agencies. Radiated emissions leave a circuit board or cable and propagate through the air as electromagnetic waves (Dahl et al., 2016). These can then be received by an unintended receiver, and interfere with its operation. An unintended receiver is any piece of equipment whose operation is affected by the unwanted emission. Conducted emissions refers to electromagnetic energy which leaves a circuit through wires or cables. Conducted emissions may cause problems directly, or may generate radiated emissions (Dahl et al., 2016). In most applications, *all* cables need to be considered for their EMI-generating ability. This includes the following cable types:

1) Signal input
2) Interface and control
3) Power input
4) Output.

EMI mitigation for each one of these cable types employs different techniques based on a number of factors. These are:

1) The frequency range of the nominal operational signals on these cables
2) The frequency range of the EMI on these cables
3) Whether the EMI is differential mode (DM) or common mode (CM)
4) The operational current and voltage on these cables
5) Any regulatory agency safety requirements that apply to the cable type.

CM and DM

When analyzing signals that may radiate from cables consisting of two or more conductors, it is critical to understand the differences between CM and DM signals.

CM signals are signals which appear on all cables (taken together) with respect to a reference (typically ground).

DM signals are signals that appear between conductors.

EMI from cables is typically generated by CM signals. This distinction is important, because different Class-D switching topologies generate differing amounts of CM noise on output cabling. This is more critical for "filterless" designs, where there is no local L-C filter to act as a low-pass filter for these switching voltages and currents.

Take for example a full-bridge switching stage topology used in a filterless design. In this case, the switching scheme utilized influences EMI, because it determines the amplitude of the DM and CM signals on the cables. The two principal modulation schemes are AD and BD. These are illustrated in Figures 15.2 and 15.3.

In AD modulation, the switching of the two half-bridges composing the full-bridge output is 180 degrees out of phase. When one output is high, the other is low, and vice versa. In BD modulation, the outputs are in-phase at idle (i.e. no differential output voltage to speaker), and as the output differential signal level deviates from idle, the two outputs of the full-bridge are no longer the inverse of each other. The principal effects on EMI are due to the differences in CM voltage signal appearing on the output cables in the two switching schemes. In Class AD, there is virtually zero CM voltage, whereas in Class BD, the CM voltage amplitude is 0.5 × Vdd at the switching frequency. This can present EMI problems. These are addressed by the addition of ferrites and more elaborate switching schemes, which retain the power-saving advantages of BD switching, but reduce the emissions energy.

The discussion above is most relevant to filterless Class-D amplifiers. However, for conventional Class-D amplifiers that employ L-C filters, the large-amplitude square-wave signal of the power switching stage is reduced to a much lower amplitude residual sinusoidal waveform (residual ripple at the Class-D switching frequency) with "spike noise" at the switching transitions of the power transistors.

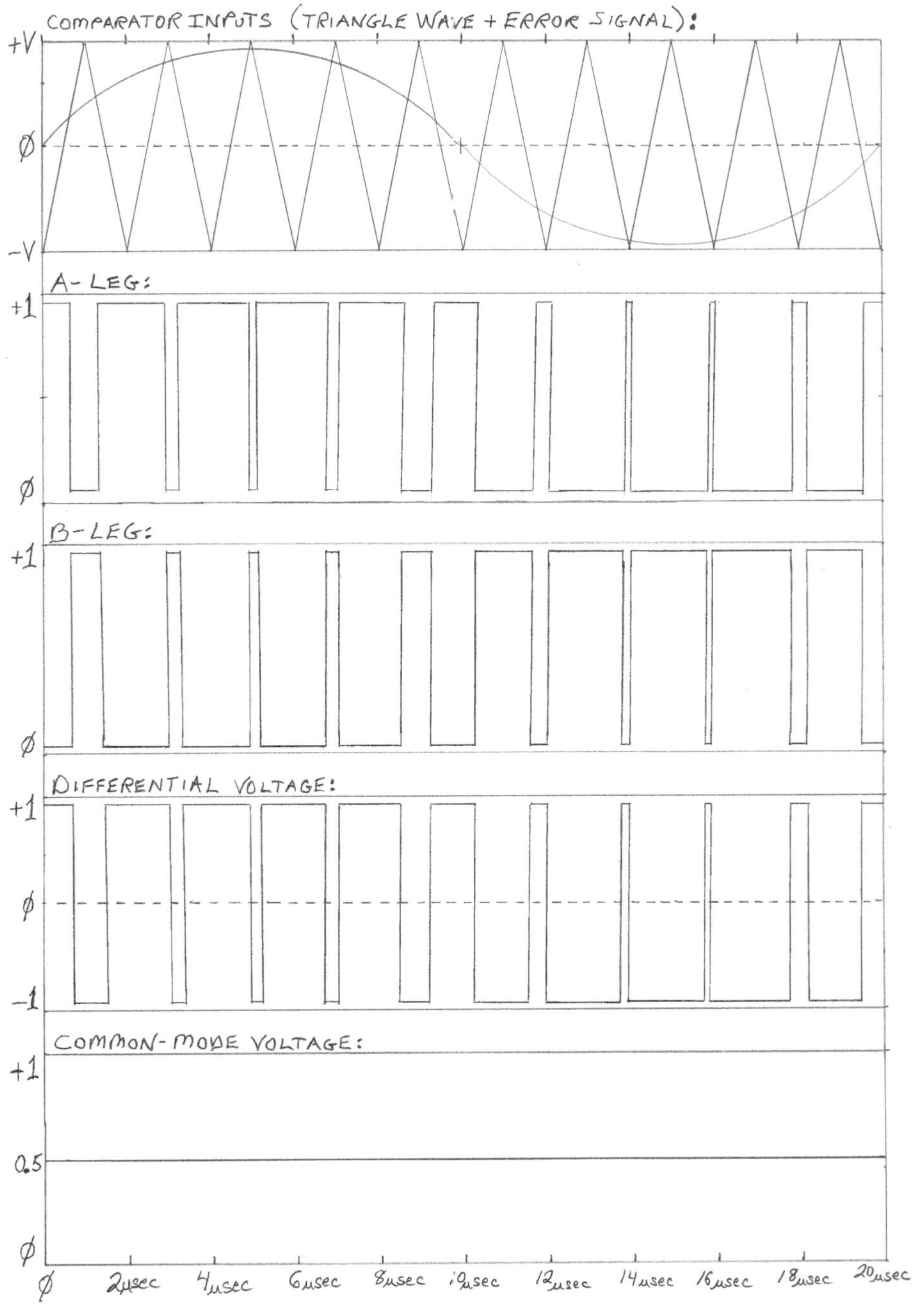

Figure 15.2 AD Modulation

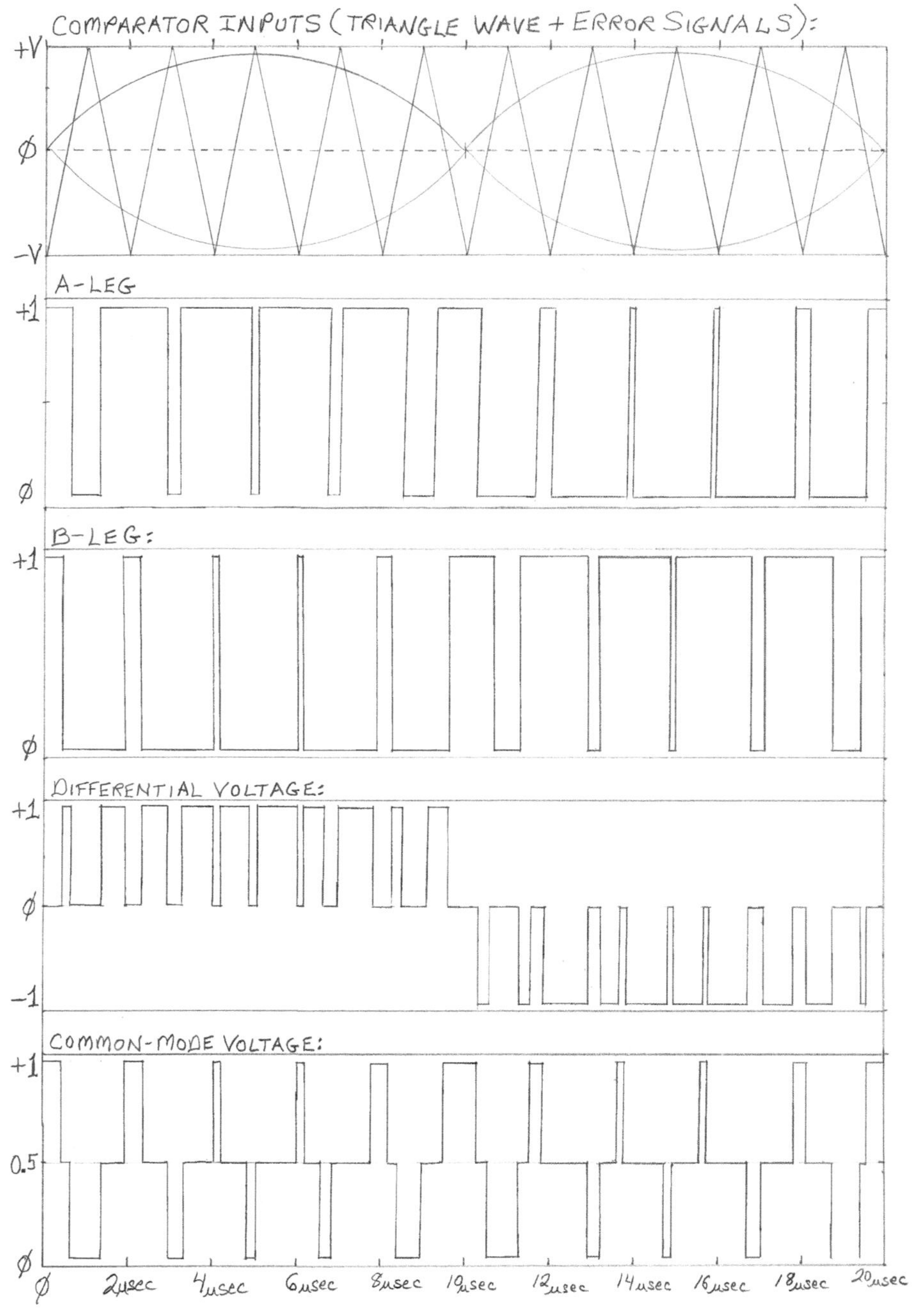

Figure 15.3 BD Modulation

This "spike noise" is much higher in frequency than the Class-D switching frequency (typically 10–60MHz). It is influenced by several factors including:

1) Switching speed of the power transistors (di/dt and dv/dt of switching transitions)
2) Reverse recovery characteristics of mosfet body diodes
3) Shunt capacitance of the Class-D filter inductor
4) ESL of the Class-D filter capacitor(s)
5) Excitation of other parasitics in the switching and output circuit traces.

These above can be mitigated using the following means:

1) Controlling the switching speed of the power transistors. This is a trade-off against efficiency, and must be done carefully. The easiest method of controlling mosfet switching speed is by adjusting the value of the gate resistor.
2) The addition of a parallel Schottky diode of sufficient voltage rating can help to steer free-wheeling current away from the mosfet body diode, in favor of the lower-voltage Schottky, which does not exhibit a reverse recovery characteristic (because it is a majority carrier device). Adding a low-ESL ceramic or film capacitor directly from the drain of the upper mosfet to the source of the lower mosfet, for (each) half-bridge stage, may help to circulate reverse recovery currents locally (in a tight local current loop), and decrease noise associated with these snappy reverse recovery currents exciting PCB parasitics, if allowed to flow outside of this tighter current loop. See Figure 15.1.
3) Use of an inductor that uses a single-layer winding, with no overlap of starting and ending winding turns, can significantly reduce shunt capacitance and raise self-resonant frequency. The designer must also be aware of capacitance of the winding turns to the core, as this represents another path for capacitive current. The addition of a coating on the core, or tape covering of the core may help to decrease winding-to-core capacitance.
4) The Class-D filter capacitor (in a L-C demodulation filter) should be a low-ESL ceramic or metalized film capacitor. In general, SMD types will have lower ESL than leaded parts.
5) Reducing inductance, and other parasitic elements, in the high dv/dt and di/dt traces is a critical part of the PCB layout process, and is covered in Chapter 16 on PCB layout.

6) Addition of an R-C series snubber from the switching node to ground (for a half-bridge topology). This R-C series snubber can be "tuned" to damp out high-frequency ringing at the switching node. However, its effectiveness is usually proportional to its power dissipation. Todd (1993) is an excellent source of design information on snubbers. Surprisingly, the power dissipated in the snubber is determined by the capacitor value, but is actually dissipated by the resistor. It is always preferable to mitigate switching node ringing by *careful layout*, rather than suppressing ringing caused by a suboptimal layout through the use of a power dissipative R-C snubber network.

Spread-Spectrum Techniques

A popular method used to reduce EMI involves spread-spectrum techniques. Fixed-frequency Class-D amplifier designs, such as the reference amplifier, use a fixed switching frequency. In this topology all switching noise and EMI will be harmonically related to the switching frequency. Emissions energy will be concentrated at specific frequencies. If instead of a fixed switching frequency, some *dither* is added to the switching frequency, the emissions energy can be spread out over a wider range of frequencies. This technique does not decrease the total emissions energy, it merely spreads it out, so that the peak, quasi-peak, and average signal levels measured by EMI measuring equipment (usually in units of dBμV) are lower than they would be without this technique. Usually the spreading is small (less than 10% of the nominal carrier frequency). Variable-frequency amplifier topologies have an EMI advantage over fixed-frequency designs for the same reason. The emissions energy is spread out over a wider range of frequencies. However, if an amplifier design is non-compliant, it is sometimes more difficult to troubleshoot the variable-frequency design, because emissions might only occur under a very specific set of conditions which may be hard to duplicate and stabilize long enough to analyze.

Radiated and Conducted Susceptibility/Immunity Tests

In addition to radiated and conducted EMI, cables must also be considered in conducted and radiated susceptibility/immunity tests. Conducted susceptibility tests employ a *clamp* that is placed around the cable-under-test. RF energy is injected into the clamp. The effect of this RF energy on the operation of the amplifier is assessed. This effect can be audible *noise* heard on attached speakers, or even operational disturbances of the amplifier, such as complete loss of audio, or momentary or complete shutdown, or damage. Similarly, radiated susceptibility tests involve subjecting the amplifier-under-test to a strong RF field. In this case, the entire setup, including the amplifier, all attached cables, and speakers (or resistive dummy loads substituted for speakers), is subjected to the RF field. The same effects of audible noise and/or operational perturbations apply as for the conducted immunity tests.

In the case of these conducted and radiated immunity tests, it is again the cables that allow the most efficient ingress of disturbing RF energy into the amplifier-under-test. In most cases, the same "fixes" employed to reduce conducted and radiated emissions from the amplifier also help to reduce susceptibility here as well.

Bypass Noise to Chassis

One of the most effective means of reducing EMI emissions from cables is to bypass noise and EMI currents to chassis ground. If a Class-D design uses a metal enclosure that is earth-grounded, this is a big help in reducing EMI. All signals on cables should be bypassed to this chassis ground. This should be done as close as possible to the egress point of the signals. The reason for this is that even a small exposed signal connection *beyond* the bypass point can be exposed to, and pick up, RF noise from radiating emitters inside the enclosure. This is why *feedthrough* components are so effective in reducing EMI. Feedthrough capacitors, for example, are basically a continuation of the shield around the signal wire, with the bypass point being internal to the feedthrough capacitor. Unfortunately, feedthrough components are generally expensive, and are not usually implemented in lower-cost consumer electronics devices. The alternative is to use low-ESL ceramic capacitors as close to the egress point as possible. Another type of capacitor construction is useful as well. This is described in Johanson Dielectrics (n.d.). This construction uses two capacitors in one small SMD package, with a center terminal. These exhibit very low inductance due to a current cancellation effect. The dual-capacitor construction makes them ideal for filtering CM noise on two traces (output and return, for example), with a very small PCB footprint, allowing them to be positioned very close to signal egress points on the PCB (i.e. connectors).

In general, the EMI-reducing effect of noise bypass capacitors can be enhanced by adding a series *ferrite* bead. It is important when adding these components to understand the frequency of the EMI noise component to be filtered, and the frequency of the operational signal on the cable. The filtering components should not degrade the operational signal. For example, if the signal to be filtered is the Class-D output signals to a speaker, these components should not rolloff the frequency response of the amplifier in the audible range (below 20KHz). In addition, they should not degrade the THD performance of the amplifier. Surprisingly, some ferrite beads, though having negligible impedance in the audio frequency range, *will* degrade THD when used to filter amplifier output signals in DM. The solution for this is to use *CM* ferrite beads. This makes sense for two reasons:

1) The EMI signal mode that causes problems in cables is generally CM and not DM.
2) A CM choke does not "see" the DM signal currents (the audio signal currents), so these do not contribute to core saturation. This allows CM chokes to achieve very high CM inductance values with relatively small cores, without saturation effects. It is the saturation effects of DM beads that cause the THD degradation when used in the audio signal path.

There are many varieties of CM chokes suitable for EMI filtering of cables, even at high current. For Class-D applications, these are principally ferrite block types and powdered iron core types.

It is critical, when applying CM chokes for EMI filtering in a Class-D design, that the currents in all wires going through the choke cancel. Figures 15.4 and 15.5 illustrate this concept. If this rule is followed, a practically unlimited number of signal and return wires can be run through a single choke core, without any DM current being available to saturate the choke. Also, since no net DM current is present, this current does not induce flux in the core that can couple to any other wires, so there is no crosstalk effect among these multiple wires even though they are going through the same CM choke. An example of this concept used to good effect is in dual-channel amplifiers, where the two output channels can be *either* configured as two separate outputs, in a stereo left/right pair (Figure 15.4), *or* configured in a Bridge-Tied-Load (BTL) configuration to drive a single speaker at higher power (Figure 15.5).

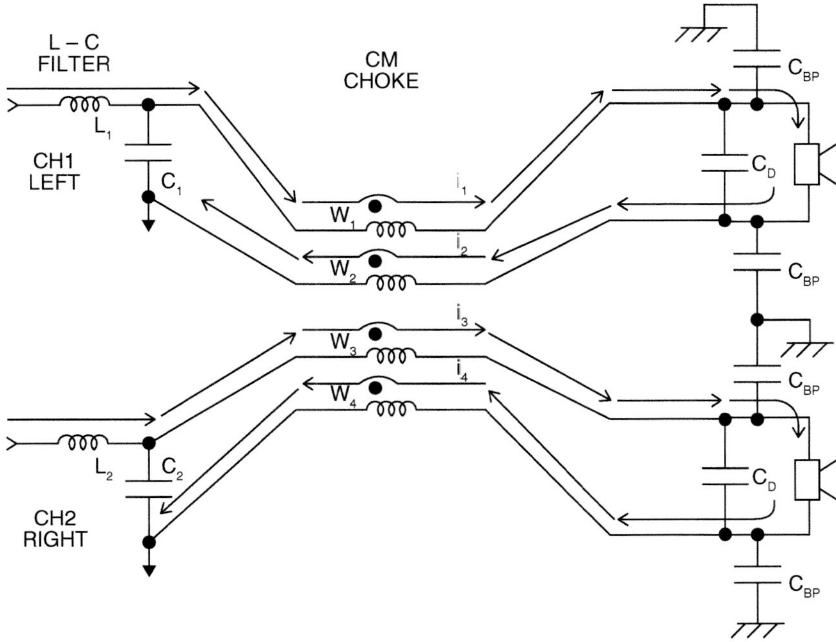

Figure 15.4 Currents in a Four-Winding CM Choke in a Two-Channel Amplifier Configuration

In Figure 15.4, the output current from CH1 (i_1) and return current from CH1 (i_2) are identical, and cancel each other in the CM choke. Likewise, the output current from CH2 (i_3) and return current from CH2 (i_4) are identical, and cancel each other in the CM choke. Note the "polarity dots" on the windings of the CM choke. The full CM inductance is realized when currents flow *in the same direction* in all windings. A CM current for the CH1/left output, for example, would flow *into* the "dot" end of winding 1 (W1) *and* winding 2 (W2). Such a current would see the full CM inductance and be attenuated. DM currents, which are the output audio signal current and its return current, flow in opposing directions through W1 and W2 windings, and therefore cancel, and generate no flux in the core. Adding the output current (i_3) and return current (i_4) for the CH2/right output, flowing through windings W3 and W4 *on the same core*, operates similarly. The currents in W3 and W4 cancel each other. Thus, even though there are four DM currents flowing through the CM choke windings, all DM currents cancel and the net DM current is 0A. Since no net DM current flows, no flux is induced in

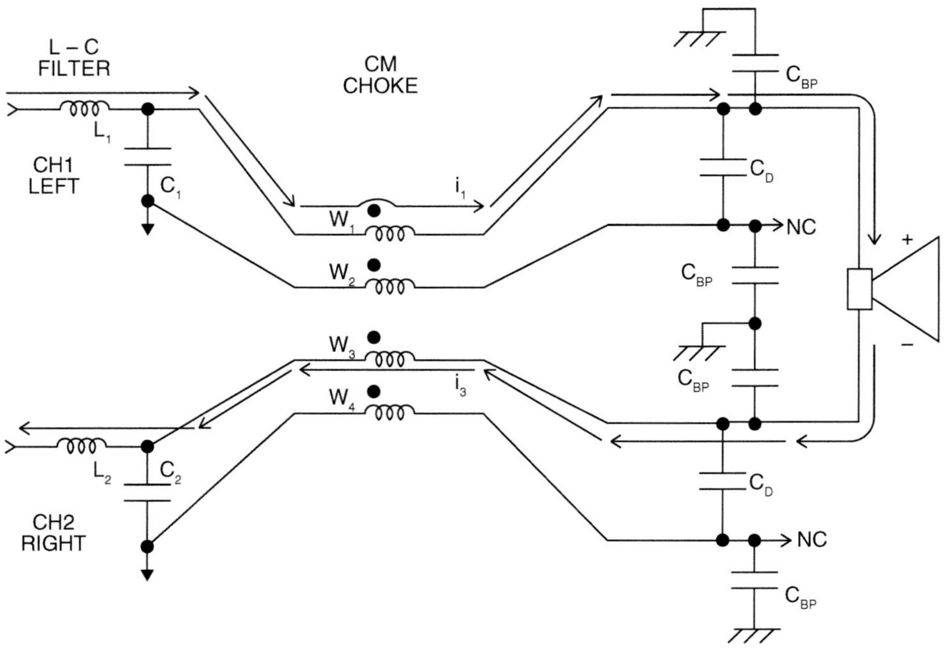

Figure 15.5 Currents in a Four-Winding CM Choke in a Single BTL-Connected Amplifier Output

the core, and there is no flux-coupling between wire pairs, and therefore no cross-talk is induced between these two separate audio channels.

Of course, in this analysis, we have not mentioned the CM noise currents. Those currents flow in the same direction on all wires, and see the full CM inductance of the CM choke. This inductance corners against the four capacitors bypassing to chassis (C_{BP}), and they are attenuated before they get onto the speaker cable where they could potentially radiate EMI. The DM output capacitors (C_D) are much larger than the C_{BP} values, and, for this simplified analysis, serve to short the two ends of windings W1/W2 and W3/W4 together, from a high-frequency CM noise standpoint.

Figure 15.5 illustrates the same concept, applied to a BTL output arrangement of the two Class-D channels. In this case, the output audio current of CH1/left (i_1) is cancelled by the return current flowing into CH2/right (i_3). There is no audio signal current flowing through windings W2 and W4, so again the net DM current is zero. Therefore, there is no flux induced in the core by these DM currents. CM noise currents in i_1 and i_3 see the full CM inductance of the choke.

Electrostatic Discharge (ESD)

Fortunately, the concept of placing low-ESL ceramic capacitors as EMI bypass capacitors for signals leaving an enclosure (chassis) on a cable, is *also* a best practice for ESD mitigation. These capacitors allow ESD currents to be bypassed as close to the chassis opening as possible. This causes ESD currents to be contained in small local loops, and not propagate further into the PCB where they can potentially disrupt circuit operation, cause transient pops and clicks, or even damage components. It is also good practice to add ESD suppression components at these same locations. The reference design utilizes these at the audio signal inputs. When selecting these components for ESD suppression at audio signal inputs, the following criteria need to be considered:

1) Breakdown voltage rating. These types of suppression components typically have a working voltage range, as well as a breakdown voltage rating, where both voltage and current are specified. Be sure that the devices will not begin to conduct over the expected operational input voltage range of the amplifier. Also be sure that the maximum clamping voltage specified for the ESD suppression device is low enough to protect any circuit components from damage.
2) Capacitance. These devices usually are available in low-capacitance varieties, so that their capacitance will not cause undesired rolloff of the audio signal.
3) Leakage current. A leakage current is usually specified at a specified voltage and temperature. Be sure to check that this leakage current will not cause undesired voltage offsets, when flowing through any input resistor networks.

These ESD components should be connected to chassis ground, which is usually configured as a separate ground strip along the edge of the PCB. This copper should be connected to chassis (enclosure) metal with as short and direct a mechanical connection (think low inductance) as possible. The suggested PCB layout in Figure 15.6 shows this chassis ground strip.

Figure 15.6 shows the EMI/ESD bypass caps connected from output connector pins to a CHASSIS-GND copper strip on the PCB. This CHASSIS-GND copper is then connected to the metal enclosure through the two mounting holes in the lower corners of the PCB.

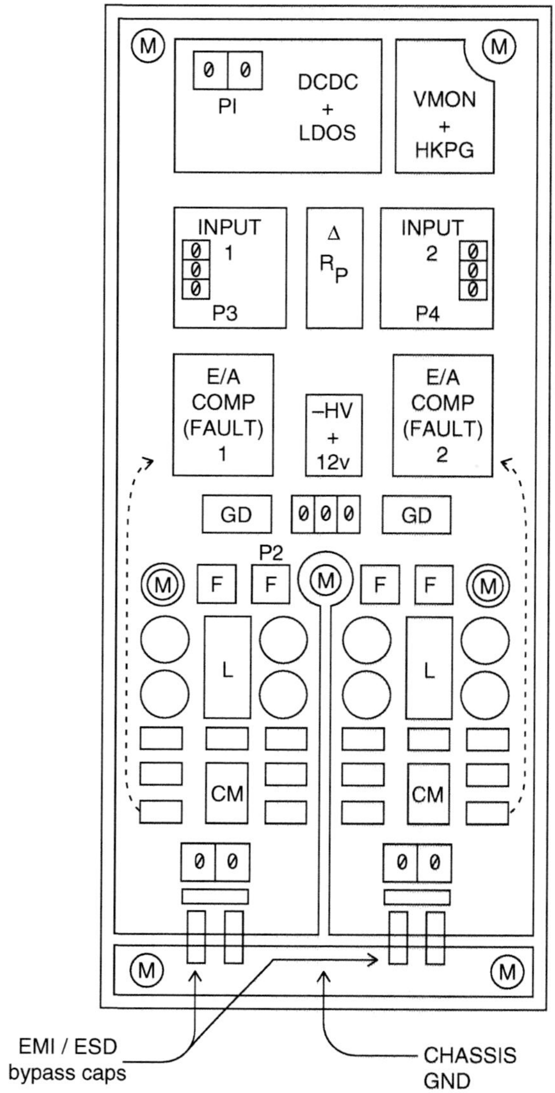

Figure 15.6 Component Placement Concept

If the design does not have a metal enclosure, then the circuit board ground should be used as the EMI and ESD bypass ground.

Bibliography

Dahl, N. J., Iversen, N. E., Knott, A., & Andersen, M. A. E. (2016). Comparison of simple self-oscillating PWM modulators. In: *Proceedings of the 140th Audio Engineering Society Convention*, Paris, France, June 4–7, 2016 (article 9562). New York: Audio Engineering Society.

Johanson Dielectrics. (n.d.). Datasheet, X2Y® filter & decoupling capacitors. Camarillo, CA: Johanson Dielectrics.

Karaca, T. D., & Deutschmann, B. (2015). Electromagnetic evaluation of Class-D switching scheme. Graz, Austria: Institute of Electronics, Graz University of Technology.

Mei, S., Hu, Y., Xu, H., & Wen, H. (2022). The Class D audio power amplifier: A review. *Electronics*, *11*(19), 3244. https://doi.org/10.3390/electronics11193244

Nielsen, D., Knott, A., Pfaffinger, G., & Andersen, M. A. E. (2009). Investigation of switching frequency variations and EMI properties in self-oscillating Class-D amplifiers. In: *Proceedings of the 127th Audio Engineering Society Convention*, New York, NY, October 9–12, 2009 (article 7907). New York: Audio Engineering Society.

Taranovich, S. (2023, August 21). Tried-and-true techniques to lower EMI in power-supply designs. *Electronic Design*.

Texas Instruments. (2013). Managing EMI in Class-D audio applications, SNAA050A. Originally published 2007, revised 2013. Dallas, TX: Texas Instruments.

Todd, P. C. (1993). Snubber circuits: Theory, design and application. Texas Instruments, Seminar 900, Topic 2. Dallas, TX: Texas Instruments.

Wyatt, K., & Sandler, S. (2022, July 28). DC-DC converters – Solid return plane or cutouts under switch node and inductor? *Signal Integrity Journal*.

Chapter Sixteen

PCB Layout and Thermal Considerations

Problems a Poor Layout Can Cause

A circuit designer may have arrived at the most elegant schematic design, with all the correct components selected, and all circuits designed to operate within datasheet specification limits. Every operating condition may have been anticipated, and the designer may feel that they have produced an optimal design. However, without careful detailed PCB layout, the odds of achieving a high-performing Class-D audio amplifier design are low. Careful PCB layout is absolutely critical in order for a Class-D amplifier to achieve high performance. Some of the typical problems associated with poor, or suboptimal, PCB layout are:

1) EMI emissions and susceptibility
2) Ringing and excessive voltage and current excursions on switching waveforms
3) Excessive voltage drop
4) Crosstalk
5) Thermal issues
6) Increased THD and noise.

This chapter will address all of these potential shortfalls and more. It will also explain how to approach the PCB layout in a logical sequence, in order to achieve the best result in the shortest time.

 DOI: 10.4324/9781003327943-16

Develop a PCB Floorplan (Placement)

Component placement is the first step, taken with an overall view of three larger factors. These are:

- Signal flow
- Noise coupling
- Thermal considerations.

These considerations determine where each circuit group should be located on the PCB. Then the designer addresses details within the functional blocks with two fundamental aspects in mind:

- Common impedance
- Current loops (Honda & Huang, 2008).

The first and most important step for PCB designers is to group components dedicated to a common purpose, such as:

- Audio input circuitry
- PWM control circuitry
- Gate driver stage
- Switching stage
- Housekeeping circuitry
- Fault processing circuitry
- Voltage regulators (DCDC + LDO).

By identifying which components belong together, the PCB designer can decide on the placement of the remaining circuitry by seeing how they interface with the already placed functional blocks, and choosing the most appropriate open areas of the board (Honda & Huang, 2008) that will result in the shortest interconnections.

Best Practices

One essential rule that PCB designers should follow can be called "proximity priority". This dictates which components get priority placement around an IC, for example. As IC packaging gets smaller, there is less overall perimeter to the package, and therefore, not all components associated with a given IC can be placed directly at the pins they are connected to. Therefore, it should be understood which components get priority placement over others.

For digital ICs in general, bypass caps should be as close as possible to Vcc and GND pins. This allows transient currents with fast edges to travel in the smallest possible current loops. Pullup and pulldown resistors, carrying low static currents, have lowest placement priority.

For analog ICs, again, bypass capacitors have highest placement priority to Vcc/ Vss and GND pins, both for transient current handling, but also for minimum power supply noise and PSRR considerations. When considering the length of current loops, it is important to consider the entire current path, including sourcing *and* return currents (i.e. from bypass capacitor to Vcc pin and from ground pin to capacitor ground pin. The connection of *both* of these pins to the PCB ground plane must also be as short and as low inductance (i.e. wide) as possible). For op-amps, the feedback network to the inverting input is an especially noise-sensitive and stability-sensitive net. Therefore, these feedback components should be placed as close as possible between the output and inverting input pins. The inverting input pin especially (being high impedance) is one of the most noise-sensitive circuit nodes. Try to minimize the copper area and capacitance of this node. As an example, if the feedback network consists of a C (or series R-C) in parallel with an R, then the C or R-C will get placement priority over the R. Undesired capacitance at the inverting node of an op-amp can yield peaking in the response, or even instability and oscillation (Ardizzoni, 2005).

Many references on PCB layout emphasize the impedance characteristics of decoupling caps. Different capacitor values and types have different self-resonant frequencies, above which the impedance becomes inductive, and noise bypassing degrades (Texas Instruments, 2016a). By being cognizant of the impedance characteristics of various bypass cap technologies, an optimal bypass network can be designed (Putzeys, 2014). In general, larger capacitances are used for low-frequency bypassing and as local reservoirs of high transient energy. Aluminum electrolytic capacitors are often used for these purposes, because of their larger capacitance values. These however, have significant series inductance (ESL), and therefore low self-resonant frequencies. To complement these, higher-value ceramic capacitors can be added in parallel. Ceramic capacitors have excellent capacitance/size ratio, and relatively high self-resonant frequencies. A rule of thumb for creating wide-bandwidth noise decoupling networks is to use parallel ceramic capacitors with values a decade apart, for example, 1μF, 0.1μF, 0.01μF. Of course, when noise bypassing at even higher frequencies is important, even smaller values of ceramic bypass capacitor can be added. Always keep in mind that the impedance characteristics of the noise bypass network (decreasing-impedance-with-frequency) should complement the falling PSRR-with-frequency characteristic of the circuit or component being bypassed. When multiple parallel supply bypass capacitors are used, the smallest-value capacitors should get placement priority next to the supply and return pins of the IC, over the next higher value. This is because the lowest-value capacitor has the highest self-resonant frequency, and bypasses the highest-frequency noise currents. To do its job effectively, it needs the lowest possible trace inductances in its bypass loop, and therefore the smallest total trace lengths. Note that the total trace lengths include any vias. Therefore, to achieve the shortest possible bypass loop, the bypass caps should be on the same side of the PCB as the component they are bypassing, thus eliminating vias (Ardizzoni, 2005).

In general, it is best practice to always be cognizant of current paths for high or fast transient currents. Analyze the entire current path, including return current paths through ground planes or ground traces, and make these current loops as small and short as possible. A good example of this is a gate-drive loop for a power mosfet. For example, the gate-drive circuit for a high-side mosfet is described in Figure 16.1. *Note* that the source node of the upper mosfet is *referenced* to the switching node, but gate current does not flow into the switching node. The gate current is confined to a loop described by the dashed arrows in the figure. This loop should be connected such that currents from other loops, such as the inductor current/output current loop, do not mix with the smaller upper gate drive loop. Note that the trace "A" does *not* carry gate-drive current. It only carries the switching current between the two mosfets and the inductor current.

Figure 16.2 shows how the gate-drive routing should *not* be done.

In Figure 16.2, the gate-drive current *and* the high-amplitude switching current are allowed to mix in trace "A". The high-amplitude switching current flowing through the impedance of trace "A" will generate a voltage that will add to or subtract from the desired V_{GS} voltage for the upper mosfet. This can interfere with the desired switching speed and timing for the upper mosfet, and can generate undesired V_{GS} voltage spikes.

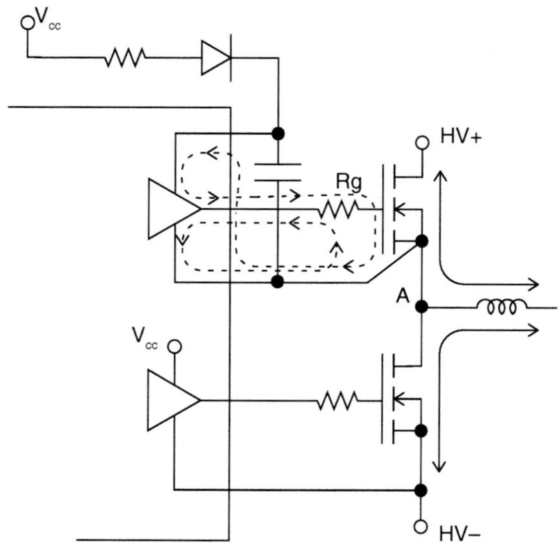

Figure 16.1 Currents Flowing in a Gate Drive Circuit with Proper Current Containment

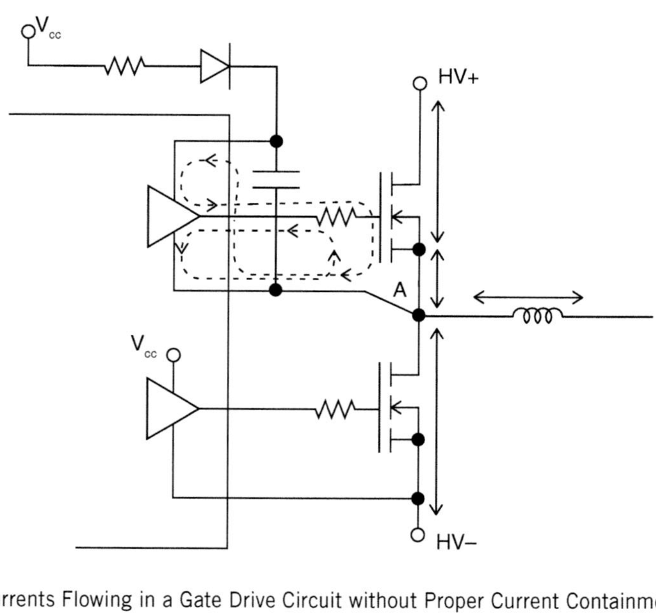

Figure 16.2 Currents Flowing in a Gate Drive Circuit without Proper Current Containment

The ideal design of the above circuit is for trace "A" to have zero length and inductance.

Switching Current Containment

Switching current containment refers to the concept of restricting high-frequency switching currents (which can cause noise and EMI) to small, defined current loops. The discussion above for the gate-drive loops is one illustration of this concept. The following discussion develops this concept further in consideration of the main high-current, high-frequency switching loops in a half-bridge Class-D design.

Figure 16.3 depicts a simplified schematic for the power stage and power supply rails. V1 and V2 depict the high-voltage power supply rails (+/−48VDC in the reference design), which for this example can be considered to be located at some distance from the Class-D power stage. The Class-D power stage consists of all components to the right of inductors L1, L2, L3. Figure 16.3 illustrates high-frequency switching current and audio frequency current flow when audio current is flowing towards

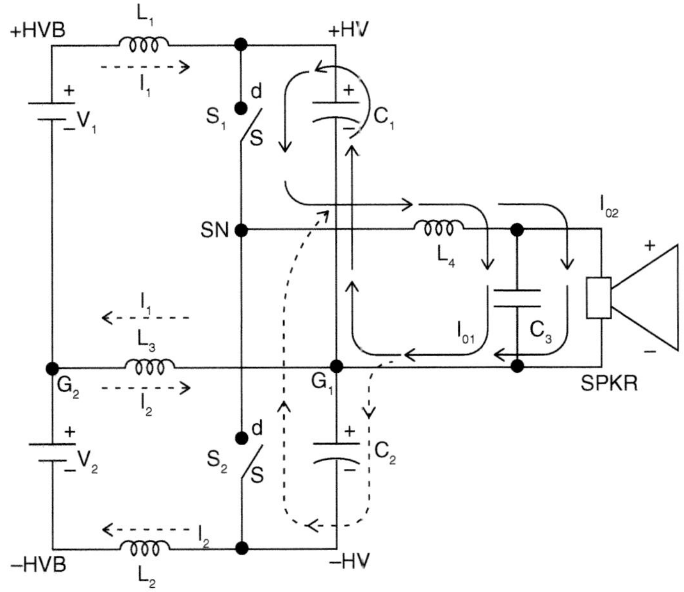

Figure 16.3 Class-D Power Stage (Half-Bridge) Where Current is Being Sourced to the Speaker through L4, Illustrating Switching Current Containment

the "+" terminal of the speaker (solid arrows). In this state, current is sourced by C1 and flows through the upper mosfet (S1) from drain to source (d-to-s). This current continues to flow through the Class-D inductor (L4). The great majority of the switching frequency ripple current flows through the Class-D filter capacitor (C3), and very little flows through the speaker due to the inductances of the speaker wire and the speaker itself (voice coil and discrete inductances in the speaker). This leaves the audio frequency output current to flow to the speaker. This same current returns to ground node G1 and to the negative terminal of C1, completing the current loop. Now, when S1 opens, current flow will continue as the dashed arrow path indicates. *Note* that current flow through S2 is source-to-drain (s-to-d), and current flow through C2 is in a charging direction, such that the voltage across C2 increases. This current causes "supply pumping" of C2. This dashed-arrow current will first flow through the body diode of S2 (during the deadtime) and then flow through the source-drain channel of S2 when it switches ON.

Now consider the case for negative audio current, with current flowing in a direction *from* the speaker "+" terminal into the output terminal of the Class-D amplifier. This is depicted in Figure 16.4. In this case, when S2 is ON (switch closed), current (solid arrows) will flow from drain-to-source (d-to-s). Current,

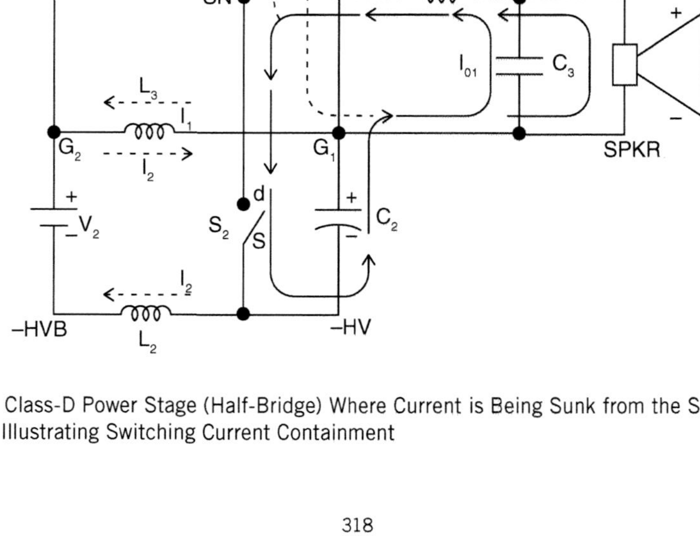

Figure 16.4 Class-D Power Stage (Half-Bridge) Where Current is Being Sunk from the Speaker through L4, Illustrating Switching Current Containment

energy, and charge will be sourced by C2. When S2 switches off, current will continue to flow in the dashed-arrow loop. *Note* that the current direction through S1 will be source-to-drain (s-to-d). This will cause C1 to charge, thus causing supply pumping of C1. This dashed current will first flow through the body diode of S1 (during the deadtime) and then flow through the source-drain channel of S1 when it switches ON.

The critical point to discern from Figures 16.3 and 16.4 is that all of the high-frequency switching current has been confined to the current loops in the Class-D power stage (solid and dashed arrows). Therefore, if the layout of the power stage is made as compact as possible with the shortest possible connections, the emission of EMI energy can be minimized.

Another critical point is that supply currents I_1 and I_2 can theoretically become DC currents if L1 and L2 values are large, and this will not interfere with the operation of the Class-D power stage (so long as energy storage capacitors C1 and C2 have large enough capacitance to support audio current demands at all audio frequencies, without excessive voltage droop). So, the switching currents have been contained or confined to the Class-D power stage. Also, currents I_1 and I_2 *cancel* in inductor L3. In fact, L3 only serves to separate G1 from G2 in the figures. It is not necessary or desirable, because it represents inductance in the ground plane. If L1 and L2 are made infinite, then there will be perfect current cancellation in the L3 circuit path. Therefore, if the copper area consisting of the area between circuit nodes G1 and G2 is the "quiet" ground plane, it will have *no* switching or audio frequency currents flowing through it, creating undesirable voltage drops.

This is an idealized concept, but it is useful in illustrating the principle of understanding which loops carry high-frequency switching currents and how they can be restricted to flow in defined areas of the PCB, without contaminating noise-sensitive areas.

In practice, however, L1 and L2 cannot be made infinite, because the high-voltage power rails will need to support audio frequency currents. The capacitance available *locally* at the Class-D stages (C1 and C2 in Figures 16.3 and 16.4) will not be large enough to completely supply the necessary audio frequency currents without excessive supply rail droop or pumping, *but*... This supporting audio current contribution from the high-voltage rails (V1 and V2 and their associated local bulk capacitances) is increasingly needed as audio frequencies *decrease*. Therefore, the low-pass characteristic of impedances L1 and L2 aligns nicely with this requirement, and allows low-frequency current support while choking off high-frequency switching currents. Another reason why this is serendipitous, is that this means there will be increasing voltage ripple in the high-voltage rails as audio frequencies decrease, and this aligns well with the PSRR characteristics of the negative feedback loop. The increasing loop gain of the negative feedback loop at low frequencies is just what is needed to combat the increasing power supply ripple voltage amplitude at low audio frequencies.

In summary, with reasonable values for C1 and C2, the current paths I1, I2 will carry high currents at *audio frequencies*, and the current cancellation in L3 will not be effective because I_1 and I_2 will flow at different times. I_1 will flow during the positive AC half-cycle of audio output current, and I_2 will flow during the negative AC half-cycle of audio current. Therefore, the impedances of the L1, L2, L3 connections will be critical. Voltage droops in the L1, L2 connections will contribute to crosstalk between multiple channels sharing the same voltage rails, and voltage droop on the ground connection G1-to-G2 will contribute to crosstalk between channels as well, if the ground connection is shared (common impedance, explained below).

Another example of current containment is in the design of the switching DCDC converters. These are the +7V and −7V converters, as well as the +12V gate drive supply converter (which is referenced to the −48V rail). These are all *buck*-type, stepdown DCDC converters. This converter topology is characterized as having "smooth" output current and pulsatile input current. However, the switching currents in these DCDC converters can be constrained to flow in small local loops, by adding an inductor in the input path to the converter. This causes the switching current to circulate locally, and filters the input current, reducing high-frequency switching current amplitude.

Functional Blocks

There are functional blocks that generate noise, and there are functional blocks that are sensitive to noise. The PCB designer should identify these blocks and decide best placement based on these characteristics, as well as mechanical and thermal requirements.

- Noise-sensitive functions:
    - The audio input circuitry
    - The PWM control circuitry, especially the triangular ramp generation circuitry.

- Noise-generating functions:
  - The gate driver stage
  - The switching stage (Honda & Huang, 2008).

Key Component Placement

Taking the above into consideration, the PCB designer starts placing key components in a given open space. Typical key components that affect overall design and should be regarded in this design stage are as follows:

- Audio signal input connectors
- Error amplifiers
- Gate driver ICs
- Power mosfets
- Bus capacitors
- Output low-pass filter components
- Power supply inlet
- Speaker output connectors (Honda & Huang, 2008).

Current Loops

Current loops that carry large alternating currents have the greatest potential to generate EMI in the circuit and deserve priority in the layout. Characterized by high peak currents and sharp switching edges (fast current rise and fall times), the currents flowing through these loops can easily inject noise into surrounding circuitry. To minimize the impact of these current loops in the amplifier, designers should minimize the loop area and total length of these current paths (Honda & Huang, 2008).

Common Impedance

Common impedance refers to the wires or traces shared by two or more separate circuits.

Some circuits require minimum impedance to the ground whereas others prefer separate routing to ground to minimize cross-couplings from adjacent circuits. A designer must minimize common impedance paths shared by noisy and quiet

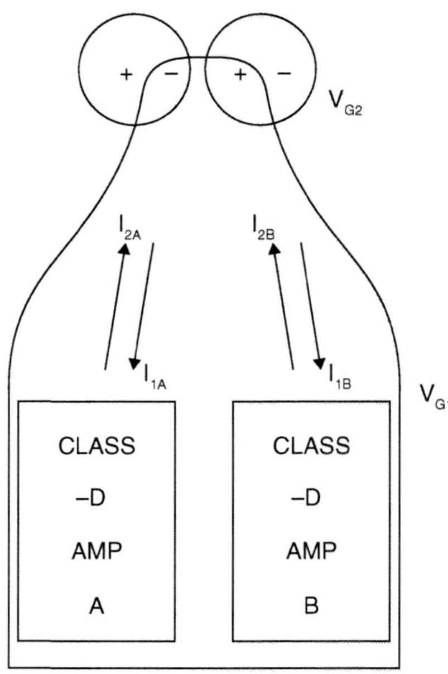

Figure 16.5 Common Impedance

circuits. The concept of "star ground" effectively prevents common impedance from affecting the performance of the amplifier. Figures 16.5 and 16.6 illustrate these concepts.

In Figure 16.5, the ground connection of two separate Class-D amplifiers, back to the bulk DC power source, is shared (*common*). The currents I_{1A} and I_{2A} supplying AMP A will create a voltage drop in the ground copper, between points VG1 and VG2. This voltage drop will affect AMP B. However, in Figure 16.6, the ground paths from the Class-D amplifiers back to the DC power source are separated. The currents I_{1A} and I_{2A} supplying AMP A will still cause a voltage drop in the ground copper, but this time it is between VG1A and VG2. The ground potential difference between VG1B and VG2 is *unaffected* by currents flowing in the separate VG1A to VG2 path.

Figure 16.7 shows another representation of the effects of common impedance connections, versus star-grounded connections.

The upper circuit in Figure 16.7 depicts a common impedance connection. AC ground return currents from CH2 (termed the "aggressor" channel), of +/−1A,

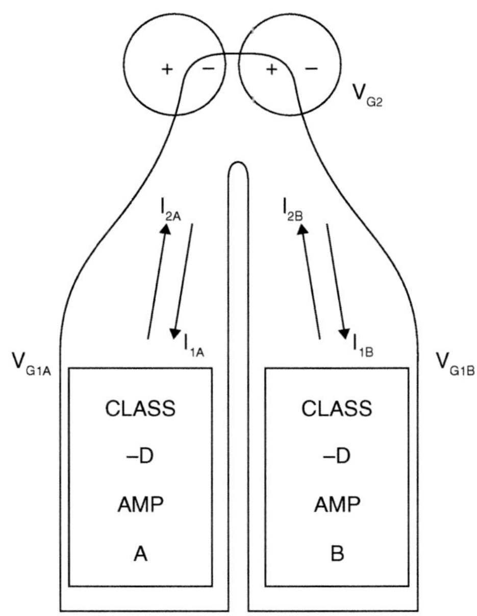

Figure 16.6 Star Ground

create a +/−10mV differential between the reference 0V ground node at the left of the diagram (to which the low-level preamp circuits are referenced) and the output return node of CH2, by flowing through the 10 milliohm ground return resistance. This +/−10mV differential voltage then becomes the "ground reference" for the error amplifier in CH1, and the output voltage of CH1 sees this error voltage amplified by the gain of the power stage (×10 in this example), producing an output error of +/−100mV.

The lower circuit in Figure 16.7 illustrates a star-grounding configuration. In this case, the AC ground return currents from each channel flow back to the 0V ground reference through independent low-resistance paths. There will still be a voltage generated at the output return node of CH2, which is +/−10mV with respect to the 0V ground reference, but this does not affect CH1 at all. CH1's power stage and error amplifier have no differential voltage imposed between its local output return node and the 0V ground reference on the left of the figure.

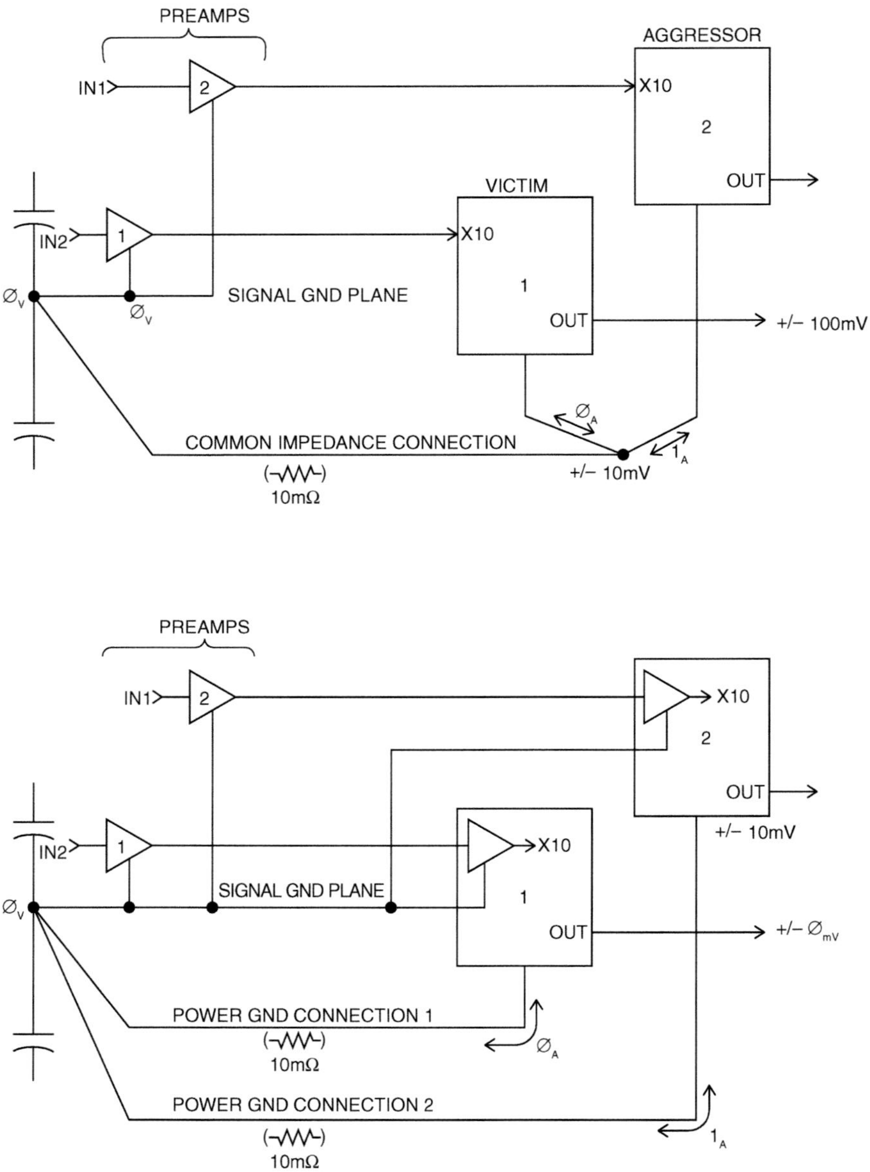

Figure 16.7 Common Impedance and Star Grounding Connection Configurations

Shielding of the Switch Node

There are differing opinions about the use of GND PLANE directly underneath the switching node. On the one hand, it *increases* the capacitance at the switching node, which causes increased CV^2f losses. But on the other hand, it has been found to decrease EMI by providing a local return for noise currents, and shielding for the E-field generated by the Class-D inductor. Wyatt and Sandler (2022) discusses this design decision in detail, with supporting measurements. These were made on a DCDC converter and not a Class-D amplifier. However, synchronous DCDC converters use the same half-bridge topology for the power switches as is used in Class-D amplifiers, and therefore the switching characteristics and critical noise issues are the same. Wyatt and Sandler found a slight advantage to using the full ground plane instead of making cutouts in the ground plane under the high dv/dt nodes. The copper area of the switch node (including component pads/leads), and traces should be minimized, and although ground shielding (on PCB layers) under the switch node has a net benefit, the separation of these layers deserves attention. It may be possible to increase the distance between these layers in a multi-layer PCB, for example. Doubling the distance between these layers will halve the capacitance.

In general, the switching node of Class-D circuits shares many of the characteristics and operating conditions of synchronous buck (step-down) DCDC converters. Therefore, many of the concepts and techniques discussed in the references regarding switch-mode power conversion, and DCDC converters in particular, are highly relevant to Class-D power stage design and layout.

In addition, it *always* makes sense to decrease the loop area "hot loop" at the switching node. These hot current loops are EMI transmitting antennae. Decreasing the loop area that these switching currents traverse, decreases their transmitting efficiency and EMI. It is also beneficial, at the same time, to decrease the *surface area* of the copper that is experiencing high dv/dt switching. Capacitance is directly proportional to surface area, so decreasing copper surface area decreases this capacitance and the associated capacitive switching losses. This large dv/dt creates a strong electric field (E-field), which has a significant noise spectrum up into the tens and hundreds of MHz (Toporski, 2022). An example of something that can be done, that is totally free, is to *swap* the positions of the bootstrap diode and any series current limiting resistor in the upper transistor gate drive bootstrap circuit. Connect the resistor to the *anode* lead (low-voltage/non-switching

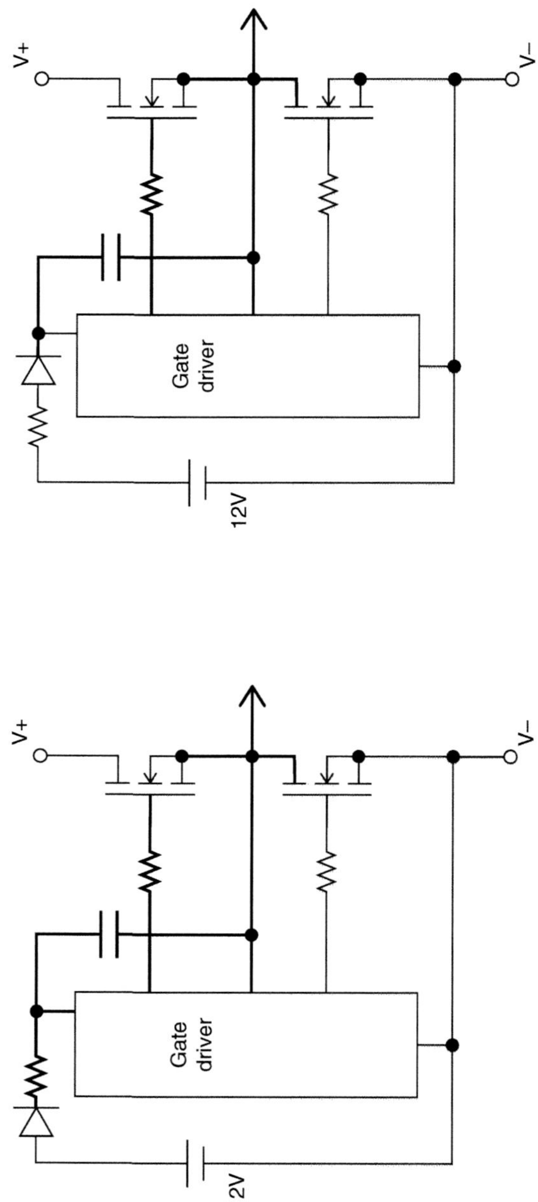

Figure 16.8 Commonly Used Bootstrap Circuit for Providing Bias Voltage for the Upper Gate Driver

terminal) of the bootstrap diode instead of to the cathode switching terminal. Figure 16.8 illustrates the benefit.

In the circuits in Figure 16.8, the *bold* traces and components experience high dv/dt switching. The circuit on the *left* includes the bootstrap resistor and its terminal pads, in the high dv/dt area. The circuit on the *right* does not, yet the functionality is identical. In this example, a simple rearrangement of the rectifier and resistor of the bootstrap circuit reduces the area of exposed copper experiencing high dv/dt.

L-C Demodulation Filter Layout

The inductor of the L-C demodulation filter opposes fast changes in current, so the current waveform does not have sharp transition edges like the switching voltage. Even though di/dt is not very large, there is still a ripple current with a strong time-varying magnetic field (B-field) at the 500kHz switching frequency (of the reference design, for example). This B-field can be a potential issue for nearby circuits that are sensitive to this frequency range. This B-field can vary greatly depending on the actual construction of the Class-D inductor (see more information in Chapter 9). In general, devices using a "gapped" magnetic core (such as a gapped-core ferrite inductor) will have higher stray B-field than distributed gap core structures (such as powdered iron toroidal cores, or composite cores). However, ungapped solenoid-style cores and air-core inductors will have the highest level of B-field emissions. These solenoid-style cores, though very inexpensive, are rarely appropriate for Class-D designs because of this B-field effect, except perhaps in extremely low-power, low-cost designs. Air-core inductors on the other hand cannot achieve the inductance usually required for the Class-D demodulation filter function, though they are sometimes used as a second stage filter (requiring lower inductance). They have the advantage of freedom from core saturation, but the disadvantage that there is no core to concentrate magnetic flux lines, so there is little containment of the B-field. Also, they have a large physical size for a given inductance value.

PCB Trace Sizing and Current-Carrying Considerations

There are useful nomographs and online calculators for trace sizing on PCBs according to various Institute for Printed Circuits (IPC) standards. These nomographs allow a calculation of copper cross-sectional area, as well as temperature rise at various current levels, differentiated between traces on external or internal PCB layers.

PCB Layer Count and Stackup

For Class-D amplifier design, as for switch-mode power circuits in general, it is advisable to use a PCB layer count of at least four layers. Since PCB layer counts are available in two-layer increments, the next lowest stackup below four layers is two layers. A two-layer PCB is a simple double-sided PCB. Routing critical circuits and proper ground plane coverage of critical areas, along with power distribution, become extremely difficult using a two-layer PCB. Using a four-layer PCB allows more degrees of freedom for component placement and routing.

One example PCB layer stackup that yields good results for Class-D amplifier design is the following:

- Top-Side = Signal Routing (with Ground fill)
- Layer 2 = FULL GROUND PLANE layer
- Layer 3 = DC POWER (with Ground fill)
- Bottom-Side = Signal Routing (with Ground fill).

Note the high degree of ground fill copper used. This has many benefits:

1) Increasing the total copper on the PCB increases the power dissipation capability and thermal mass of the PCB. This helps in the heatsinking of surface-mounted device (SMD) components.
2) Provides shielding of traces as both EMI radiators and receivers.
3) Lowers overall ground plane impedance, which will decrease I^2R losses, as well as noise voltages.
4) Copper weight balance across the PCB is maintained. This aids in PCB manufacture and minimizes warpage.

The ground areas on all layers should be connected together with vias. It is also prudent to *stitch* the ground layers along the four sides of the PCB, close to the edges of the PCB, using stitching vias, as an added EMI reduction step.

For even better performance and ease of layout (increasing degrees of freedom in placement and routing), the use of a six-layer PCB should be considered. The PCB layer stackup below illustrates the benefits:

- Top-Side = Signal Routing (with Ground fill)
- Layer 2 = FULL GROUND PLANE layer
- Layer 3 = DC POWER (with Ground fill)
- Layer 4 = DC POWER (with Ground fill)
- Layer 5 = FULL GROUND PLANE layer
- Bottom-Side = Signal Routing (with Ground fill).

This stackup allows an extra layer to ease DC power connection and therefore decrease I^2R losses. It also provides a full ground plane layer immediately adjacent to the top and bottom signal layers. This shortens current loops, improving bypassing, and lowering noise, thus improving signal-to-noise ratio (SNR).

Other Routing Considerations

Avoid long parallel traces routed on the same layer and in close proximity. These will experience inductive coupling (crosstalk) (Ardizzoni, 2005).

Avoid parallel traces routed on top of each other on adjacent PCB layers. These will experience *both* capacitive and inductive coupling (Ardizzoni, 2005).

When traces must cross, try to do so at a right angle (orthogonal) to minimize both inductive and capacitive coupling. A ground plane layer in between these (if possible) will act as an electrical shield, and will minimize the capacitive coupling between these traces.

A discussion of controlled impedance lines which maintain a characteristic impedance, and are routed as microstrip or stripline, is beyond the scope of this treatment, since signal integrity considerations for Class-D amplifiers rarely have a requirement for controlled impedances within the Class-D design itself. However, other signal processing circuits which share the same PCB as a Class-D design may employ controlled impedances, and these may dictate the PCB layer stackup requirements. Therefore, the Class-D amplifier designer may indirectly need to be cognizant of these controlled impedance considerations.

Johnson (1993) is an oft-cited reference for digital signal integrity.

Thermal Considerations

It is important to understand that for most modern low- to moderate-power audio Class-D amplifier designs, the PCB will serve the role of heatsink for many SMD devices. In order to accomplish this effectively, the copper surface area around power-dissipating SMD components must be maximized. This is another reason for filling exposed areas on the top and bottom layers with copper. These layers are exposed to air and allow surface copper to convect heat away from the PCB.

The DirectFET packages used in the reference design (IRF6775) are purpose-designed to benefit from PCB cooling. In addition, their flat profile allows easy addition of a heatsink to their top side if additional heatsinking is necessary.

Note: The metal case of the DirectFET package is electrically connected to the DRAIN terminal. Therefore, an insulator is required to maintain electrical isolation between the drain and a metal heatsink. This allows the heatsink to be connected to ground, which is beneficial for EMI, allowing it to act as a shield. In many cases, a pliable gap pad is used as the insulator. The gap pad allows efficient heat transfer between the case of the DirectFET and the metal heatsink, while still maintaining electrical isolation. Always check the dielectric voltage breakdown rating and final thickness of the compressed gap pad or insulator to be sure it is appropriate for the application.

Two thermal designs will be presented. In the first example, a heatsink will be used, in combination with a gap filler pad acting as a Thermal Interface Material (TIM). In the second example, no heatsink will be used, and the PCB will be the primary means of conducting and dissipating heat from the power dissipation source.

Thermal Design 1 (Using Metal Heatsink)

Figure 16.9 illustrates the mechanical arrangement for the thermal example to be analyzed.

Three parameters which must be known in a design that uses convection cooling, instead of forced air (fan) cooling, are the power (in watts) that must be dissipated, the maximum junction temperature to which the power device junction will be allowed to rise, and the maximum *local* ambient temperature in which the heatsink will operate. This example will use a single power dissipation

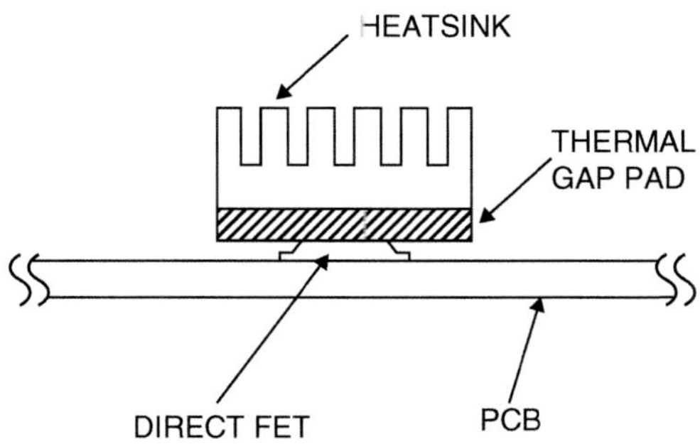

Figure 16.9 Heatsinking Arrangement for Top-Cooling of a Surface-Mounted Power Device (DirectFet)

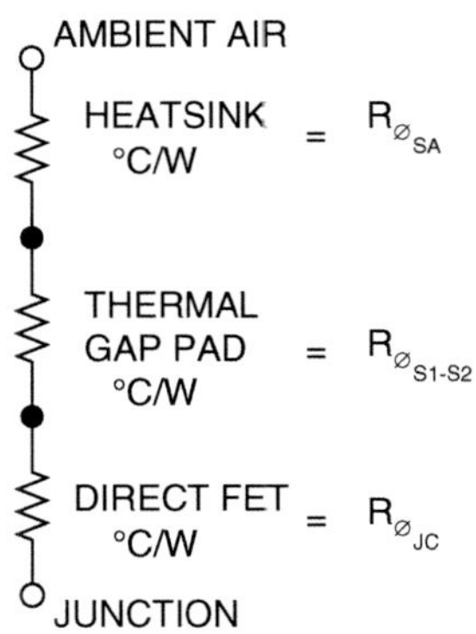

Figure 16.10 Thermal Design Example Using a Series Connection of Thermal Impedances, Depicted as Series-Connected Resistors

source for simplicity; however, in an actual Class-D design, it is more likely that a single heatsink will be used to cool multiple (at least two) devices.

For this example, the power dissipation of the DirectFet will be assumed to be 5W. The maximum junction temperature of the DirectFet will be limited to 110 °C. This will provide a comfortable safety margin for the device's absolute maximum junction temperature specification of 150 °C. The maximum local ambient air temperature in which the heatsink will operate will be 45 °C.

The second step is to draw a "resistance ladder" for the thermal impedances involved in the heat flow path, from the semiconductor junction (heat source) to ambient air (the "ultimate heatsink"), as per Figure 16.10.

Figure 16.10 includes the following thermal resistances arranged in a series ladder:

- Thermal resistance from junction-to-case for the DirectFet = 1.4 °C/W max (from datasheet)
- Thermal resistance of the TIM (thermal gap pad), from its lower surface (surface S_1) to its upper surface (surface S_2). This will be calculated below.
- Thermal resistance of the heatsink, from its lower surface (mounting surface) to ambient air (from datasheet).

The dimensions used for thermal resistance will be degrees C/watt = °C/W.

In order to calculate the thermal impedance of the *gap pad*, the surface area of the heat-conducting path must be known. In this case, the *top* of the DirectFet package will make contact with the lower surface of the gap pad. This is surface S_1. The area is (from the datasheet mechanical dimensions) 6.25mm × 4.80mm = $30\text{mm}^2 = 30 \times 10^{-6}\text{m}^2$. The compressed *thickness* (t) of the gap pad is used (see Figure 16.11).

EQ. 16.1 is used to calculate the thermal impedance of the gap pad, using the mechanical dimensions and the *thermal conductivity* (K) specification from the datasheet (in W/m-K). The length and width dimensions used will be those for

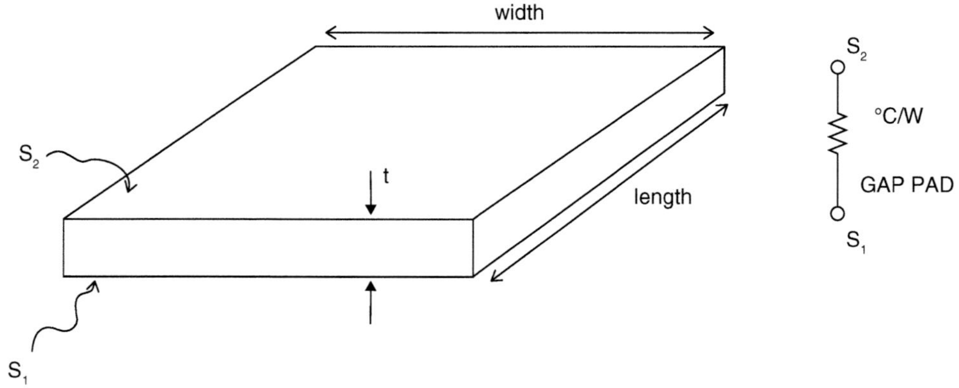

Figure 16.11 Mechanical Dimensions of a Thermally Conductive, Electrically Insulating Gap Pad

the top of the DirectFet package. A compressed thickness of 0.020" = 0.508mm = 0.000508m. All dimensions will be in meters.

Gap pad thermal resistance Θ = t/(length × width)(K) = 0.000508 / $(30 \times 10^{-6}\text{m}^2)(5)$ = **3.39 °C/W** EQ. 16.1

Note: For these thermal calculations, °C and °K are used interchangeably, because we are only concerned with *delta* degrees (which are the same in both units), and not absolute temperatures (which have a 273-degree difference), where 0 °C = 273.15 °K. The variable "K" in EQ. 16.1 is thermal conductivity and is 5 W/m-°K for the gap pad used in this example (Bergquist 2018).

These calculations, so far, yield the following results:

Temperature *delta* from junction-to-case for the DirectFet = 5W × 1.4 °C/W = 7 °C

Temperature *delta* across the compressed gap pad thickness = 5W × 3.39 °C/W = 16.95 °C

The *total* allowed temperature delta between the ambient air temperature and the maximum junction temperature is 110 °C − 45 °C = 65 °C.

Therefore, the total temperature delta between the heatsink mounting surface and ambient air needs to be

(65 °C − (7 °C + 16.95 °C)) = 41.05 °C

This means a heatsink with the following maximum thermal resistance from "sink to ambient" ($\Theta_{S\text{-}A}$) must be selected:

$\Theta_{S\text{-}A}$ = 41.05 / 5W = 8.21 °C/W

Note: Since the DirectFet package has such a low mounted profile, a soft conformable thermal gap pad can be used to advantage. If the correct uncompressed thickness is selected for such a gap pad, it can compress and conform to the top of the DirectFet case, and yet still make contact with the PCB surface immediately surrounding the DirectFet package. This will greatly increase the surface area used in the calculations above and reduce the thermal impedance of this interface. Figure 16.12 illustrates this configuration.

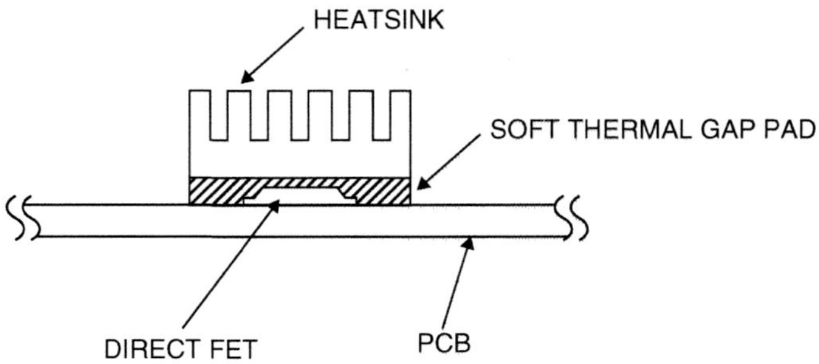

Figure 16.12 Surface-Mounted Power Component (DirectFet) Covered by a Conformal Gap Pad and Heatsink

Thermal Design 2

The second thermal design does not use a dedicated heatsink. Instead it relies on conduction of heat from the heat source (power semiconductor) into the PCB, and conduction of heat from the PCB to the surrounding ambient air, or to an enclosure, and then (ultimately) to the surrounding ambient air.

For this case, the parameter that matters for the DirectFet is the theta junction-to-PCB = $\Theta_{J\text{-}PCB}$.

For the IRF6775, this is specified as 1.4 °C/W typical.

Therefore, for 5W of power dissipation in the DirectFet, the junction will be (5W × 1.4 °C/W) = 7°C hotter than the *local* PCB surface temperature. It is important to consider the maximum operating temperature of the PCB material. Most FR4-based PCBs can handle operating temperatures of at least 90 °C. Certain PCB types can handle higher operational temperatures, so be sure to check the maximum operating temperatures for the PCB material that will be used in the design. For this example, the maximum allowable PCB operating temperature will be 90 °C. As a result of this limitation, the maximum temperature of the DirectFet junction will be (90 °C + 7 °C) = 97 °C.

The design challenge now will be to keep the PCB temperature below the 90 °C limit. This can be achieved as follows:

1) Use multiple mechanical mounting points (PEM® PCB fasteners) for conducting heat from the PCB to an enclosure.

2) Use as much copper (in the PCB) as possible, for heat spreading. This is especially true for the outer PCB surfaces, which will be exposed to ambient air.
3) Use of *thermal vias*: These are used to conduct heat from the copper that is immediately in contact with the power/heat source, to other PCB layers. For example, if the power semiconductor is mounted on the top side of the PCB, the immediate copper around the device can have *multiple* thermal vias that conduct the heat to copper areas on other layers and particularly to the bottom PCB layer.
4) Use a thermal gap pad on the exposed top and (usually) bottom of the PCB to conduct heat to an enclosure over a wider surface area, thus resulting in lower thermal resistance.

Note: There are PCBs available with metal substrates, which feature excellent thermal resistance, both through and across the PCB, but these do not usually find usage in commercial Class-D audio amplifiers, due to their relatively high cost compared to conventional FR4-based PCB materials.

Heatsinking of Through-Hole Components

So far the design examples above have focused on surface-mounted semiconductors. These are certainly the more popular types for Class-D design, for low- to moderate-power designs. However, for very high power designs, through-hole power semiconductors are still used. These are typically in TO-220 or TO-247 plastic power packages. These are conveniently mounted to heatsinks, using either screws or clips. Most such devices (mosfets) have the DRAIN exposed at the rear mounting surface of the device, and thus require a TIM to provide electrical insulation between the device mounting surface and a metal heatsink, which is usually at circuit ground or chassis ground potential.

Thermal Considerations for Other Components

As part of the complete thermal design of the amplifier, it is important to know which components to keep away from heat-producing components, or hot areas of the PCB. These heat-sensitive parts are principally Aluminum Electrolytic (AE) capacitors, crystal oscillators, film capacitors, and some types of ceramic capacitors.

AE capacitors have lifetime ratings that are highly temperature dependent. A rule of thumb is that lifetime doubles with each ten-degree C decrease below the rated operating temperature. For example, a typical AE capacitor might have a datasheet lifetime specification of 5000 hours at 105 degrees C. This can be extrapolated to:

10K hours at 95 °C = 1.14 years (at 8760 hours/year)
20K hours at 85 °C = 2.28 years
40K hours at 75 °C = 4.56 years
80K hours at 65 °C = 9.12 years
160K hours at 55 °C = 18.24 years

Note that the temperature listed above is the local ambient air temperature to which the AE capacitor is exposed.

Crystal oscillator datasheets should be consulted, as there are some parts that only have an 85 °C rating.

Some ceramic capacitor dielectrics (such as X5R) only have an 85 °C rating. It is advisable, whenever possible, to use ceramic capacitors of X7R or NP0/C0G dielectrics, which have temperature ratings of at least 105 °C.

Though 85 °C may seem at first to be a sufficient maximum temperature rating, it is possible to exceed that temperature at certain parts of the PCB that are in close proximity to the mosfet power switches and Class-D inductor, especially when PCB copper is being used to conduct heat away from the packages of these devices.

Thermal Sensing

This conduction of heat away from high-power SMD devices using PCB copper presents an opportunity. By placing an SMD thermistor in close proximity to the power device, and ideally soldered to the same copper as the power device, the temperature of the power device can be easily sensed by control circuitry. This thermal information can be used for various actions, such as disabling the amplifier, in the case of an overtemperature condition (OT protection), or decreasing output power, and/or providing a visual indication (via an LED) of an OT condition to the user. It can even serve to control a fan's speed, if one is used.

Example Layout

The example layout in Figure 16.13 is just one possible layout that might be used for the reference amplifier. It incorporates many of the suggestions in this chapter; however, it is important to recognize that PCB layout involves compromise. Each layout task comes with its own set of constraints. There may not be as many degrees of freedom in the layout as the circuit designer might desire. For example, due to cost and size considerations, the number of PCB layers and the PCB shape and size may already be determined by other product considerations, and the Class-D amplifier designer will have to design the layout within those constraints. The example layout in Figure 16.13 assumes a simple rectangular shape, and optimizes circuit placement, signal flow, and connector number and position. It utilizes the concepts of "current containment", common impedance, and ground plane discussed in this chapter.

Figure 16.13 shows suggested placements for the various circuits which comprise a Class-D amplifier, as well as an outline of the *ground plane* layer.

Features

1) By using a solid ground plane, with a cut between the two output channels, travelling back to power connector P2, the high output audio currents in the two channels will not experience common impedance, and voltage drops in one channel will not crosstalk to the other channel.
2) High-amplitude, high-frequency switching currents in the Class-D half-bridge stage will remain local among the power mosfets (F), the rail capacitors, and the output L-C filter in each channel.
3) The switching DCDC that generates the +12V rail (referenced to the −HV rail) is situated in the center of the PCB with a "U-shaped" plane cut around its perimeter. This contains the switching currents of this converter within its confines, and prevents them from polluting the ground plane. The ground plane above P2 is therefore free of high-frequency switching currents.
4) The output voltage feedback signals for each channel (dotted traces) run back to the error amplifier in each channel without having to cross over any noisy high-frequency switching signals. Differential output voltage sensing (Vout+ and *local* Vout− = output ground return) could optionally be used here.

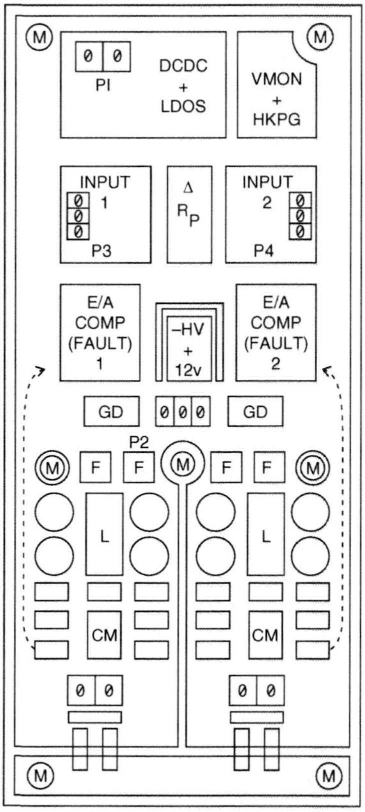

Figure 16.13 Placement and Layout Example for the Reference Class-D Amplifier Design

5) Noise-sensitive circuits such as input circuits and the ramp generator circuit are placed far from noisy high-power switching circuits.
6) Separate input connectors (P3, P4) and low-voltage (+12V) power connector (P1) allow greater placement and routing flexibility, decreasing crosstalk between channels.
7) The four mosfets (F) and three mounting holes (M) are arranged to allow the attachment of a heatsink and gap pad.

Bibliography

Ardizzoni, J. (2005). A practical guide to high-speed printed-circuit-board layout. *Analog Dialogue*, *39*(09), 1–6.

Bergquist (2018, October). Datasheet, TGP 5000 Gap Pad.

Duraij, M., Iversen, N. E., Petersen, L. P., & Boström, P. (2015). Self-oscillating 150 W switch-mode amplifier equipped with eGaN-FETs. In: *Proceedings of the 139th AES Convention*, New York, NY, October 29–November 1, 2015 (Article 9378). New York: Audio Engineering Society.

Honda, J., & Huang, C. (2008). PCB layout with IR Class-D audio gate drivers. Infineon (International Rectifier) Application Note AN-1135. El Segundo, CA: International Rectifier.

Johnson, H. (1993). *High Speed Digital Design: A Handbook of Black Magic* (1st ed.). Englewood Cliffs, NJ: Prentice Hall.

Miraglia, C. (n.d.). Compression characteristics of thermal interface gap filler materials. Carteret, NJ: Fujipoly America.

Miraglia, C. (n.d.). Fundamentals of heat transfer in thermal interface gap filler materials. Carteret, NJ: Fujipoly America.

Porley, M. A., & Chesser, K. (2022). What are the basic guidelines for layout design of mixed-signal PCBs? *Analog Dialogue*, *56*(3), 1–4.

Putzeys, B. (2014, April 22). The G word, or how to get your audio off the ground. *EDN*.

Sun, J., Jiang, L., & Zhang, H. (2022). How to optimize switching power supply layout by minimizing hot loop PCB ESRs and ESLs. RAQ Issue 207. *Analog Dialogue*, *56*(4), 1–6.

Texas Instruments. (2011, May). Understanding thermal dissipation and design of a heatsink. Texas Instruments Application Note SLVA462. Dallas, TX: Texas Instruments.

Texas Instruments. (2016a). LC filter design. Texas Instruments Application Report SLAA701A. Originally published October 2016, revised November 2016. Dallas, TX: Texas Instruments.

Texas Instruments. (2016b). Datasheet, TPA3251 175-W stereo, 350-W mono PurePath™ ultra-HD analog-input Class-D amplifier, SLASE40D. Originally published May 2015, revised April 2016. Dallas, TX: Texas Instruments.

Toporski, T. (2022). Switch node layout considerations for EMC. Kirkland, WA: MPS.

Wyatt, K., & Sandler, S. (2022, July 28). DC-DC converters – Solid return plane or cutouts under switch node and inductor? *Signal Integrity Journal*.

Chapter Seventeen

Power Supply Considerations

Block Diagrams of ACDC and DCDC Power Distribution

The discussion of power supply considerations is aided by first considering an overall block diagram of the power distribution architecture used in the reference design Class-D amplifier (see Figures 17.1 and 17.2).

While illustrating the detailed circuit design of the reference Class-D amplifier *is* a goal of this book, illustrating the detailed circuit design of the power supply is not. The reasons for this are twofold. First, it necessarily limits the scope of the book to Class-D amplifier design, which, as justified previously, is due to the scarcity of coverage of the subject in book form, as well as the complexity and volume of the material to cover. Second, although there are specific topologies and architectures of power supply design that complement Class-D amplifier design more than others (these will be discussed), in other ways, the design of the power supply architecture is not exclusive to Class-D design, and therefore, a detailed discussion of it does not promote understanding of Class-D design. It is a goal of this book to limit the subject matter to topics specifically relevant to Class-D design, and provide less coverage of items that are more generically applicable to other types of amplifier design. This is the case for power supply design.

The discussion of power supply considerations is divided into two subject areas: ACDC and DCDC converters and linear regulators (LDO = Low DropOut types and standard dropout types).

DOI: 10.4324/9781003327943-17

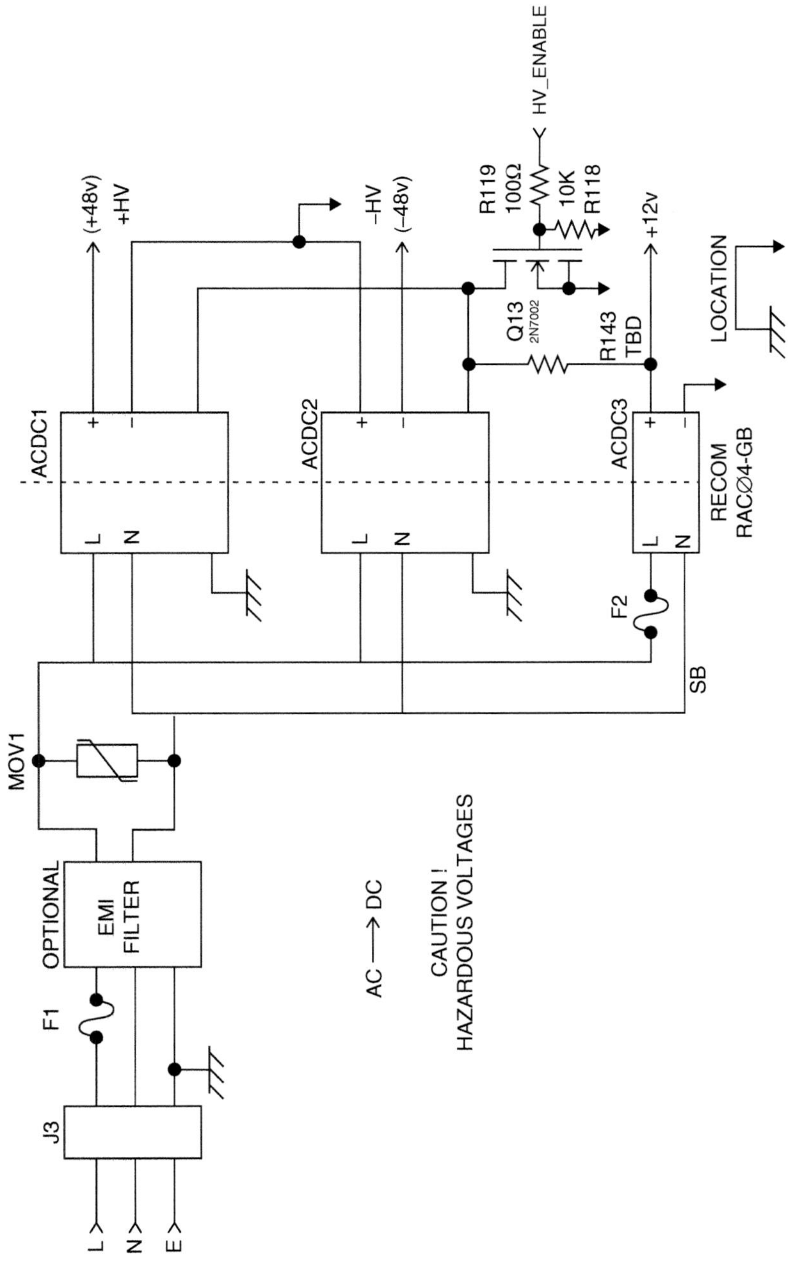

Figure 17.1 Block Diagram for the ACDC Converter Supplies

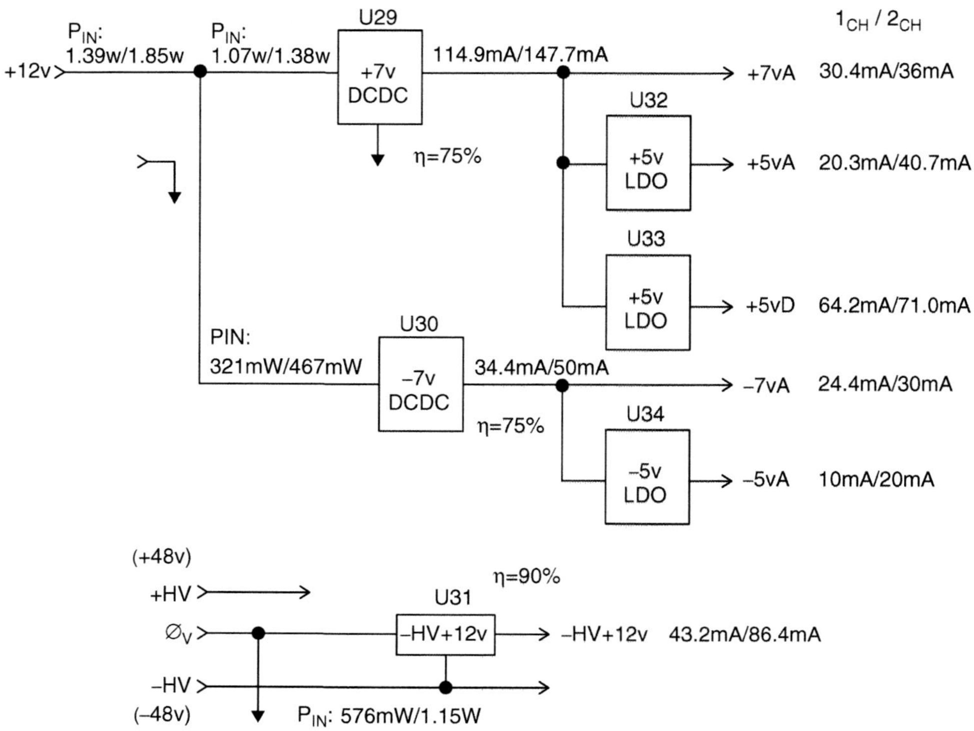

Figure 17.2 Block Diagram for the Low-Voltage DCDC Converter Circuits

ACDC Converters

In the reference design, there are three separate ACDC converters. ACDC1 and ACDC2 are identical, and are single-output, 48VDC output types. The output of ACDC1 is wired to produce +48VDC output polarity, whereas the output of ACDC2 is wired to produce a −48VDC output polarity. This configuration was chosen because single 48VDC output types are widely available, whereas *dual*-output ACDC supplies capable of producing positive and negative 48VDC outputs are extremely rare and expensive. Another benefit of the wide availability of the single-output 48V ACDC is that it is available in very small form factors. A 2×4" open-frame form factor was chosen for the reference design. In addition, the ACDC selected has a secondary-side REMOTE ON/OFF feature, which is used to enable/disable the +/−HV Class-D amplifier rails.

Specifying Original Equipment Manufacturer (OEM) ACDC Converters for Audio Amplifier Applications

It is important to understand the power demands of audio amplifiers, in order to align these requirements with the common specifications of original equipment manufacturer (OEM) ACDC converters. The main issue here is that *audio* power demand is very dynamic, whereas OEM ACDCs are specified for constant power output. Another way to state this is that audio content has a high peak-to-average power ratio. When dealing with voltage and current waveforms, the term *crest factor* is appropriate. This is defined as the peak-to-RMS value of these waveforms. However, when dealing with power, peak-to-average is the correct terminology because power is not correctly expressed as having an RMS value.

Of course, it is difficult and inaccurate to try to come up with a single peak-to-average value for audio power, since audio content has tremendous variety. However, it is commonly accepted that music has a peak-to-average ratio of between 6 and 24dB depending on the type of music.

$$\text{Peak-to-Average ratio (in dB)} = 10 \log (\text{Peak Power (W)}/\text{Average Power(W)})$$

The peak unclipped amplifier output power (for one channel) is 100W. If the least dynamic ratio of 6dB is used, the *average* amplifier output power will be 25W. If the most dynamic ratio of 24dB is used, the *average* amplifier output power will be 0.4W. (*Note* the "10 log" factor used for *power*, as opposed to the 20 log factor used for *voltage* or *current*.) As can be seen, this is an extremely wide range of power.

A simplification that is useful here is to understand that the ACDC power supply needs to be able to support the peak output power demand of the amplifier, but only for short intervals. There is much debate in the audio industry as to how long that "interval" should be, again due to the great variety in music content (and audio content in general, including speech etc.). Luckily, most realistic intervals used for this specification are relatively short, compared to the thermal time constants of ACDC converters. What this means is that an ACDC converter can be selected with an output power specification sufficient to support the *peak power* demand of the amplifier, but, from a *thermal* standpoint, it need only support the *average power* demand of the amplifier.

Using the reference design as an example, each amplifier channel can produce 100W peak power. If the efficiency of the amplifier is 90%, then the ACDC power supplies must supply 100W/0.9 = 111W to each amplifier (55.5W for the +48V rail and 55.5W for the −48V rail). Peak power demand will be equal from the two amplifier voltage rails.

Now, it is important to consider what happens when low-frequency audio output power is produced. Consider the output voltage and current levels in Table 17.1, when the amplifier is producing 100W output power.

Table 17.1 illustrates that the peak output power delivered by the amplifier is 200W, when it is producing 100W *average* output power. If the amplifier is 90% efficient, the power supply will have to supply the amplifier with 222.2W peak to support this level of peak audio output power. This peak power is required from the positive rail supply when positive audio output voltage/current is delivered to the load, and from the negative rail supply when negative audio output voltage/current is delivered to the load. Thus, if there were no large rail capacitance present to store energy, the power supply would need to deliver these peak currents. When there is rail capacitance present, the peak current demand is *shared* between the rail capacitance and the power supply. If the power supply is current-limited, it will supply its maximum current during peak demands, and the rail capacitance will supply the remainder. During these periods the rail voltage will droop below its regulated voltage as the capacitance discharges. This rail voltage droop increases as the audio frequency decreases, because the rail capacitance discharges for a longer time interval during each audio signal period.

Table 17.1 Voltage, Current, and Power Output Parameters for a 100W Class-D Amplifier Driving an 8-Ohm or 4-Ohm Load

Parameter	*100W into 8-Ohm Load*	*100W into 4-Ohm Load*
RMS Voltage	*28.28V*	*20.00V*
Peak Voltage	*40.00V*	*28.28V*
RMS Current	*3.54A*	*5.00A*
Peak Current	*5.00A*	*7.07A*
Peak Output Power	*200W*	*200W*
Peak Input Power @ η = 90% (Required from 48V ACDC)	*222.2W*	*222.2W*

A simplified/idealized numerical example will help to explain this more clearly. Suppose a power supply can deliver exactly 111.1W at +48V (as the positive rail supply). A Class-D amplifier operating at 90% efficiency can deliver 100W, with this input power limit.

100W represents 20V instantaneous into a 4-ohm load
100W represents 20VRMS into a 4-ohm load
20VRMS = 28.28Vpeak

This means that for a *sinusoidal* audio output voltage waveform, the output power will exceed 100W *between* the points on the sine wave depicted in Figure 17.3.

The shaded areas on the 20Hz sinusoidal output voltage waveform in Figure 17.3 are intervals during which the output power exceeds 100W, and therefore the input power exceeds 111.1W. During these intervals, the rail capacitance must supply the necessary extra input power for the amplifier. For a 20Hz signal, these intervals are (18.75msec − 6.25msec) = 12.5msec during the positive half-cycle and 12.5msec during the negative half-cycle. This is exactly *one-half* of the total signal period (50msec for 20Hz). It can be understood therefore that the rail voltage droop will increase as the audio output signal frequency decreases.

A note about amplifier output power ratings: Amplifier power ratings are often given in Watts RMS. Strictly speaking this is incorrect, because power (in Watts) is not expressed as an RMS value. Instead, RMS refers to the output voltage and current whose product equals this power. The output voltage and current *are*

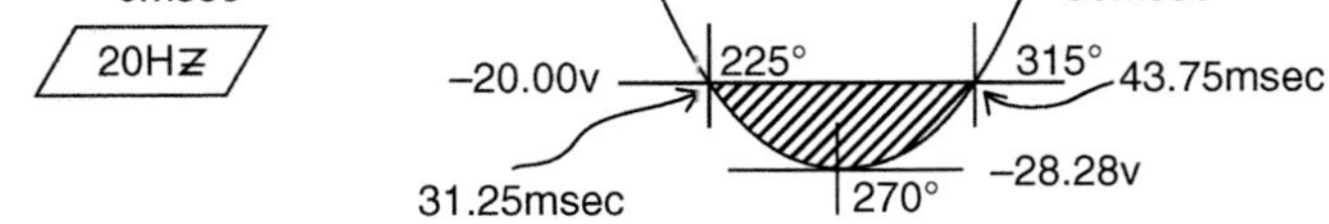

Figure 17.3 20Hz Sinusoidal Output Voltage Waveform, with Areas Marked Where Power Exceeds 100W

RMS quantities. Output power in Watts is actually correctly referred to as average output power.

Supply Rail Pumping

Supply rail pumping is a phenomenon common to half-bridge Class-D topologies. It arises from the fact that the inductor used in the L-C filter stores energy on each switching cycle (at the high Class-D switching frequency). This energy is sourced from one of the power supply rails, and returned back into the opposite power supply rail. The effect is exacerbated as the audio output frequency decreases and the output power increases. The interval during which *net* energy is pumped into one of the rail capacitors lengthens as the audio frequency decreases, and the voltage on that rail capacitor increases or is "pumped" up. This excess energy in the pumped-up rail capacitance is then discharged when the audio output signal reverses polarity, and the opposite rail capacitance is "pumped" up.

Honda and Cerezo (2003) and Maxim Integrated (2010) provide more information on supply rail pumping, and EQ. 17.1:

$$\Delta V_{RAIL\text{-}MAX} = V_{RAIL} \,/\, (2\pi^2)(F_{SIGNAL})(R_{LOAD})(C_{RAIL}) \qquad \text{EQ. 17.1}$$

$\Delta V_{RAIL\text{-}MAX}$ occurs when:

1) Rail capacitance (C_{RAIL}) is minimum (2200µF)
2) Audio signal frequency (F_{SIGNAL}) is minimum (20Hz)
3) Load (R_{LOAD}) is lowest impedance (4 ohms for the reference amplifier)
4) Output signal amplitude is $(2/\pi)(V_{RAIL}) = (0.637)(48V) = 30.56$Vpeak (Maxim Integrated, 2010).

Note: 30.56Vpeak represents 21.61VRMS = 116.7W into 4 ohms. This is slightly higher than the 100W rated output power into 4 ohms for the reference amplifier, but this value will be used to overestimate rail pumping for a more conservative design.

Plugging the values above into EQ. 17.1, and using a value of 48V for V_{RAIL}, for the reference design, yields $V_{RAIL\text{-}MAX}$ of 17.27V. This means that under these worst-case conditions, the rail voltages can *pump up* to (48V + 13.82V) = +/−61.82Vpeak (63V rated rail capacitors will be used).

Note: EQ. 17.1 assumes that there is no voltage *droop* in the power supply rails, because the power rails can deliver the required peak currents. Therefore Δ (delta) V_{RAIL} consists *only* of the rail voltage *increase* due to "pumping".

LTspice can be particularly useful in simulating supply rail pumping, because the effect of current limiting in the positive and negative rail supplies can be simulated, and the effect on the resulting supply pumping and droop can be simulated. This is a great help in selecting the optimum power supply (current and power limitations) and rail capacitance.

Figure 17.4 is an LTspice simulation under the following conditions:

1) Vout = 28.28Vpeak into 4 ohms (100W output power)
2) Audio output frequency is 20Hz
3) +/−48VDC rails (regulated, sourcing-only, through diodes, cannot sink pumping current)
4) 2200μF on each power rail; +48VDC and −48VDC.

The simulation shows that the 100W output power level (+/−28.28Vpeak into 4 ohms = sinusoidal trace) produces rail pumping peak voltages of approximately +/−60Vpeak (uppermost and lowermost traces). The power rails are delivering approximately 6Apeak (only positive rail current is shown in Figure 17.4).

Note: For the simulation in Figure 17.4, the rail power supplies are *not* current-limited. They are delivering the required 6Apeak current. For a 48VDC rail, this represents (48V × 6A) = 288Wpeak. If the power rail cannot supply that level of peak current, it will be drawn from the rail capacitance and the rail voltage will droop.

Figure 17.5 is another simulation. But this time, the rail power supplies *are* current-limited. This simulation is meant to show the effect of using +/−48V rails with lower output power ratings. In this case, there is still rail pumping, but there is also rail droop during the time intervals when the required rail current exceeds the maximum output current capability of the rail power supply. During those intervals the rail voltage will droop, as the output capacitance supplies the current deficit between what the amplifier demands and what the rail power supply can deliver. For this simulation, the output power of each of the rail power supplies was limited as closely as possible to 100Wpeak.

1) Vout = 26.5Vpeak into 4 ohms (87.78W audio output power)

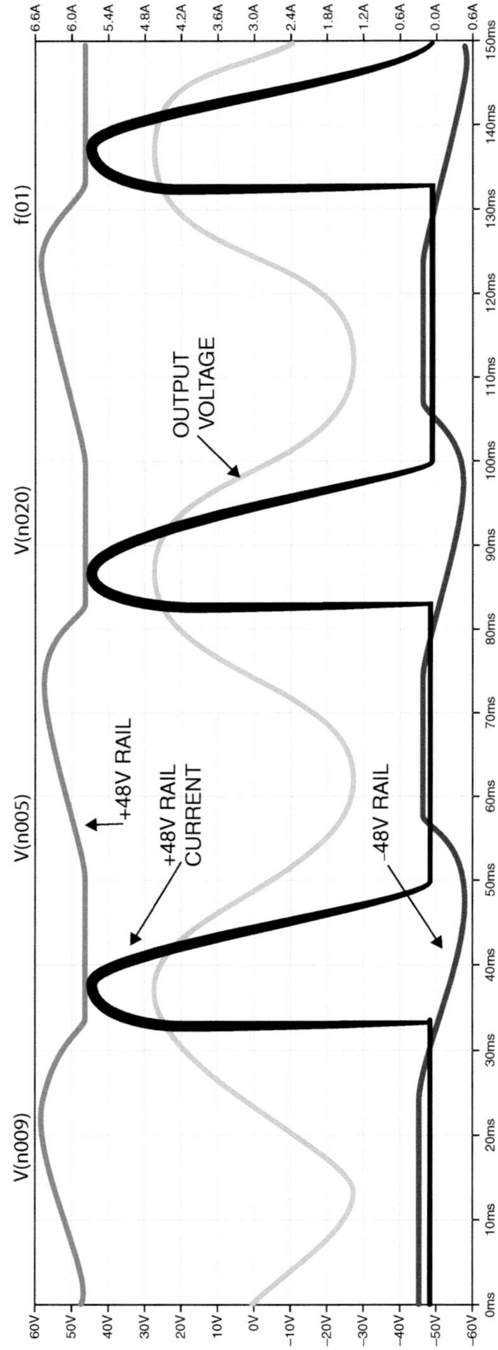

Figure 17.4 LTspice Time-Domain Simulation for 100W Amplifier Output into a 4-Ohm Load, at a Frequency of 20Hz, Power Supplies Not Current-Limited

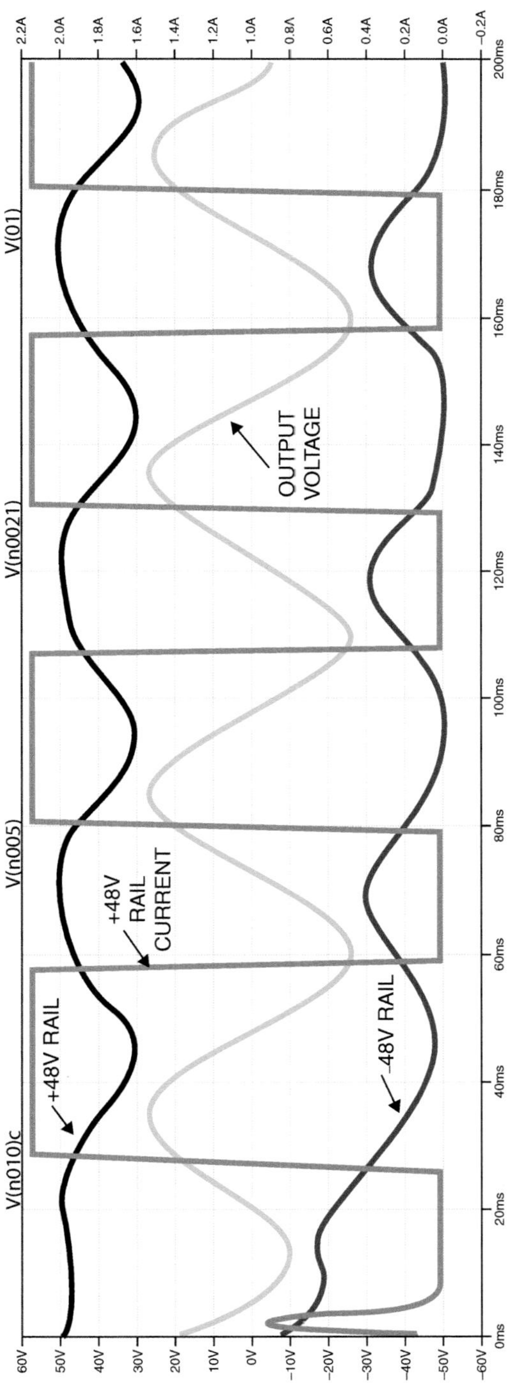

Figure 17.5 LTspice Time-Domain Simulation for 100W Amplifier Output into a 4-Ohm Load, at a Frequency of 20Hz, Power Supplies Current-Limited at 2.17A Each

2) Audio output frequency is 20Hz
3) +/−48VDC rails (current limited at 2.17A each = (2.17A × 48V = 104Wout). Rails source current-only, through diodes, and cannot sink pumping current)
4) 2200μF on each power rail; +48VDC and −48VDC
5) Rail voltages = 30.8Vmin (due to droop) / 50.0Vmax (due to pumping).

Power Rail Pumping: Half-Bridge versus Full-Bridge Class-D Topologies

The discussion of supply rail pumping above is relevant to half-bridge Class-D topologies that use split-rail, (positive and negative) power supply rails. Supply rail pumping does *not* occur in full-bridge Class-D topologies. In full-bridge topologies that use only one supply rail, current, energy, and charge are drawn from that (one) supply rail during *both* positive and negative signal polarities. This is one of the great benefits of the full-bridge Class-D topology. This is illustrated in Figure 17.6.

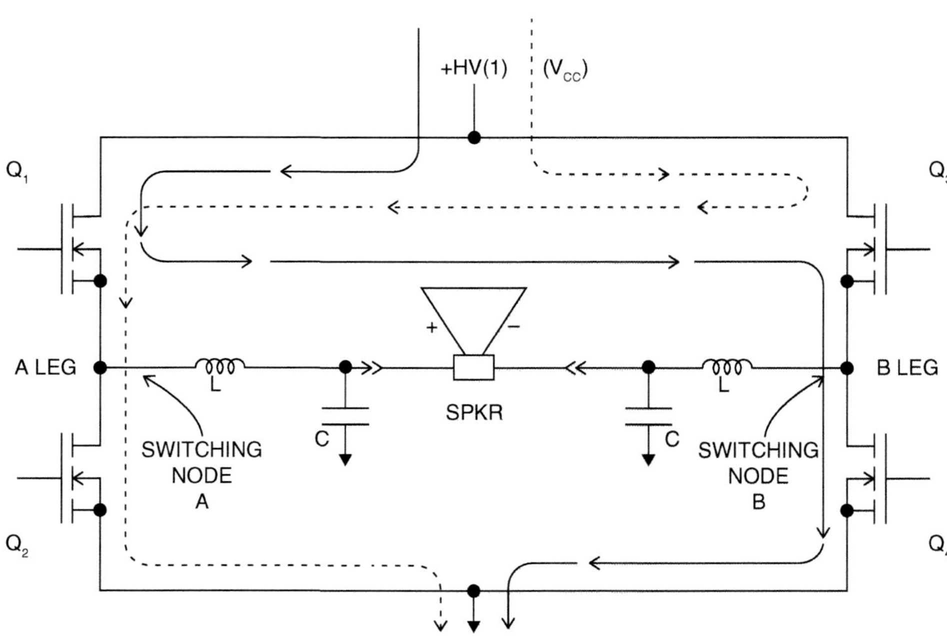

Figure 17.6 Current Flow in a Class-D BTL Power Stage, Using a Single Power Supply Rail

Output Connection Phasing to Reduce Power Rail Pumping

There is also a technique that can be used to reduce power rail pumping for stereo amplifier channel pairs utilizing the half-bridge Class-D topology with positive and negative power supply voltage rails (split rail design). It requires that *pairs* of Class-D amplifiers be configured with opposite output signal phasing, as per Figure 17.7.

Figure 17.7 illustrates a very versatile circuit topology for reducing supply rail pumping in split-rail designs (positive and negative voltage power supply rails). In one case, it consists of two separate amplifier channels, configured as a *stereo*

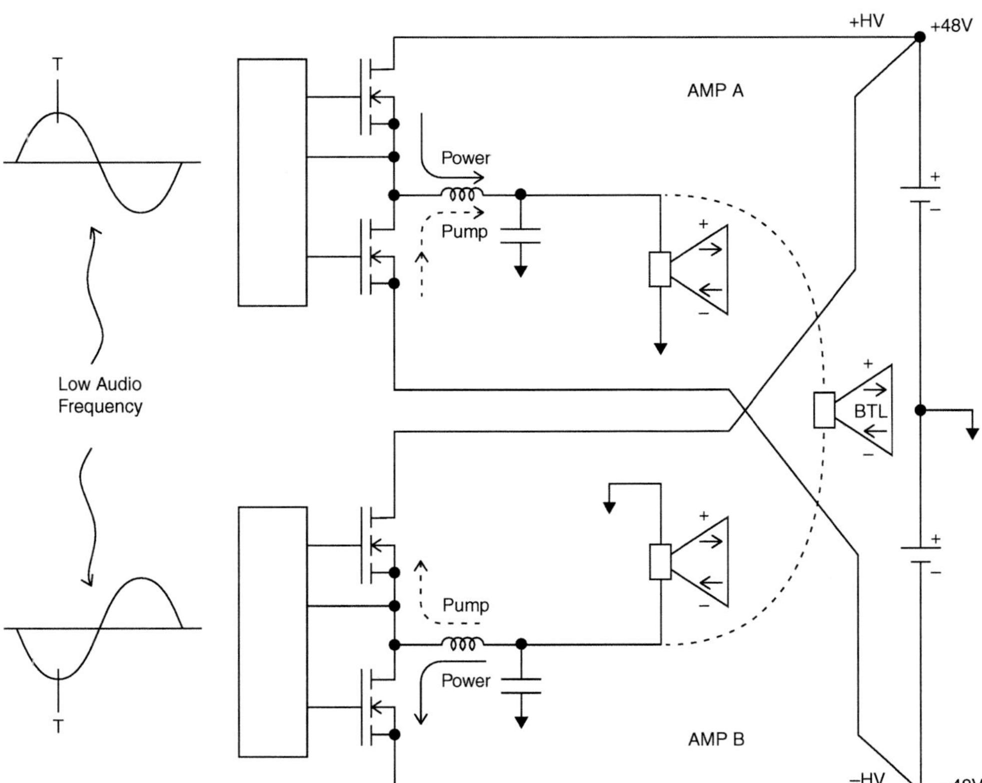

Figure 17.7 Connections of Two Separate Class-D Amplifiers, Utilizing Half-Bridge Power Stages, and Shared Positive and Negative (Split Rail) Power Supplies, with Phasing Arranged so as to Minimize Supply Pumping

pair, each having its own audio input signal. Pumping mitigation relies on *two conditions*. The first is the fact that low-frequency (*bass*) audio content is usually shared approximately equally and identically between left and right channels, because low frequencies lack directionality and do not define the stereo image. There is, however, left/right separation of higher-frequency audio content in left/right stereo pairs. The second condition is that the audio input signal supplied to channel B is inverted in phase compared to channel A's audio input signal. Then, if the operation of the circuit at time "T" is analyzed, it is seen that AMP A has a positive output voltage and current. It will source current from the positive rail, and pump current into the negative rail. Meanwhile, AMP B has a negative output voltage and current. It will source current from the negative rail, and pump current into the positive rail. In this way, the two amplifiers, working together, mitigate supply pumping.

It is important that the connection polarity to the speakers match that shown in Figure 17.7, in order for the audio output from the speakers to be in-phase (i.e. identical speaker-cone movement direction). As the figure shows, positive current flowing *into* the "+" terminal of the speaker will cause the cone to move *out*.

If this scheme is used for stereo pairs, the input signals will be two *different* signals, with similar low-frequency content, and having opposite phases. However, if the signals were exactly *identical*, with inverted phases, the outputs of the two channels could be connected as "BTL" = Bridge-Tied-Load", to a single speaker. This arrangement can then supply *four times* the power to the load (because it can supply twice the voltage across the load). In this scenario, supply pumping is eliminated, by the same mechanism discussed above.

It will also be recognized that if, instead of an analog input signal, the modulator of the upper amp (A) used a PWM input stream, and the modulator of the lower amp (B) used the same PWM stream inverted, then this would be a full-bridge topology with AD modulation.

ACDC Power Supplies Optimized for Class-D Audio

Mendenhall (2005) is a valuable source for information on power supplies suitable for audio amplifier applications. Typically, a purpose-designed dual-output ACDC switching power supply is optimum for powering a half-bridge Class-D amplifier such as the reference design. The ACDC power supply should be extremely efficient, compact, and lightweight, to complement these characteristics of the

Class-D amplifier. However, ACDC power supply design is outside of the scope of this book.

Housekeeping ACDC

In addition to the two separate ACDC converters which produce the +/−HV (+/−48VDC) amplifier rails, under enable control, there is also a low-power ACDC producing +12VDC, which is always on whenever AC input power is present. This +12VDC is used by the DCDC converter section of the power distribution circuitry, to create the point-of-load (POL) rails. The selection of +12V as the main DCDC distribution rail is based on a number of factors:

- Availability of 12V ACDC converters
- Availability of 12V-input DCDCs
- Efficiency of DCDCs with +12V input
- Optional usages of the +12VDC rail.

Low-power ACDC converters with +12VDC output are readily available, in a great variety of sizes, and output power ratings, and in PCB-mount packages.

There is a large variety of DCDC converter designs that operate with a 12VDC input, making the component selection and design of the downstream POL DCDC converters simpler.

The efficiency of the downstream DCDC converters is important in achieving overall high efficiency in the reference design (measured from AC input to Class-D amplifier output). Therefore, efficiency of these DCDC converters should be optimized. The efficiency of these types of DCDC converters is often specified for +12V input, which again makes selection easier.

The reference design is proposed as a convection-cooled design, requiring no fan. Fans can produce ambient audible noise which adds to the ambient noise floor in the listening environment, detracting from the achievable real-world signal-to-noise ratio (SNR) of the design. Therefore, it is preferable if an amplifier design can operate without fan-cooling. Also, as a mechanical, moving device, fans do have wear out mechanisms and limited lifetime. That said, however, the addition of a low-noise, high-reliability, high mean-time-to-failure (MTTF) fan can allow operation at higher ambient temperatures, and/or operation in enclosed spaces, where some ventilation holes are blocked (top and bottom holes, for example), and perhaps even higher power in a smaller overall enclosure. If a fan is desired (for these reasons), +12V is a convenient voltage for fan power. Fans powered by +12V are readily available, in great variety, again optimizing the selection process.

Low-Voltage Power Distribution

In addition to a detailed schematic page for the three DCDC converters, and a separate page for the three LDOs (Low DropOut voltage regulators), there is also a *block diagram* (Figure 17.2) which details the power distribution architecture for the low-voltage power.

Output noise and power conversion efficiency are the two most critical specifications of interest in the low-voltage converter section of the reference design.

The +5VD rail powers the digital circuitry, as well as any analog circuitry that is relatively insensitive to noise (such as the housekeeping voltage-monitoring comparators). A DCDC converter is well suited for powering this rail.

The +7VA and −7VA rails are analog rails which power the op-amps in the small-signal section of the reference design. Here, noise *is* critical, and must be balanced against the need for high power conversion efficiency. The DCDC converters that produce these rails are designed with low output noise in mind.

The +/−7VA DCDC converters are followed by very low noise LDO voltage regulators. These create +/−5VA for the PWM comparator, which is very noise-sensitive. These LDOs operate with only 2V input-output headroom, which yields relatively high efficiency, considering they utilize a linear voltage pass regulator topology.

Efficiency (LDO) = Vout/Vin = 5V/7V = 71.43%

The last DCDC converter to discuss is the converter that creates the +12V gate supply voltage for the power mosfets of the Class-D half-bridge stage. This +12V output ("−HV+12V") is a separate 12V output, which is referenced to the −HV rail. This rail is not to be confused with the main +12VDC rail, which is referenced to circuit ground (0V).

This converter is a simple *buck* converter topology, creating 12V from a 48V input, except in this case, this DCDC converter is referenced to the −48V rail, and its Vin node is connected to GND. Its 12V output is referenced to the −48VDC rail.

DCDCs; Detailed Circuit Discussion

Referring to *Low Voltage 1* page of the reference schematic, U29 is the BUCK DCDC converter which produces the +7VA output from the +12VDC input (from the housekeeping power supply). The reasons that U29 was selected were the following:

1) Small package size
2) Low cost
3) Forced Continuous Mode operation (synchronous rectification)
4) High fixed switching frequency for small component sizes
5) Low output current capability.

This converter utilizes a synchronous rectification topology, meaning that there are two mosfets arranged in a half-bridge configuration. This arrangement in a step-down (buck) converter is advantageous because the lower mosfet replaces a rectifier, allowing for much lower ON-state voltage drop (lower power loss) and bidirectional current flow. This latter mention of bidirectional current flow is critical, because it allows a converter to run in *continuous mode*, even down to zero output current. This will be explained in more detail now.

Continuous Conduction Mode (CCM), or Forced PWM (FPWM) in this context, means that inductor current never stays at 0A. Instead, inductor current has a smoothly ramping characteristic, with a controllable (small) peak-to-peak value. Usually, the peak-to-peak ramping current (referred to as ripple current) is a small percentage of the total DC output current capability of the buck converter. The smaller this peak-to-peak ripple current is, the less the potential for noise it can produce. In contrast to continuous mode, discontinuous mode implies that inductor current is discontinuous, having periods where it reaches and remains at 0A. Sometimes this is referred to as the inductor "running dry". Discontinuous mode is characterized as having much more dynamic inductor current waveforms, in the shape of a sawtooth, with high peak-to-average ratios, and usually having more ringing and potential for noise production. Synchronous buck converters allow current to flow in both directions through the output inductor; therefore, even with 0ADC output current, the inductor ripple current can still have a smoothly ramping characteristic, even smoothly ramping through 0A, flowing in both a positive and a negative direction. In a non-synchronous buck converter, the lower mosfet is replaced with a rectifier. This rectifier now blocks reverse current from flowing through the inductor, and the inductor current becomes discontinuous when the DC (average) value of the output current drops below one-half of the peak-to-peak ripple current amplitude. U29 was chosen because it has forced continuous mode operation. This means inductor ripple current will be well controlled and have a consistent peak-to-peak value over the full range of output load currents. It is important that the maximum load current rating for the buck converter exceed the maximum expected load current in operation. However, it is advantageous to choose a buck converter that has a peak load current capability that does not *far exceed* the expected maximum load current. The reason for this concerns how the device will operate in overload. If the buck converter output is short-circuited, for example, the switched currents through the mosfets internal to U29 will rise to the internal overcurrent trip level. This overload current will also flow through the output inductor. It is important that the output inductor be able to handle this overload current level without saturating. If the output inductor saturates, the current will rise much faster than the normal di/dt ramping current through the inductor, and the internal overcurrent protection circuitry in the controller IC may not be fast enough to protect the device in this condition. Therefore, it is best to check the buck controller datasheet for the overcurrent trip threshold (sometimes specified in the buck converter datasheet as peak inductor current or peak switch current). Be sure to select an inductor that can handle this overload current level without saturation. *Note:* Since inductor saturation characteristics vary widely, depending on the type of inductor chosen and the core material, it is reasonable for an inductor with a "soft saturation" characteristic to be driven into partial saturation, of perhaps 30% during this condition. The controller will generally still be well protected because this type of inductor will still have sufficient inductance, even in this partially saturated condition, to slow the rate-of-rise of fault current, such that the controller's overcurrent protection circuitry will react fast enough to protect the buck controller from damage.

Figure 17.8 illustrates the continuous mode circuit operation, showing the small ripple current it produces in the output inductor, as well as how 0ADC output current can be supported in forced CCM operation.

It is worthwhile to note that while the output current of the buck converter has low-ripple continuous current flow, the *input current* waveform of the buck DCDC is discontinuous. This means the 12V housekeeping rail has the potential to be noisy.

Schottky diode D6 is an important addition to the +7VA output. It prevents this rail from being pulled below ground (negative) by the −7VA output, as the two rails turn on. Whenever bipolar power supplies power a load that is strung between positive and negative rails (without a center ground connection point), such as op-amps, it is possible for one of the rails to drag the other rail to the opposite polarity, if it starts to ramp up before the opposite rail. In some cases, this inverse polarity on the opposite rail can prevent its DCDC from starting up, and the reverse polarity on the power rail is most likely in violation of the absolute maximum voltage characteristics listed on the datasheet for many of the components powered by that rail. Therefore, this reverse polarity condition at start-up must be prevented. Schottky diode D7 serves a similar role for the −7VA rail, but has an additional function discussed in the next section.

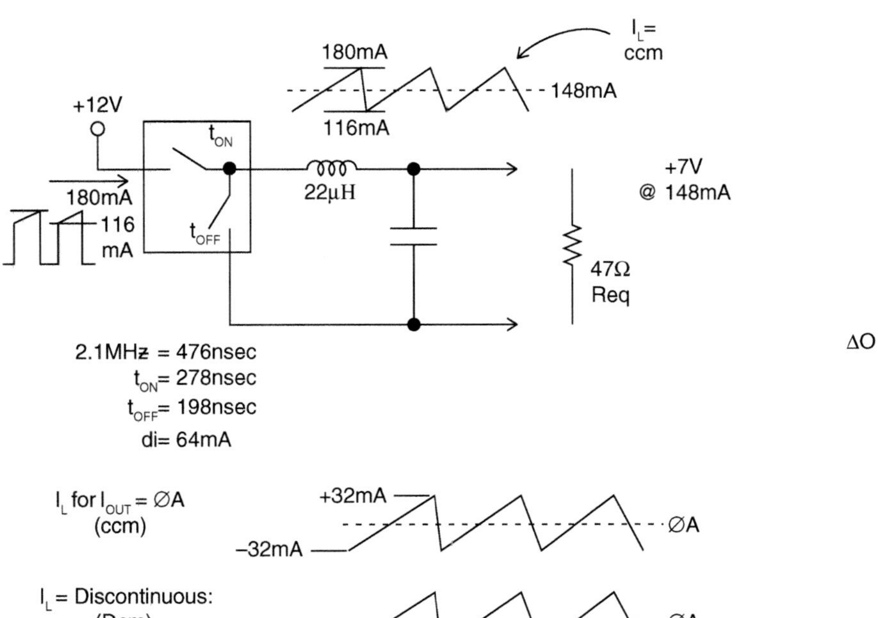

Figure 17.8 Buck DCDC Converter Output

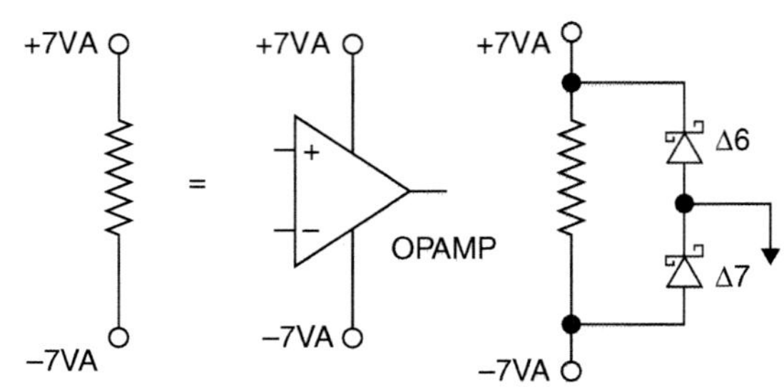

Figure 17.9 Connection and Use of Polarity Protection Diodes in the Reference Amplifier Design

Figure 17.9 illustrates the usage of polarity protection diodes D6 and D7.

−7VA DCDC

The −7VA DCDC at first looks identical to the circuitry that produces the +7VA output, but on careful inspection, it can be seen that the output inductor is connected to ground, and what is normally the return net is now the −7VA output net. This DCDC converter topology is called Inverting Buck-Boost. The circuit topology is still continuous mode. The inductor current is continuous. The waveforms look like a buck converter operating with a 19VDC input and producing a 7V output. The addition of Schottky diode D7 also facilitates start-up of DCDC converter U30. This is illustrated in Figure 17.10. When +12V is first applied to U30, its internal CONTROL circuitry is biased by start-up current flowing in the path depicted. So, at initial startup, the −7VA rail is pulled up as high as +0.5V, where it is clamped by D7. The internal control circuitry of U30 "sees" a start-up bias voltage of (12V − 0.5V) = 11.5V. U30 starts PWM action, and the −7VA rail ramps up, due to current (I_{SF}) flowing from L3 through the Synchronous Mosfet (SF), each time the Synchronous Mosfet switches on. Once the −7VA rail has ramped up to be more negative than −0.5V, D7 will be off, and the −7VA rail will then continue ramping up to −7V, and the input voltage "seen" by U30 will ramp up from 11.5V, at initial start-up, to a final value of (12V + 7V) = 19VDC.

Figure 17.11 illustrates current waveforms in each branch of the +7VA DCDC stepdown continuous mode buck topology. Buck converters are characterized as having relatively low output current ripple at the switching frequency. However, their input current waveforms feature high-amplitude switching currents (also

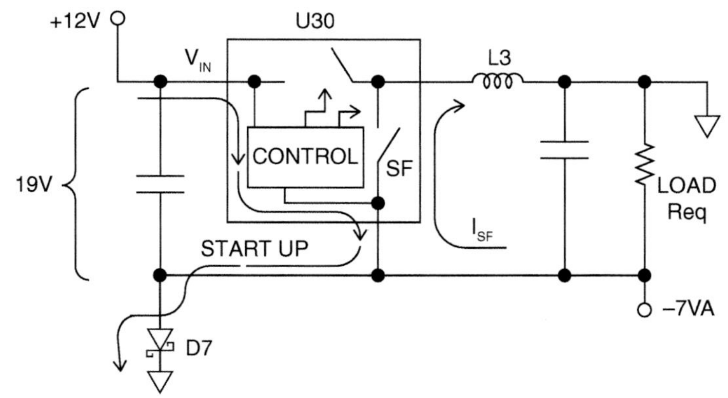

Figure 17.10 Inverting Buck-Boost DCDC Converter Topology that Produces the –7VA Rail in the Reference Design, Illustrating Current Flow during Start-up

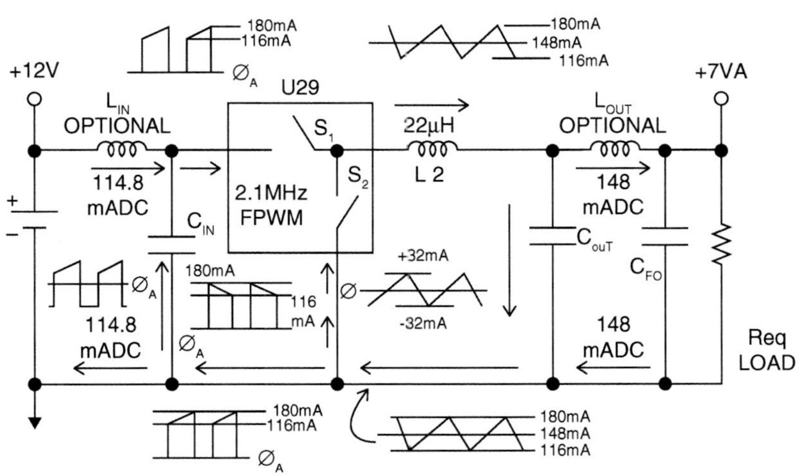

Figure 17.11 Buck DCDC Converter Topology that Produces the +7VA Rail in the Reference Design, Illustrating which Circuit Branches Have Pulsatile Current Flow and which Have Lower Peak-to-Peak Values of Ripple Current

at the switching frequency). The purpose of Figure 17.11 is to illustrate the characteristics of the current in each of the circuit branches in the buck converter. By understanding this, the effects of these switching currents can be minimized. Note particularly the return current branches on either side of U29's ground pin. The branch on the right has triangular ripple current, and the branch on the left has pulsatile switching current. It is advantageous (from a noise reduction standpoint) to stop these currents from flowing in the ground plane of the PCB, where they can create voltage differentials that will affect other circuits.

Figures 17.11, 17.12, and 17.13 illustrate one approach to achieve that objective. When *optional* input (L_{IN}) and output (L_{OUT}) chokes are added, the input and output currents are low-pass filtered, and become essentially DC currents. Next, if the switching-current circuit branches discussed above are shortened to a single point connection (input capacitor return terminal, output capacitor return terminal, and U29 ground terminal), all connected at a single point in the layout, then these switching currents will not circulate in the ground plane. Figure 17.12 illustrates this schematically. Figure 17.13 illustrates one possible way of implementing this scheme in a PCB layout.

The last of the DCDCs to be discussed is U31, which produces the 12V output that is referenced to the −48VDC amplifier rail. Similarly to U29 and U30, U31 operates in forced continuous mode, with its stated noise advantages and continuous inductor current over the full output current load range. One difference in the selection of U31 is that it can operate with an input voltage as high as 75VDC, whereas U29 and U30 are limited to 36VDC input.

All the selected DCDC converters operate at a fixed frequency. This is helpful in that it makes filtering easier, since the noise source frequency is known and not varying with load condition. It also means each DCDC has a "signature" (exact fixed frequency and duty cycle), so that if there are noise or EMI problems, their source can be found via the frequency and duty cycle of the offending noise. This is predicated on the fact that even though different fixed-frequency DCDCs may

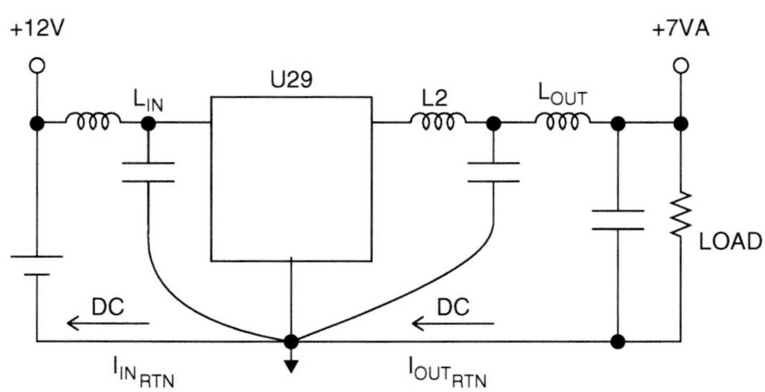

Figure 17.12 Buck DCDC Converter with Optional Input and Output Inductors Added to Locally Contain Switching Currents

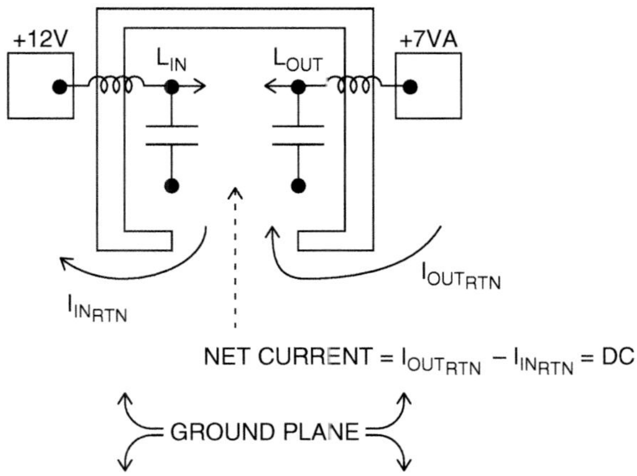

Figure 17.13 One Possible Layout Method to Implement the Switching Current Containment Scheme Illustrated Schematically in Figure 17.12

be resistor-programmed to run at the same switching frequency, they do not use crystal-controlled oscillators, so their actual switching frequencies will be slightly different, and distinguishable during troubleshooting.

LDOs

Even though forced-continuous-mode buck converters were chosen for low-noise DCDC conversion, following these up with LDOs to produce the very low noise rails (+5VA and −5VA) provides even greater noise suppression.

The LDOs chosen for U32 and U34 were selected for their low output noise characteristics, superior power supply rejection ratio (PSRR), and high input voltage rating. U32 produces the +5VA output, whereas U34 produces the −5VA output. Both LDOs operate with 2V (nominal headroom). This amount of voltage headroom is sufficient to allow the LDO circuitry to operate optimally with excellent PSRR, while at the same time not being too excessive and wasting power. LDO power dissipation (Watts) is Iout × (Vin − Vout).

The same low-noise LDO that was chosen for the noise-sensitive +5VA output was chosen for U33, the +5VD output, even though low noise is not a critical requirement for this digital power supply rail. The reasons are that the current demand of the +5VD rail is low, so the same low-current LDO type could be used, simplifying the BOM (Bill of Materials).

Bibliography

Analog Devices. (2019). Datasheet, ADP7182, −28 V, −200 mA, low noise, linear regulator. Rev. M. Wilmington, MA: Analog Devices.

Analog Devices. (2024). Datasheet, ADP7142, 40 V, 200 mA, low noise, CMOS LDO linear regulator. Rev. I. Wilmington, MA: Analog Devices.

EOS. (n.d.). Datasheet, 225 Watt industrial, 4EM-20-320, 39-DE60-46750-002 / D10. Mumbai, India: EOS Power India Pvt Ltd. (Datasheet for 225W/48V/2×4" ACDC power supply.)

Honda, J., & Cerezo, J. (2003, October 8). Class-D audio amplifier design tutorial. El Segundo, CA: International Rectifier.

Maxim Integrated. (2010, April 16). Avoid overvoltage stresses by minimizing power supply pumping on single-ended output, Class D audio amplifiers. Application Note, Tutorial 4260. San Jose, CA: Maxim Integrated Products, Inc.

Maxim Integrated (Analog Devices). (2020). Datasheet, MAX17760, 4.5V to 76V, 300mA, high-efficiency, synchronous step-down DC-DC converter, 19-100795. Rev. 0. 4/20. San Jose, CA: Maxim Integrated Products, Inc.

Mendenhall, E. (2005). SMPS for audio power electronics. *Switching Power Magazine*.

RECOM. (n.d.). ACDC Datasheet RAC04-GB, 4 watt single output EMC Class B. Gmunden, Austria: RECOM Power GmbH. (12V ACDC housekeeping power supply.)

Texas Instruments. (2018). Datasheet, TPS560430 Simple Switcher® 4-V to 36-V, 600-mA synchronous step-down converter, SLVSE22B. Originally published September 2017, revised June 2018. Dallas, TX: Texas Instruments.

Appendix: Current Demand for Reference Design DCDCs and LDOs

This appendix lists detailed current demand for each output, as well as efficiency and power loss for all DCDC converters and LDOs in the low-voltage power distribution circuitry.

Underlined values = must multiply by two for *two-channel design*.

+7V (Total includes +5V_LDO below) 1 CH = 50.73mA, 2 CH = 76.66mA
U1 Preamp = 2.8mA
U2 E/A = 2.8mA
Ramp Gen Mirror = 12mA
U21 Buffer = 4mA
U18 Servo = 2.8mA
Ramp Reset = 6mA
Subtotal: 1 CH = 30.4mA, 2 CH = 36mA

+5V_LDO 1 CH = 20.33mA, 2 CH = 40.66mA
U3 COMP = 10mA
U28 Hall Sensor = 8.33mA
Fault Divider = 2mA

−7V (Total includes −5V_LDO below) 1 CH = 34.4mA, 2 CH = 50mA
U1 Preamp = 2.8mA
U2 E/A = 2.8mA
Ramp Gen Mirror = 12mA
U21 Buffer = 4mA
U18 Servo = 2.8mA
Ramp Reset = 0mA
Subtotal: 1 CH = 24.4mA, 2 CH = 30mA

−5V_LDO 1 CH = 10mA, 2 CH = 20mA
U3 COMP = 10mA

+5V_DIG 1 CH = 64.17mA, 2 CH = 70.97mA
U10 LVC1G57 = C_{PD} = 22pF = 100uA
U11, U12, U13, U14, U15, U16, U17, U25 LVC1G57 = C_{PD} = 22pF = 800uA Tot.
U4 Si8244 = 4mA
Q1/U8 2.5VREF = 3mA
U9 Vdet = 1mA

LV_OK = 20mA
U6, U7 LM339 = 4mA Tot.
Q11, Q12 LATCH = 5mA
U24, U26, U27 LM393 = 2.4mA Tot.
U19, U20 CLK DIST LVC1G17 = C_{PD} = 26pF = 200uA Total
U22 VHC393 = *per counter* = Icc(OPR) = (C_{PD} × Vcc × F_{IN}) + Icc($_{quiescent}$)/2 and C_{PD} = 23pF

Counter 1 = 16MHz_{in} = 1.86mA
Counter 2 = 8MHz_{in} = 940uA
Counter 3 = 4MHz_{in} = 480uA
Counter 4 = 2MHz_{in} = 250uA
Counter 5 = 1MHz_{in} = 135uA
Total = 3.67mA

U23 OSC = 20mA.

−HV+12V 1 CH = 43.2mA, 2 CH = 86.4mA
U4 Si8244 = Using power dissipation equation for DirectFet IRF6775:

Pd = Qg × Vg × Freq, where Qg (IRF6775) = 25nC Typ., 36nC Max.
Pd = 216mW/FET = 18mA/FET (from +12V)
Active supply current for Si8244 (at 500KHz) = 3.6mA Typ.

Chapter Eighteen

70V and 100V Constant-Voltage Amplifiers (High Impedance)

Definition and Nomenclature

Constant-voltage amplifiers are *not* limited to use of Class-D topologies. They can be constructed using other amplifier topologies, such as Class-AB. The reason that a chapter is devoted to them in this book on Class-D audio amplifier design is that the Class-D topology provides some benefits for constant-voltage operation that other amplifier classes lack. The information in this chapter will be limited to the particular aspects of constant-voltage amplifier design that are unique to Class-D design.

Unfortunately, the term "constant voltage" has caused much confusion for this class of amplifiers. The output voltage of these amplifiers is not constant. It is an analog signal of continuously varying amplitude, and, in the case of an audio amplifier, is a varying analog audio signal with dynamic range (i.e. decidedly *not* constant). What the term is intended to mean is that, if set up properly, the constant-voltage amplifier will produce a peak output voltage signal conforming to one of the constant-voltage standards. The principal standards in use today are 70.7V for the USA (often abbreviated to 70V) and 100V for Europe. This means a 100W rated 70V constant-voltage amplifier will produce 70.7VRMS (+/−100Vpeak) at 100W. This also means that it will produce a peak output current of 100W/70.7VRMS = 1.414ARMS = +/−2.00Apeak when driving a load impedance of 70.7V/1.414ARMS = 49.98 ohms.

Similarly, a 100V constant-voltage amplifier rated at 100W output power will produce 100VRMS (+/−141.4Vpeak) at 100W. This also means that it will produce

DOI: 10.4324/9781003327943-18

a peak output current of 100W/100VRMS = 1.00ARMS (+/−1.414Apeak) when driving a load impedance of 100V/1.00ARMS = 100 ohms.

Typical electromagnetic loudspeakers are usually designed to have a nominal impedance of between 4 and 8 ohms. Since this is a relatively low impedance, amplifiers designed to drive such loudspeakers directly are termed low-impedance amplifiers, or "Lo-Z". Conversely, amplifiers designed to drive constant-voltage loads are termed high impedance or "Hi-Z", because as noted above, the loads driven by these amplifiers, especially lower-power amplifiers, have impedances significantly higher than 4 or 8 ohms.

Before we leave this introductory section, it will be helpful to describe the voltage/current/power operating conditions for constant-voltage amplifiers in a little greater detail, to get a better understanding of their use conditions (Bohn, 1997; Giddings, 1990; Crown Engineering Staff, 2005).

Constant-voltage amplifiers were born out of the need for efficient power distribution over distance. This is key to understanding their operation. The problem they address is the same electrical power distribution problem addressed by electrical power utilities in the distribution of AC power over long distances, from the power generation facility (hydroelectric, coal, or nuclear power station) to the end user (a home or business). Whenever electrical power is distributed over long distances, there will be distribution losses in the connecting cables due to I^2R heating losses in the distribution cables. In order to minimize these distribution losses, the choices are *either* to decrease the resistance of the distribution cables (R), by increasing the diameter of the copper conductors, which increases cable weight and cost, *or* to decrease the current (I). Here it is important to recognize that power is the product of current times voltage (I × V), and 100W of power could be distributed in any number of combinations of I × V. For example:

100W = 100V × 1A

100W = 10V × 10A

100W = 1V × 100A

Obviously, the lowest-current approach would incur the lowest I^2R power losses in the distribution cabling.

In the area of AC power distribution for power generation stations, the distribution voltage can be as high as 1100VAC. With such a high distribution voltage, each amp of current carries 1100W of power.

Constant-voltage amplifiers are used to power distributed speakers that may be arranged at great distances from the amplifier. For example, these could be multiple ceiling speakers in a large conference room, or hallways of a hotel, or many loudspeakers arranged throughout a large stadium or racetrack.

In these application examples, the 70V constant-voltage amplifier is set up such that it will produce 70.7VRMS when driven by the peak input signal, and will produce its rated output power under that condition. At this point, it is important to understand that the loudspeakers themselves have the same low-impedance voice coils as loudspeakers driven by low-impedance amplifiers (4–8 ohms); the *difference* is that there is a *step-down* transformer usually associated with the 70V loudspeaker. This transformer will step-down the 70V distributed audio signal to a voltage appropriate for the speaker's voice coil. Since constant-voltage applications are essentially *power distribution* systems, architects of these systems deal in terms of power. 70V speakers are selected based on their power ratings, and usually come with adjustable *taps*, for selecting a power level (and therefore a volume level). Suppose for example, that a system requires twenty 10W ceiling speakers for a conference room and two 25W speakers at the entrance foyer. The designer adds up the total power (250W), and then adds some headroom. Bohn (1997) describes a "very seasoned rule of thumb" of 1.5 times the calculated power, to account for headroom. So in this case, a constant-voltage 70V amplifier rated for at least 375W would be selected.

It should be recognized that the load impedance driven by constant-voltage amplifiers can have a very wide range. For example, for the case above, if all the speakers were connected, and the power tap in each speaker were set for the power ratings mentioned above, the amplifier would need to deliver 250W at 70.7VRMS. The load impedance (Z) "seen" by the amplifier in this case would be:

$$Z = V^2/P = 70.7^2/250\text{W} = 20 \text{ ohms}$$

However, suppose only a *single* 10W ceiling speaker were connected. This might be the case when the system is being installed and tested, for example. In that case, the amplifier sees a load impedance of:

$$Z = 70.7^2/10\text{W} = 499.8 \text{ ohms}$$

This vast difference in load impedance is mentioned here, because it demonstrates that a Class-D amplifier employed in a Hi-Z, constant-voltage application must be stable (from a feedback compensation standpoint) and meet audio performance specifications (frequency response flatness etc.) over this wide range of load impedance, basically, from rated power load impedance, to no load ($Z = \infty$).

Note: Readers may recognize that low-impedance amplifiers also need to be stable at no load ($Z = \infty$), as it is very common for a low-impedance amplifier to be running with speakers disconnected, either intentionally or not. For example, a multi-channel amplifier might be running with all of its output amplifier channels enabled, but with speakers only connected to two outputs. The subtle difference here, however, is that the Lo-Z amplifier's audio performance is only tested over the range of (low) load impedances for which it was designed (for example, 4- or 8-ohm loads, or at most 2- to 16-ohm loads). Its audio performance at no-load is rarely tested, as that is not a relevant operating condition for a Lo-Z amplifier. In contrast, a Hi-Z constant-voltage amplifier could be called on to operate with a single speaker (as mentioned above) and therefore a relatively high impedance load, and will still need to meet its rated audio performance specifications under that loading condition.

Constant-Voltage Amplifiers Employing Output Step-up Transformers

One of the ways that a constant-voltage audio distribution system can be realized is by using step-up transformers at the amplifier output. When step-up transformers are used, a conventional Lo-Z amplifier can drive 70V or 100V loudspeakers by using an output transformer to step-up the low-voltage output to the 70V or 100V constant-voltage level. This voltage is then distributed the required distances to the 70V or 100V speakers, where it is again stepped-down to the appropriate voltage for the speaker's voice coil. Lo-Z Class-D and Class-AB amplifiers can both be employed in this way. However, this scheme suffers from insertion loss at *both* ends of the distribution system (i.e. in both the step-up and step-down transformers). Insertion loss is the term used to describe the power dissipated or I^2R losses in the internal transformer windings (Bohn, 1997), as well as core losses in these transformers. There are other disadvantages to using step-up output transformers, which will be discussed. These are weight, size (volume), cost, load regulation, and frequency response (range and flatness).

Transformers must be designed to support a Voltage × Time product, without saturation. When an AC voltage is applied across the primary winding of an ideal transformer, the current that flows is simply the secondary winding current (load current) times the turns ratio of the transformer (Number of Secondary Turns / Number of Primary Turns). However, a real transformer has a non-ideal term which is the magnetizing current. This is current that flows in the primary winding, that is not reflected in the secondary current. The amplitude of this magnetizing current is inversely proportional to the primary inductance of the transformer primary winding, and is proportional to the voltage applied across the primary winding *and* the time interval over which such voltage is applied (thus the standard inductance relationship):

$$di = (V_L)(dt)/L \qquad \text{EQ. 18.1}$$

In order to keep magnetizing current low, so that it is negligible compared to the reflected load current (secondary winding current), either primary inductance must be *increased*, or the time that the voltage is applied to the primary winding must be *decreased*. In addition to this relationship, there is also a finite maximum Voltage × Time product which the primary inductance can support without saturating. Core saturation is characterized by a sharp decrease in permeability of the core material, and therefore an associated sharp increase in primary magnetizing current. It is to be avoided. In order to increase the V × T product that can be supported by the primary winding, while keeping primary magnetizing current low, and without core saturation, it is necessary to increase primary winding inductance. This is accomplished by increasing transformer core size and permeability, and increasing the number of primary turns. All of these steps *increase* the physical size and weight of the transformer.

In an audio application, the *time* constraint is fixed by the desired low-frequency response of the system. For example, if low-frequency response extending down to 20Hz is desired, then the primary winding voltage will be applied for half a cycle at 20Hz, or $T_{20Hz}/2$ = 25msec. If this results in too large a transformer core, to accommodate the V × T product at 20Hz, one common approach used in constant-voltage systems is to limit the low-frequency response by the use of a *high-pass filter.* Typically such filters have a 2-pole response (12dB/octave) and corner frequencies of between 60 and 80Hz. The use of such a filter can significantly reduce the size of the step-up transformer in constant-voltage systems.

Another characteristic of step-up transformers is *load regulation.* This is defined as the change in output voltage with load current, expressed as a percentage. Typically transformers driven with a fixed primary voltage will produce a defined output voltage at NO-LOAD, and a reduced output voltage at full load. The difference can be as high as 25% or 30%. This is significant, and must be taken into account when designing such transformer-coupled systems.

Frequency response can also vary over the audible range. In addition to the low-frequency rolloff described above, there is typically high-frequency rolloff as well, due to capacitances in the transformer windings.

Although the disadvantages of step-up transformers have been emphasized here, there is one important characteristic of transformer-coupled systems that is desirable, and sometimes required, and that is *isolation*. An output step-up transformer provides a galvanically isolated high-voltage output, which is sometimes required in certain countries and in certain installations.

Direct Drive

A more efficient method of generating the Hi-Z output voltage is by direct drive. This means using a constant-voltage amplifier purpose-designed, with an output stage capable of directly producing the 70.7V or 100VRMS output voltage. It is here that Class-D amplifiers shine, especially when compared to Class-AB amplifiers.

The efficiency of Class-AB amplifiers is highly dependent on the difference between the supply rail voltages and the output voltage, since the power loss in the amplifier is mainly due to the product of the output current times the voltage drop across the power output transistors. In contrast, Class-D amplifiers have much less of an efficiency dependency on this relationship. In fact, increasing the rail voltages for Hi-Z operation will only increase the switching loss component of power dissipation, and as discussed in other chapters this can be optimized through various means at the designer's disposal. Capacitive switching losses will also increase in the Class-D design, due to the high voltage swings, so the designer must be diligent in power switch selection. Gate charge and C_{OSS} (energy and time related) of candidate power switches must be carefully reviewed.

Because Class-D amplifier efficiency is less dependent on supply rail voltages (as compared to Class-AB amplifiers), it is possible to design Class-D amplifiers that can directly drive *either* Hi-Z or Lo-Z loads without the use of step-up transformers. This type of amplifier output is extremely versatile and popular among commercial amplifier applications. This type of amplifier provides great flexibility for installers of commercial audio systems. For example, a multi-channel version of this amplifier can be configured such that some of its channels power Hi-Z ceiling speakers (in a conference room, for example) while two of its channels power Lo-Z speakers on either side of a podium or dais. An alternate type of multi-channel commercial amplifier that is popular addresses this need by having a fixed number of dedicated Lo-Z and Hi-Z channels. Though this can meet the same need, it is less flexible than a multi-channel amplifier that has all its channels capable of driving *either* Lo-Z or Hi-Z loads.

Output Stage Configurations for Hi-Z and Full-Range Loads using Direct Drive

Figure 18.1 shows a typical *half-bridge* arrangement for a Class-D output stage. In order to drive Hi-Z outputs, the power stage must be able to produce the output voltages in Table 18.1, and the power switches selected (mosfet, GaN, etc.) must be capable of withstanding the operational voltages listed.

The half-bridge power stage in Figure 18.1 has the advantage of low parts count, but has the disadvantages of requiring very high voltage power switches (mosfet, GaN, etc.), and suffering from supply rail pumping under high audio output power and low audio frequency conditions (discussed in Chapter 17).

Figure 18.2 shows typical *full-bridge* arrangements for Class-D output stages. In order to drive Hi-Z outputs, the power stage must be able to produce the output voltages in Table 18.2, and the power switches selected (mosfet, GaN, etc.) must be capable of withstanding the operational voltages listed.

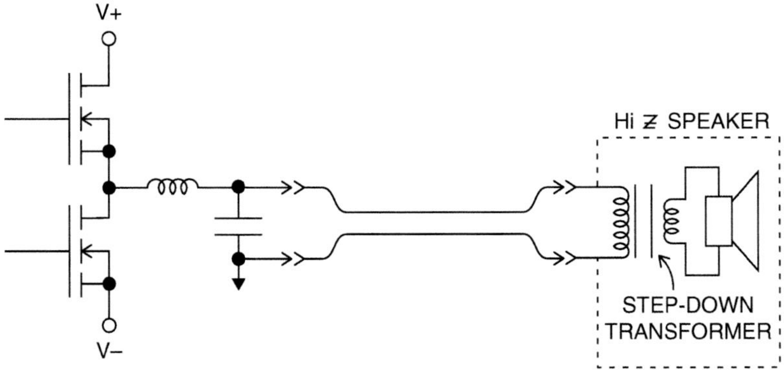

Figure 18.1 Constant-Voltage Amplifier Output Consisting of a Half-Bridge Class-D Power Stage Driving a Hi-Z Speaker that Incorporates a Step-down Transformer

Table 18.1 Peak Voltages for a Half-Bridge Class-D Power Stage Directly Driving 70V and 100V Hi-Z Loads

Hi-Z Standard	*Peak Output Voltage*	*Power Supply Rail Voltages*[a]	*Operational Power Switch Peak Voltage*	*Minimum Power Switch Voltage Rating*[b]
70.7V	+/−100.0V	+/−111.1V	222.1Vpeak	300V
100V	+/−141.4V	+/−157.1V	314.2Vpeak	400V

a Assumes maximum PWM duty cycle attainable is 95%.

b Rule of thumb used is that maximum operational voltage does not exceed 80% of power switch voltage rating. Designer must be cognizant of peak voltages (V_{DS}-peak) generated in the design due to ringing and overshoot. Standard available mosfet voltage ratings are used.

Table 18.2 Peak Voltages for a Full-Bridge Class-D Power Stage Directly Driving 70V and 100V Hi-Z Loads, Using Split (Bipolar) DC Power Supply Rails

Hi-Z Standard	*Peak Output Voltage*	*Power Supply Rail Voltages*[a]	*Operational Power Switch Peak Voltage*	*Minimum Power Switch Voltage Rating*[b]
70.7V	+/−100.0V	+/−55.55	111.1V	150V
100V	+/−141.4V	+/−78.55V	157.1V	200V

a Assumes maximum PWM duty cycle attainable is 95%.

b Rule of thumb used is that maximum operational voltage does not exceed 80% of power switch voltage rating. Designer must be cognizant of peak voltages (V_{DS}-peak) generated in the design due to ringing and overshoot. Standard available mosfet voltage ratings are used.

The power switching stage on the right of Figure 18.2 uses a single supply rail. This eases power supply design or selection. It must be noted that this stage presents a *quiescent* (IDLE) voltage of one-half of the supply rail voltage, at both output (speaker) terminals, when the amplifier is switching at 50% duty cycle (no output audio signal) condition. This must be reviewed in light of any applicable safety standards which the amplifier must meet for a given application.

This single-rail, full-bridge (BTL) configuration has the advantage that it is not subject to supply rail pumping under low-frequency, high-output-power conditions. It has the disadvantage that it does require level-shifting of the signals in the feedback audio path, since the output voltage signal is offset by one-half of the supply voltage.

The stage on the left of Figure 18.2 uses split, or bipolar, power supply rails (positive rail and negative rail). This increases the complexity and cost of the power supply design or selection, but has the advantage of producing a 0V

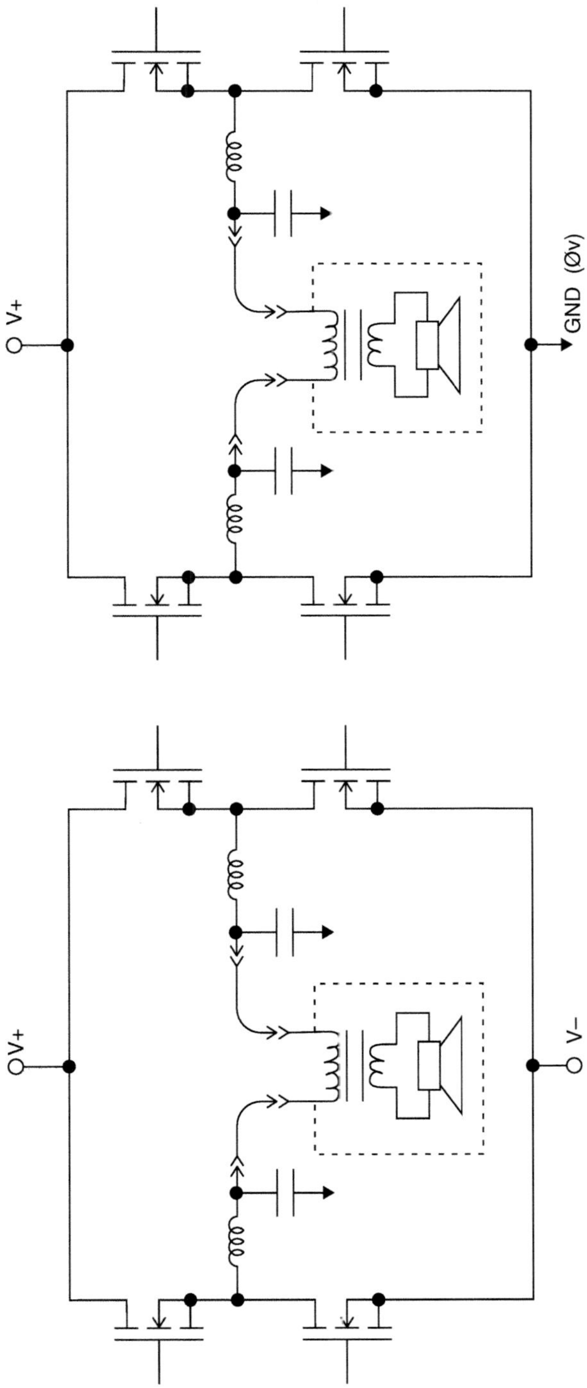

Figure 18.2 Two Examples of Full-Bridge Class-D Power Stages Driving a Hi-Z Load: One Stage is Powered by a Single DC Power Supply; the Second is Powered by Positive and Negative Polarity (Split) DC Power Supplies

quiescent (idle) voltage at both speaker terminals when the power stage switches are switching at 50% duty cycle. Therefore, the audio signal can be ground-referenced (0V-referenced) throughout the entire audio signal path, including the feedback path. It also has the advantage of not incurring supply rail pumping, because power is drawn from *both* rails during *both* polarities of audio output signal.

When deciding on a power stage configuration for a given application, there are many things to consider. These include the complexity and cost of the power supply, as well as the requirements (ratings) of the power switches and L-C filtering components.

For the power switches, a higher V_{DS} voltage rating (in the case of mosfets) is usually associated with increased $R_{DS}ON$. It is almost always beneficial from a cost and performance standpoint to use the lowest-voltage mosfet that can safely be used in the application.

It must also be noted that the full-bridge configurations allow for a reduced voltage rating for the components selected for the L-C filtering and EMI suppression at the output. For example, with supply rails of +/−78.55V, capacitors and transient voltage suppressor (TVS) devices having 100V ratings can be selected. This must be weighed against the requirements of a *half-bridge* design, where higher voltage ratings for these parts are required, but half as many parts are required.

See Chapter 8 on Class-D output configurations (half-bridge and full-bridge).

Overcurrent Protection for Direct-Drive Hi-Z/Lo-Z Outputs

Class-D outputs that are capable of directly driving Lo-Z or Hi-Z loads present a problem regarding overcurrent protection. The following example will illustrate this.

A 100W amplifier, driving a 100V constant-voltage load, only needs to produce 1ARMS of output current (+/−1.414Apeak).

This same amplifier driving a 4-ohm (Lo-Z) load needs to produce 5ARMS (+/−7.07Apeak).

If effective overcurrent protection is to be applied, and a peak overcurrent trip threshold of >7.07Apeak is selected, this could result in a maximum output power of 100V × 5ARMS = 500W, if the amplifier produced its full output voltage of 100VRMS at its rated output current of 5ARMS. This is a severe over-power condition.

Some methods of addressing this issue are to incorporate selectable overcurrent protection based on the driven load impedance, or output power limiting. The first can be user selectable, or automatic, if load impedance is sensed (monitored). The second can also be automatic if load impedance is sensed (monitored). Both of these approaches require that *both* output voltage and current be monitored. Another approach which can be effective is to sense power supply rail current. Since the power supply rails are a constant DC voltage, the current demand on a rail will be fixed for a given level of output power.

Selectable or Variable Supply Rail Voltages for Hi-Z/Lo-Z Amplifiers

Although Class-D amplifiers can operate much more efficiently than Class-AB amplifiers when the DC power supply rails are varied over a wide range (power supply rail insensitivity), it is still advantageous from a switching loss standpoint to operate with the lowest possible rail voltages sufficient for producing the required maximum output voltage. To this end, some Class-D designs utilize power supplies with multiple supply rails, or variable or switchable rail voltage. These rails can be variable or switched, on a per-amplifier-channel basis, or for all channels simultaneously, depending on the complexity of the power supply design. When an amplifier channel is configured as a Lo-Z channel, it can utilize lower-voltage power supply rails. If that channel is switched into Hi-Z mode, the power supply rail voltages will also switch to the required higher rail voltage level. In this manner, the amplifier can always operate at maximum efficiency.

Bibliography

Bohn, D. (1997). Constant-voltage audio distribution systems: 25, 70.7 & 100 volts. RaneNote 136. Mukilteo, WA: Rane Corporation.

Brown, R. E. (2009). *Electric Power Distribution Reliability* (2nd ed., p. 5). Boca Raton, FL: CRC Press.

Crown Engineering Staff. (2005, October). Guide to constant-voltage systems. Los Angeles, CA: Crown Audio. (Includes useful system design equations, as well as an interesting history of constant-voltage audio distribution systems.)

Davis, D., & Davis, C. (1987). *Sound System Engineering* (2nd ed.). Carmel, IN: Sams, 1987.

Giddings, P. (1990). *Audio System Design and Installation*. Carmel, IN: Sams. (Excellent reference on constant-voltage systems.)

Infineon. (2018). User guide, IRAUDAMP23, 500W X 2 channel constant voltage Class D audio amplifier using the IRS2452AM. Munich, Germany: Infineon.
Infineon. (n.d.). Audio amplifier solutions. Munich, Germany: Infineon. https://www.infineon.com/cms/en/applications/consumer-electronics/home-entertainment/audio-amplifier/professional-audio/#!designsupport

Chapter Nineteen

Multi-Channel Designs

Interactions between Channels in a Multi-Channel Amplifier Design

When more than one channel of Class-D amplification is employed on a single printed circuit board (PCB), an entirely new set of design considerations and concerns is presented. Most of these are due to the interaction that occurs between channels. This interaction occurs both directly and indirectly. An example of direct interaction is when noise created by the activity of one channel (switching noise, for example) directly affects the operation of the second channel. An example of indirect interaction is the interaction that occurs when one channel's demand on the power supply rails causes an effect on these rails (such as increased voltage ripple). Then that increased power supply rail voltage ripple affects the performance of the second channel, in consequence of its finite power supply rejection ratio (PSRR).

Most interaction effects between channels in a multi-channel amplifier design fall under the general category of "crosstalk". Simply stated, crosstalk means any signal induced into one channel by the operation of another channel. Some types of crosstalk and their coupling mechanisms are listed below:

1) Power supply induced crosstalk
2) Capacitive coupling
3) Electromagnetic coupling due to induced currents in the PCB
4) Coupling due to leakage fields from inductors

DOI: 10.4324/9781003327943-19

5) I × R induced voltage disturbances (common impedance coupling)
6) Within a "multi-gate" component.

This coupling can occur at various points in the signal path of the Class-D amplifier, from the input section, all the way to speaker output connections. It is very important to scrutinize all the potential coupling mechanisms that can occur in a multi-channel design during the design process. It is difficult to find and mitigate crosstalk in a finished design. It is much preferred to anticipate crosstalk mechanisms and eliminate them as early as possible in the design process. Some of these mitigation techniques are discussed in other sections of this book, such as in the chapter on PCB Layout (Chapter 16).

Power Supply Induced Crosstalk

When a single power supply is used to power multiple Class-D audio amplifier channels, there is the opportunity for crosstalk to be imparted through the common power supply connections.

Via Common High-Voltage DC Power Supply Rails

Taking the reference amplifier design as an example, it is possible to power multiple Class-D audio amplifier channels from the same positive and negative DC power supply rails. The reference design utilizes a half-bridge Class-D power circuit topology. For the purpose of this crosstalk discussion, we will refer to the active channel, whose activity is producing the crosstalk perturbation, as the "aggressor", and the channel(s) which are affected by the crosstalk perturbation, as the "victim" channel(s). The aggressor channel draws current from the positive and negative rails, as the dynamic output audio power demands of the channel change. These dynamic current demands will cause deviations in the output voltage of each power supply rail. "Load regulation" is a power supply specification that describes the change in output voltage with load current. An ideal power supply should function as an ideal voltage source. This is defined as a source whose output voltage does not change at all, as the output current changes from 0A up to the rated output current limit of the power supply.

Note: In other sections of this book, it is shown that the power supply rails supplying the half-bridge power circuits should ideally not only source current, but also have the capability to sink current (bidirectional output current capability), to avoid power supply "rail pumping". Most often this function is provided (to some extent) by the large value of rail capacitance. However, it can also be provided by power supply topologies that can both source and sink current. Synchronous rectification is one circuit technique that can allow a power supply to not only source, but also sink, some amount of output current.

However, all practical power supplies fall short of this ideal voltage source performance, and therefore the power supply output voltages *do* change with current demand. Therefore, when activity of the aggressor channel produces power supply rail voltage fluctuations on a shared (common) power supply rail, this voltage disturbance will be impressed on the half-bridge circuitry of the victim channel. The output voltage disturbance seen in the victim channel due to this crosstalk path will be determined by the PSRR of the victim channel. The half-bridge Class-D amplifier topology has inherently *zero* power supply rejection without the use of negative feedback. Therefore, in an open-loop (no feedback) design, this crosstalk path will be unattenuated and produce significant crosstalk in the victim channel(s). For Class-D amplifier designs employing feedback, the output perturbations caused by this crosstalk path will be attenuated by the feedback loop gain. The following relationships are important to understand. Typically, there are large-value capacitors used as charge reservoirs on each of the rails used to power the power switching stage of the Class-D amplifier. Peak-to-peak voltage ripple on these rails will increase with output power, and decrease with signal frequency. Unfortunately, audio output power demand typically increases with lower audio frequency content (bass), and therefore exacerbates the voltage ripple on the power supply rails. However, in a stable feedback design, the feedback loop gain is also highest at low frequency, providing the most correction and crosstalk-induced-error mitigation. Feedback loop gain decreases as frequency increases, but so does the voltage ripple, due to the rail capacitors. So, to determine the amount of crosstalk, it is important to know the peak-to-peak magnitude and frequency of the induced voltage ripple, as well as the feedback loop gain at that frequency. For example, if the peak-to-peak ripple magnitude is 1V at 1000Hz, and the feedback loop gain at 1000Hz is 60dB, then the 1V ripple will be reduced to 1mV at the victim channel's output.

Note: See also Chapter 17 for a detailed discussion of wiring of audio amplifier channel-pairs 180 degrees out of phase, to reduce supply pumping.

Capacitive Coupling

Capacitive coupling is a crosstalk mechanism whereby noise is transmitted from the aggressor channel to the victim channel by a discrete or parasitic capacitance.

A potential capacitive coupling path for crosstalk is through parasitic capacitances in a PCB layout, or in multi-circuit connectors and wire harnesses. In these cases, trace separation is the easiest fix. Always be aware of separation both laterally, on a single PCB layer, and from layer to layer on the PCB. When traces are on top of each other in a multi-layer PCB, there will be capacitance through the PCB material, determined by the thickness of the separating layers and the dielectric constant of those insulating layers.

Examine the PCB for copper traces or planes that experience high dv/dt. The circuit nodes with the highest dv/dt and highest switching voltage amplitudes are nodes associated with the Class-D switching stage. These include the switching node, as well as the gate drive circuitry of the upper mosfet. Remember that the switching node extends to the bootstrap rectifier as well as the output inductor and Zobel network components. The output inductor has significant surface area, and so can generate a strong E-field due to high dv/dt (at least at the terminal and associated winding turns attached to the switching node). Again, distance and separation between circuits, and away from these high dv/dt nodes, is the easiest and most economical tool to use for mitigating crosstalk. The other technique to use is shielding. High dv/dt nodes can be shielded. The shield should return capacitive currents to the circuit or channel in which they were generated (usually to that channel's ground return plane).

A diagnostic signature of a capacitive crosstalk path is that it will show crosstalk *increasing* with signal frequency (due to a capacitor's decreasing impedance with frequency). Another diagnostic characteristic is that capacitive crosstalk will be sensitive to signal voltage amplitude, and insensitive to current level. So, for example, if the amount of crosstalk stays the same when the aggressor channel is loaded or unloaded, then the crosstalk path is likely to be capacitive.

Electromagnetic Coupling

This type of crosstalk mechanism is induced through currents flowing in PCB traces, inductive components, connections, and wire harnesses. When a current flows through any conductor, a magnetic field is created, with orientation described by the "right-hand-rule". If another trace is close by, currents will be induced in that trace. The diagnostic signature for uncovering this crosstalk mechanism is that crosstalk will be seen to increase with the output current of the aggressor channel. The solution for this type of crosstalk is, again, trace separation, and/or careful analysis of current loops. It is always beneficial to keep current loops small. This is important for both the aggressor and the victim (emitter and receiver) loops. If the source and return currents in a loop are kept as close as possible (and/or twisted, in the case of wires), the smallest possible loop is achieved. This advocates for very tight PCB layout (within a Class-D channel), with adequate separation between channels.

Coupling Due to Leakage Fields from Inductors

An ideal inductor will have no external magnetic leakage field. However, all practical real-world inductors do, and the strength of the leakage field is very much determined by the construction of the Class-D inductor and the materials used. For example, gapped ferrite cores are used in many styles of Class-D inductor. Typical ferrite core shapes for this style of inductor are PQ and RM, which are easily PCB-mounted and come in a variety of sizes. These core shapes have an intentional *air gap* which determines the A_L value of the core, and thus the value of the wound inductance. However, this air gap creates a fringing flux which emanates from the structure. The strength of the magnetic field (B-field) decreases rapidly with distance, which again argues for separation between circuits as a common mitigation strategy. This stray leakage magnetic field can be picked up by trace loops (described above) and appear as crosstalk. A signature of this type of crosstalk mechanism is that it will increase with output current and power, since both the operating magnetic flux density of the ferrite inductor will increase with output power, and so will the associated leakage field. Mitigation techniques for this type of inductor are to use a flux band (sometimes called a "belly-band") around the ferrite structure. This is a single turn of copper sheet material wrapped around the outside of the ferrite core. It serves to create a shorted turn to the external leakage flux, thus causing this external flux to couple to it and induce a current locally in this shorting band, instead of radiating out away from the inductor, where the stray B-field can induce crosstalk. Another idea is to use multiple smaller gaps in the ferrite structure, instead of a single large gap. These smaller gaps each produce a smaller-amplitude fringing flux. Toroidal

Class-D inductors are another style of inductor commonly used. Toroidal cores are commonly understood to have very low magnetic flux leakage. This is especially true when compared to gapped ferrite inductors. However, these toroids typically have very low permeabilities. This means the ratio of their permeability to magnetic flux, compared to that of air, is low, so they do not contain the magnetic flux as well as a higher-permeability material. Thus, they do produce some external leakage flux. That said, there are some toroidal Class-D ferrite inductors that do employ a discrete gap. These are made from high-permeability ferrites, but the physical gap produces the same fringing flux and external leakage flux problems discussed above for the PQ and RM core shapes. When a toroidal core is used without a gap, its low permeability is achieved through the use of a "distributed-gap" material. This material is made up of a mixture of magnetic and filler materials in precise combinations to achieve the desired permeability, as well as other material characteristics like saturation flux density and hysteresis power loss. Due to the distributed-gap nature of the material, these toroids produce a leakage flux field as well, but it is much lower than the external field produced by a single large gap. In summary, it is always important not to ignore this leakage flux, and to separate Class-D inductors from multiple channels, or orient them alternately at right angles (Adire Audio, n.d.) (90 degrees), to reduce this potential crosstalk mechanism.

The leakage flux produced by the Class-D inductor can induce currents in nearby traces or even in a nearby inductor. One style of Class-D ferrite inductor uses a dual-inductor construction (Coilcraft, 2018). This is convenient when a bridged-output topology is used, and two inductors are required. In that case both inductors are used in the same Class-D channel, so crosstalk is not an issue. However, this style of inductor *can* be used successfully in a multi-channel design, because the two gapped ferrite inductors present two separate flux paths in the dual-inductor structure. Coilcraft (2018) describes a low-crosstalk implementation of such a dual-inductor. However, in general, inductors in close proximity should always be examined for their potential for crosstalk.

I × R Induced Voltage Disturbances (Common Impedance Coupling)

This is the most common crosstalk mechanism, and is caused by currents from different channels causing voltage drops in traces used by other channels. It is specifically to prevent this type of coupling that "star grounding" is so often recommended as a layout technique. Star grounding allows the ground return path for each circuit to be independent. This is best explained by illustration. Figure 16.7 (in Chapter 16 on layout) depicts a scenario where a common impedance connection is present. This is a ground return connection shared by Channel 1 and Channel 2, which has an impedance of 10 mohm (10 milliohm). In the example, Channel 1 is the victim channel, and is at idle, producing 0V output. The output power of the aggressor channel (Channel 2) causes a bidirectional current of +/−1A in the common impedance connection. This causes a voltage disturbance of +/−10mV at the local ground reference for CH1 and CH2. The preamps are referenced to the signal ground plane, but their output signal is received at each Class-D channel with respect to its local ground reference, which is modulating at +/−10mV. To make matters worse, each Class-D modulator and power stage has an effective gain of 10V/V. Therefore, the voltage received from the preamps is amplified by 10× at the output of the Class-D amplifier channels. This means that Channel 1, which is at idle and expected to produce 0V output, will instead produce +/−100mV output, solely due to crosstalk induced by Channel 2.

The lower section of Figure 16.7 (Chapter 16 on PCB layout) illustrates the effectiveness of a star grounding scheme. This example also includes separate signal and power ground connections, which is another important PCB layout technique. In this example, the aggressor channel (2) still produces the same +/−1A return current. But instead of this current flowing through a common impedance connection to the "0V" return point, it flows through a dedicated independent connection. The same +/−10mV disturbance is created at the local ground reference of Channel 2 with respect to the 0V reference, but this disturbance has *no effect* on Channel 1, which continues to produce 0V, unaffected by the activity in Channel 2. The voltage disturbance caused in Channel 2's ground connection *only* affects the output voltage of Channel 2, at a 1:1 amplification factor (instead of 10:1), and can be reduced by feedback loop correction.

It should also be noted that the signal ground plane allows all the low-level, low-power circuits to share a common ground plane. Since the currents produced by these small-signal circuits are so low, they do not induce significant noise voltages in the common signal ground plane. Therefore, the signals produced by the preamps are received by the local receivers in each channel referenced to the same 0V signal ground plane.

Crosstalk in a Multi-Gate Component

Note: In this section, the term "gate" is used in the generic sense, to mean a single functional unit, such as a single op-amp or comparator. In some applications, it may be convenient and/or economical to use multi-gate components. These can be op-amps or comparators, which are typically available as singles, duals, and quads, multi-gate logic ICs, and gate drivers. The downside to the use of these multi-gate components is that they present another avenue for crosstalk between channels, that must be carefully analyzed and considered. Op-amps are typically well specified for crosstalk between multiple op-amps in the same IC package. However, other devices, such as comparators, are not. In fixed-frequency designs (such as the reference design) that use a triangular ramp signal, it is recommended to use a single high-speed PWM comparator per channel, instead of attempting to share a dual comparator between two channels. The reasons for this are:

1) It will allow for a tighter, more optimal layout of the individual channels, with the shortest possible trace lengths.
2) The comparator input is one of the most noise-sensitive areas of the design, and care must be taken to reduce noise and other possible causes of jitter in that area. Having a second comparator in the same package will inevitably produce switching transients on the shared power supplies of the comparator, and may contribute to switching threshold jitter.

Gate drivers are also best used in a one-per-channel implementation as well. The reason is (as in Chapter 16 on PCB Layout) that a tight layout in the power switching section, with the smallest possible current loops and trace lengths, will preclude using a multi-channel gate-driver in a single package that is shared between multiple amplifier channels.

Performance Considerations in Multi-Channel Designs

Multi-channel designs make it necessary to add an extra level of complexity to the performance specifications of a Class-D amplifier. One example is in output power specifications. In many commercial Class-D multi-channel designs, the maximum *total* output power specification for the product (all channels driven into a specified load impedance) falls short of the algebraic sum of the output powers of the individual channels. For example, an 8-channel amplifier may have a specification of 100W per channel into 8 ohms. It might then be assumed that the total combined output power capability of this amplifier would be 800W (8 × 100W/channel). However, the *total* output power specification (all channels driven) will often be far lower than this 800W. The manufacturer has many reasons for this, both practical and logical. From a practical standpoint, an 800W power supply will be much larger and more expensive than a 400W supply (for example). It will also tend to have greater electromagnetic interference (EMI) emissions and lower efficiency at low output power conditions. Also, it must be understood that audio output power demand is very dynamic. Audio source material (music and voice) signals have a wide dynamic range, with high peak power and relatively low average power. It is precisely for this reason that safety regulatory agencies typically use the $1/8^{th}$ power loading condition for thermal testing of audio amplifiers. For these reasons, the optimal power supply design for a Class-D audio amplifier is one that can provide high peak power for short durations, and lower continuous power. The thermal design of the power supply (including any heatsink sizing, and sizing of magnetic components like transformers and inductors, or other components with high thermal mass) will be dictated by the desired continuous power rating. The semiconductor selection (peak voltage and current limits) will be dictated by the desired peak power rating. This approach makes for an optimized power supply design, avoiding unnecessary over-design.

Shared Circuit Allocation

A multi-channel Class-D amplifier design affords the possibility of "sharing" some circuits among multiple channels, to save cost and complexity.

Power Supplies

The most obvious circuitry that can be shared is the power supplies. This includes both the ACDC supply and any DCDC converters generating low DC voltages. As mentioned earlier, the downside to sharing the power supplies is the potential for channel-to-channel crosstalk through the common power supply connections. Conversely, the upside for using separate supply is the potential for virtually eliminating this type of crosstalk. It is quite common for a single ACDC supply and DC low-voltage rails to be shared among multiple channels. Dedicated AC supplies per channel are used in some very high power amplifiers. This design approach usually also uses separate AC line cords for each supply. Another downside to the use of separate AC supplies, aside from the higher cost, is the increased leakage current that will occur due to the separate EMI filters required for each of the ACDC power supplies. This leakage current, principally due to the "Y"-type safety capacitors used in the EMI filters, must be kept below limits specified in the applicable safety standards for each target country (such as UL62368 for the USA).

Ramp Generator

Referring to the reference amplifier, it is possible to use a single ramp generator circuit in a multi-channel Class-D fixed-frequency amplifier design. In this case, the final ramp output is fanned out to the multiple channels. This approach does however provide another crosstalk path between channels. If noise is induced on the ramp signal by one channel (termed the "aggressor" channel in crosstalk parlance), it can cause noise and distortion in the "victim" channel. The triangular ramp signal connects to one input of the high-speed PWM comparator in each amplifier channel. With some comparators, it is possible for noise to couple out of the input pins during switching transitions. This can either be actual switching noise, or a slight change in input impedance, which introduces non-linearity. This switching noise and/or non-linearity can be a source of channel-to-channel crosstalk. Some type of isolation is necessary to reduce this potential crosstalk path. Two different isolation methods are shown in Figure 19.1.

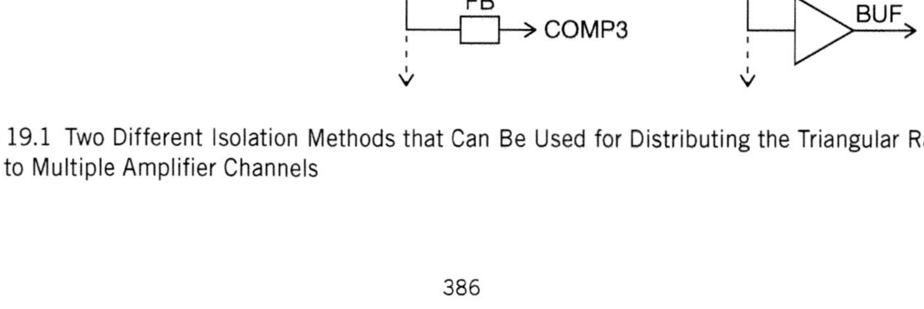

Figure 19.1 Two Different Isolation Methods that Can Be Used for Distributing the Triangular Ramp Signal to Multiple Amplifier Channels

Separate ferrite beads added in the ramp signals to each channel (as per Figure 19.1) can help reduce crosstalk. However, as discussed in Chapter 4, it is critical that the inflection points of the ramp signal be very sharp (i.e. not "rounded" by bandwidth limiting of the signal). This means that the impedance of the added ferrite beads cannot be too high. A typical value of approximately 22 ohms is prudent to achieve even better crosstalk immunity, at the expense of increased complexity. A separate high-speed analog buffer (such as MAX4200) can be added to drive the ramp signal to each channel. Figure 4.2 (Chapter 4) shows an example of a possible discrete high-speed analog buffer circuit that can be employed in each amplifier channel, to buffer the ramp feeding the PWM comparator in each amplifier.

Bibliography

Adire Audio. (n.d.). Crosstalk in inductors. Thousand Oaks, CA: Adire Audio. www.adireaudio.com

Birch, T. H., Ploug, R., Iversen, N. E., & Knott, A. (2012). Investigation of crosstalk in self-oscillating switch mode audio power amplifier. In: *Proceedings of the 132nd AES Convention*, Budapest, Hungary, April 26–29, 2012 (Paper 8631). New York: Audio Engineering Society.

Coilcraft. (2018, March 1). Inductors for Class-D amplifiers. Document 1404. Cary, IL: Coilcraft Inc.

Ploug, R. O., & Knott, A. (2012). Minimizing crosstalk in self oscillating switch mode audio power amplifiers. In: *Proceedings of the AES 48th International Conference on Automotive Audio*, Munich, Germany, September 21–23, 2012 (Article 3-2). New York: Audio Engineering Society.

Score, M., Darling, T., & Dapkus, D. (1999). Class-D audio amplifiers: System considerations. In: *Proceedings of the 107th AES Convention*, New York, NY, September 24–27, 1999 (Article 5041). New York: Audio Engineering Society.

Chapter Twenty

Audio Performance Measurements and Instrumentation

Overview

Class-D audio power amplifiers are evaluated using the same tests as are used to evaluate other amplifier types. As the adoption of Class-D audio amplifiers increases, and they are used in applications traditionally dominated by linear amplifiers, performance comparisons will always be made, as indeed they should be. Therefore, audio amplifiers must be evaluated using the same tests and same performance criteria.

That said, it is important to be cognizant of the following questions:

1) What characteristics of Class-D audio amplifiers affect audio test equipment setup and measurement?
2) Which measurements are particularly valuable in evaluating performance of Class-D audio amplifiers?
3) What equipment is necessary for more reliable and accurate measurement of Class-D audio amplifier performance?

This chapter will address the above questions, and will not cover in detail the routine tests performed on all amplifiers, as these are covered well in other texts (e.g. Cordell, 2019) and references.

Hofer (2003) discusses the characteristics of Class-D amplifiers (generalized as "switch-mode power amplifiers") that influence or interact with audio test

DOI: 10.4324/9781003327943-20

equipment. The switching nature of the power switches in the power stage of a Class-D amplifier gives rise to considerable amounts of high-frequency energy and fast voltage and current slew rates (SRs). While the L-C filter employed in Class-D designs attenuates the ultrasonic energy content in the amplifier output, there is still residual high-frequency energy present. This is predominantly due to the imperfections, non-linearities, and parasitics of the L-C filter. These are, principally, shunt capacitance of the Class-D filter inductor, and ESR/ESL of the Class-D filter capacitor, all of which degrade high-frequency filtering performance.

The input stage and signal processing stages of precision audio analyzers use operational amplifiers (op-amps) with modest SRs. These can exhibit non-linear behavior if driven (at the analyzer input) by the extremely high voltage SRs present in Class-D amplifier output noise. The non-linear behavior of these input stages, and signal processing op-amps, will cause measurement error. Hofer (2003) makes the point that this modest SR of analyzer input op-amps is an unavoidable tradeoff. It is necessary in order to achieve the high precision and extremely low noise that are required in audio performance measurements. It is necessary, therefore, to include a passive low-pass filter between the output of the Class-D amplifier-under-test, and the audio analyzer input. The purpose of this filter is to attenuate the fast-slewing artifacts that can cause non-linear behavior in the analyzer. Audio Precision (www.ap.com) offers various models of passive filter that are designed specifically for Class-D amplifier measurements. These are AUX-0025, AUX-0040, and AUX-0100. These models differ in channel count and measurement frequency range (Hofer, 2003).

Note: When measuring the outputs of *filterless* Class-D amplifiers, extra external filtering is required before the outputs of these types of Class-D amplifiers can be connected to precision audio analyzers. These filterless Class-D amplifiers lack the internal L-C demodulation filter common to conventional Class-D amplifiers. Instead of a low-level (1–2VRMS) residual ripple output waveform, these filterless (or filter-free) Class-D amplifiers have high-amplitude switching waveforms at their outputs. Palmer (2019) provides guidance on external R-C filters which should be used when measuring outputs from these types of amplifiers using audio analyzers.

Characterizing Class-D Outputs in the Time Domain

When the output voltage waveform of a conventional (i.e. *not* filterless) Class-D amplifier is displayed on an oscilloscope, it is normal to see residual ripple at the switching frequency of the amplifier (300KHz–1MHz typically). This output waveform is typically sinusoidal, but could be triangular or trapezoidal, depending on the circuit details of the output L-C filtering employed. The amplitude of this ripple is typically in the range of 0.5–2VRMS, and may have residual noise spikes, synchronous with the switching transitions of the power switches in the half- or full-bridge power stage. The frequency components in these noise spikes can have a very wide frequency spectrum, typically extending from approximately 20MHz to hundreds of MHz, depending on switching speed and parasitics. It is for this reason that it is best that these noise spikes be filtered, and prevented from coupling onto any connected cables, in order to prevent EMI problems. Half-bridge Class-D amplifiers utilizing split power supply rails (positive and negative voltage rails, as in the reference design) will have idle (quiescent = no audio signal) output voltage centered at 0VDC, with minimal DC offset (typically less than +/−100mVDC). Class-D amplifiers utilizing a full-bridge power switching stage will have quiescent (no audio signal) output voltage centered at the midpoint of the power supply rail voltage (for single-supply systems, which is most typical). Both output terminals (each connected to one of the half-bridges comprising the full-bridge stage) will be centered at this same midpoint voltage. However, if the full-bridge power stage utilizes split supplies (as would be the case if two half-bridge amplifier channels were wired in BTL (full-bridge) mode), then each output terminal would be centered at 0V, since these are essentially two half-bridge outputs running off split supply rails.

Noise

Noise measurement is an excellent first measurement to make for Class-D amplifiers, because the noise floor will influence many subsequent measurements such as signal-to-noise ratio (SNR), total harmonic distortion plus noise (THD+N), power supply rejection ratio (PSRR), crosstalk, etc.

For unweighted noise measurements, it is common practice to use either 20KHz or 22KHz as the high-frequency limit.

A-weighted noise measurements are meant to mimic the frequency response of human hearing. In this way they are intended to be better predictors of how noise will sound to a listener. The use of A-weighted values in product literature and specifications is popular because they yield lower values, and so are preferred for product marketing literature. In North America, the A-weighted response with RMS detection is the most popular, whereas in Europe, the CCIR-468 response with quasi-peak detection is preferred. Cordell (2019) also advocates the use of an AES17 filter to reduce out-of-band noise that has been exacerbated through the use of oversampling processes, such as sigma delta converters (see Chapter 21 on Alternate Modulation Schemes).

Fast Fourier Transform (FFT)

A Fast Fourier Transform (FFT) noise measurement can provide an effective way of detecting whether noise is white (spread evenly over the measurement range) or concentrated at discrete frequencies, such as AC-LINE-related hum or buzz artifacts. This type of measurement can be extremely helpful in uncovering noise sources during amplifier design or troubleshooting.

THD+N

THD+N is one of the most useful and revealing measurements that can be made on an audio amplifier, because it measures multiple distortion products simultaneously. Typically, this measurement is made using a high-precision, extremely low-noise signal from the audio analyzer, driving the amplifier-under-test. The output signal of the amplifier is then processed by the audio analyzer using a notch filter to remove the precision test tone. What remains are all the distortion products plus noise produced by the amplifier-under-test. This measurement is extremely versatile, in that it can be made over many varied operating conditions. For example, THD+N can be measured at specific frequencies over the 20Hz–20KHz audio bandwidth (BW), or swept for a full-range measurement (THD+N versus frequency). It can also be made over the full output power range of the amplifier (THD+N versus power).

THD+N versus Power

The most common THD+N versus power test performed on audio amplifiers is run using a 1KHz test tone. It typically displays two main characteristics. The first is a rather abrupt increase in THD at high power. This is usually due to voltage clipping, as the amplifier reaches its output voltage limits. The second characteristic is a gradual (usually linear on a log power axis) increase in THD+N as output amplitude *decreases*. The explanation for this effect is simple. At lower output power levels, the fixed-amplitude output *noise* voltage represents an increasing percentage of the total output voltage, as output power level decreases.

It is also useful to run THD+N versus power tests at other test frequencies besides 1KHz. When low test frequencies are used, they can uncover issues such as distortion effects at high amplitude due to supply ripple or pumping. Tests run at high frequencies uncover the effects of decreasing negative feedback factor as frequencies increase. This is a consequence of the open-loop gain decreasing with increasing frequency (a necessity for a stable feedback loop). The lower feedback factor and loop gain means less ability for the feedback loop to correct for open-loop distortion mechanisms in the design.

Typically, industry specifications will show THD versus power performance graphs at 1KHz, 100Hz, and 6.67KHz. The last may seem like a strange test frequency, but this is the highest test frequency for which the third harmonic can be measured using a 20KHz filter BW in the analyzer.

THD+N versus Frequency

Another useful test is THD+N versus frequency. For this test, the amplifier can be set to any power level, and its distortion versus frequency can be measured. As mentioned above, increasing THD+N as frequency increases is due to decreasing feedback factor and loop gain, and therefore a decreasing error-correcting ability of the feedback loop to correct open-loop sources of distortion as the test frequency increases. If distortion is seen to rise at lower frequencies, this can be due to AC-coupling capacitors having a poor voltage coefficient of capacitance. This means the capacitance of these capacitors can *change* as the voltage across them changes. This is an important specification for AC-coupling capacitors. Ceramic capacitors using an X5R or X7R dielectric exhibit this characteristic. Film and aluminum electrolytic capacitors (polarized, or non-polar) do not. At higher frequencies, the voltage across these capacitors does not change; however, at low frequencies, the voltage across these AC-coupling capacitors can change over a period at the test signal frequency. If the capacitance value is sensitive to applied voltage (voltage coefficient of capacitance), then the changing capacitance over the test signal period will cause distortion, which is not corrected by the feedback loop if the AC-coupling capacitor precedes the amplifier's closed feedback loop (such as for input circuitry preceding the power amplifier circuitry).

Hofer (2003) suggests that it might be interesting to look for distortion at subharmonics of the switching frequency lying within the 20Hz–20KHz measurement BW. For example, if a Class-D amplifier has a 250KHz switching frequency, the 12th and 13th sub-harmonics are 20.833KHz and 19.23KHz, respectively. For the reference amplifier, the lowest sub-harmonics that would fall into the audio signal BW would be the 24th, 25th, and 26th at 20.833KHz, 20.000KHz, and 19.23KHz, respectively.

FFT Analysis

Though THD+N is a powerful and useful measurement, it weighs all THD components within the prescribed measurement BW equally. Thus, it provides no information as to the frequency of the THD components. FFT analysis on the other hand provides this frequency-specific information, because it displays the spectrum of the measured THD+N. This information also helps in correlating the measured spectrum to audibility. Distortion products can be classified as harmonic (multiples or submultiples of the test frequency), or non-harmonic, such as AC-line-related hum and buzz frequency components. Harmonic distortion products can be further quantified by their order. This is useful because, for example, third-harmonic distortion products are generally considered to be more audibly objectionable than second-harmonic distortion products. When selecting an FFT measurement, it is important to be aware of the window function being used. Different window functions have different "side-lobe" characteristics, which affect the accuracy of distortion measurements (Hofer, 2003).

InterModulation Distortion Measurements

InterModulation Distortion (IMD) results when two or more audio tones beat with one another in a non-linear device. New undesirable tones can be produced by the intermodulation of these audio tones. The mechanism that is primarily responsible for producing this IMD is amplitude modulation (AM). The sidebands (new tones) that are created in this way are at the sum and difference frequencies of the two original audio tones (Audio Precision, n.d.).

IMD measurements (Cordell, 2019; Hofer, 2003; Audio Precision, n.d.; Elliott, 2019) measure the non-linearity in an amplifier's response to two or more test signals. There are thus an infinite number of IMD tests that can be performed, as the frequency and relative amplitudes of these two or more tones can be varied. The basic thinking is that an ideal amplifier should produce these two or more tones perfectly, and the reproduction of any tone or tones should not be affected in any way by the amplifier's reproduction of any other tone or tones. This seems at first to be a very straightforward requirement, and rather essential, considering that an audio power amplifier produces many different frequencies simultaneously. IMD testing can produce test products that are not harmonically related and can reveal non-linearities that are audibly more objectionable than harmonically related distortion products. It is for this reason that many believe that IMD tests yield a better indication of audio amplifier sonic performance than THD+N tests.

SMPTE IMD

This test looks for undesired modulation in a high-frequency (HF) tone, due to a larger-amplitude low-frequency (LF) tone. The tone frequency used varies internationally. In the USA, 60Hz and 7KHz, in a 4:1 amplitude ratio, is common. In Class-D amplifiers, SMPTE IMD is usually a good measure of the linearity of the pulse width modulator. The modulation components of the higher-frequency signal appear as sidebands, spaced at multiples of the lower-frequency signal. The amplitudes of the sidebands are root mean square summed and expressed as a percentage of the higher frequency's level. It is helpful to consider the operating conditions exercised by this test, in the time domain. The low-frequency signal's large amplitude causes it to exercise a large operating dynamic range in the input circuits and modulator. The high-frequency signal then evokes a response from these circuits under the different operating conditions produced by the low-frequency excitation. Another way to picture this is as follows.

Suppose the high-frequency signal produces a 10% modulation change. If this occurs near the zero-crossing of the low-frequency signal, i.e. where the IDLE duty cycle of the modulator would be 50% in the absence of signal, the high-frequency signal will cause a modulation of the duty cycle from 45% to 55%. Now, however, if the low-frequency signal moves the duty cycle up to 75%, the high-frequency signal will then cause a modulation of the duty cycle from 70%

to 80%. If there are any non-linearities in the input and signal processing circuits, or in the modulator itself (such as an incremental modulation of the gain at each operating point), these will show up as IMD distortion products in this test. This test is sensitive to crossover distortion and clipping, as well as low-frequency thermal distortion, which is caused by thermal effects that change the gain of the amplifier. In this case, the thermal time constant would need to be short enough such that thermal effects could change the gain of the amplifier over the period (or half-period) of the low-frequency waveform (for example, over 25msec = one half-period at 20Hz). An advantage that SMPTE testing has in detecting low-frequency distortion mechanisms is that the distortion components occur at a high frequency. The loop gain is usually lower at high frequencies, and therefore there is less loop gain available to reduce this distortion through negative feedback. It is therefore more easily detected and measured (Hofer, 2003).

Twin-Tone IMD

Note: Other IMD tests that are under the general category of twin-tone IMD tests are CCIF (Cyclic and Commutated Intermodulation Distortion, or Composite Second-Order Intermodulation Filter) and DFD tests (Difference Frequency Distortion).

These tests consist of equal-amplitude sine waves spaced relatively close together in frequency. The advantage of these tests is that the system is sharply BW-limited, so that high-frequency nonlinearity is more easily revealed. Many non-linear distortion products will fall in-band with twin-tone testing, without significant attenuation or error. Even-order IMD products indicate asymmetrical non-linearity, whereas odd-order IMD products indicate symmetrical non-linearity. A common source of symmetrical non-linearity is the inductor in the L-C demodulation filter of a Class-D amplifier. As discussed in Chapter 9 on L-C demodulation filters, inductors can exhibit non-linearities, especially inductors employing an iron or ferrite core, which can exhibit non-linear hysteresis properties, or even partial saturation. These twin-tone tests are quite demanding for audio amplifiers, because they require high linearity and/or high loop gain error correction at high frequencies. And this is precisely where feedback loop gain is necessarily lower, in order to meet the criteria for a stable closed-loop system.

Here again, the conflicting design requirements that make Class-D amplifier design challenging are apparent.

Dynamic or Transient IMD Measurements

Transient InterModulation distortion (TIM) (Otala & Leinonen, 1977; Cherry, 1979) occurs in amplifiers that employ negative feedback, and have signal delays that make the amplifier incapable of correcting distortion when exposed to fast, transient signals. Amplifiers that employ feedback (such as Class-D amplifiers) may have multiple gain stages comprising the closed feedback loop. Each of these gain stages has its own open-loop gain characteristics. It is possible, during a transient, for one or more of these stages to saturate or clip. Once that occurs, the entire loop is basically "open", and unresponsive to further changes in the input signal, until such time as the saturated stage returns to linear operation. A TIM test measures the amplifier's response in this transient situation. This test is important because it can reveal characteristics of the amplifier that affect its real-world sonic performance. As part of Class-D amplifier design, it is important for the designer to probe the outputs of each stage of the Class-D amplifier during transient testing. This can be very revealing. The designer can see which stages saturate first, what the saturation limits are, how long it takes for each saturated stage to recover, and whether a given stage can benefit from additional circuitry that could help speed recovery by preventing hard saturation. SR (slew rate) limitations can also be revealed. As long as TIM is kept in mind, and the stage limitations are carefully monitored during the design process, it is still possible and beneficial to use significant amounts of negative feedback in Class-D design, without incurring TIM penalties.

A Discussion about SR (Slew Rate)

A Class-D audio amplifier must exhibit an adequate SR for high-fidelity audio reproduction. The SR of a sinusoidal waveform has its *maximum* at the zero-crossing, and is expressed as:

$$SR = 2\pi f(Vpeak) \qquad \text{EQ. 20.1}$$

where f is the audio signal frequency, and Vpeak is its peak voltage level.

This equation returns SR in units of Volts-per-second. Most audio amplifier specification datasheets express SR in units of Volts-per-microsecond (μsec). To convert from Volts-per-second to Volts-per-μsec, simply divide by 10^6.

For example, the reference amplifier produces 100WRMS.

100WRMS into 8 ohms = 28.28VRMS = +/−40Vpeak

A 20KHz output signal at this amplitude has an SR of:

$$SR = 2\pi(20{,}000Hz)(40V) = 5.03 \times 10^6 V/sec = 5.03V/\mu sec$$

In order to achieve this output *voltage* SR across an 8-ohm load, the amplifier must be capable of a *current* SR of 5.03V/8 ohms = 628.3mA/μsec.

Checking SR demand into a 4-ohm load:

100WRMS into 4 ohms = 20VRMS = +/−28.28Vpeak

A 20KHz output signal at this amplitude has an SR of:

$$SR = 2\pi(20000Hz)(28.28V) = 3.55 \times 10^6 V/sec = 3.55V/\mu sec$$

In order to achieve this output *voltage* SR across a 4-ohm load, the amplifier must be capable of a *current* SR of 3.55V/4 ohms = **888.4mA/μsec**.

An important consideration in Class-D amplifier design that should not be overlooked, is the effect of the demodulation inductor on output *SR*. This inductor puts an upper limit on achievable SR. This can best be explained as follows.

The SR of current through an inductor is defined as di/dt. Conveniently, the common inductor equation expresses di/dt as a function of the voltage across the inductor (V_L) and the inductance value (L):

$$di/dt = V_L/L \qquad \text{EQ. 20.2}$$

Consider the inductor, as connected in the reference schematic (L1A). One terminal of the inductor is connected to *either* the positive supply voltage (+40V) or the negative supply voltage (−40V), depending on which mosfet switch is turned on. The second terminal of the inductor can be considered to be the amplifier output voltage for this discussion. Therefore, the voltage across the inductor will vary as the output voltage changes. For simplicity, it is assumed here that the duty cycle of the modulator can reach 0% and 100% at the extremes (when called for by the feedback loop). This means that the switch-node end of the inductor can be assumed to be at either +40V or −40V, when calculating maximum possible SR.

Suppose for example that the output is at 0V. The maximum positive SR and negative SR will be the same:

$$di/dt = 40V/22\mu H = 1.818A/\mu sec$$

Now suppose the output is at +20V. In this case, V_L is 40V − 20V = 20V, and therefore the available positive SR has been halved to:

$$+di/dt = 20V/22\mu H = +909mA/\mu sec$$

While the negative SR has increased, because the negative voltage available across the inductor has increased: $V_L = -40V - 20V = -60V$.

$$-di/dt = 60V/22\mu H = -2.73A/\mu sec$$

This has some relevance to IMD testing, because if a high-frequency tone (requiring high-current SR) is required, while the amplifier output is already biased away from 0V, and towards either the positive or negative rail (by a low-frequency intermodulating tone, for example), the available maximum output current SR will be reduced. The effect of this "dynamic" SR reduction on real-world audio performance is complex, but is worth considering. In reality, this maximum signal SR is only required for a 20KHz spectral component at maximum signal amplitude. This would have to occur simultaneously with the modulating signal's offsetting of the amplifier output signal away from 0V.

The important takeaways from this discussion are:

1) The demodulation filter inductor affects output current SR.
2) As load impedance decreases, maximum required amplifier output current SR increases.

3) As amplifier output voltage deviates from 0V, available maximum output current SR decreases in the direction of the voltage offset, and increases in the direction opposite the voltage offset.
4) Item (3) above is relevant in a discussion of overcurrent protection and response time for OCP circuits, because a short circuit to one or the other voltage rails will exhibit higher current SRs than a short circuit to ground = 0V (for a half-bridge amplifier topology).

Frequency Response

It is important to measure frequency response in Class-D amplifier designs, because many design details affect this performance parameter. As mentioned in other chapters, *open-loop* Class D designs don't have negative feedback available to correct for an under- or over-damped response of the L-C demodulation filter, and can have significant frequency response deviation from flat response at higher frequencies. This also applies to closed-loop designs employing negative feedback, if the feedback is taken *before* the output L-C filter. This is sometimes the approach taken for some self-oscillating designs, and designs which seek to avoid the complexity of adding negative feedback after the output L-C filter (with its two-pole (minimum) phase shift). *Note*: The reference design does take loop feedback *after* the output L-C filter.

Frequency response is typically measured at 1W output power level, but it is useful and relevant to performance with real-world audio reproduction if the frequency response is also measured at other power levels, up to clipping. It is quite possible for clipping to occur at different power levels at different frequencies. However, a THD+N versus frequency sweep at different power levels will be a more sensitive test for this effect.

Damping Factor

Damping factor (DF) is a measure of the ratio of the load impedance to the output impedance of the amplifier, and is touted as a measure of how well the amplifier can "control" the load. It is informative to look into this in greater detail. Figure 20.1 illustrates the voltage divider formed between the output impedance of the amplifier (Z_{OUT}) and the load (Z_{LOAD}). It is easy to see that if the output and

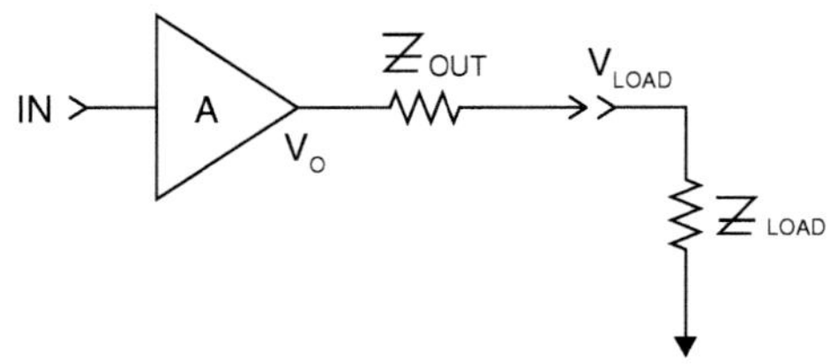

Figure 20.1 Voltage Divider Formed by the Output Impedance of the Amplifier and the Load Impedance

load impedances were fixed quantities, the voltage impressed across the load by the amplifier would be a lower amplitude, but would not vary with frequency. However, if the load impedance were dynamic, and changed with frequency (as electrodynamic and electrostatic speakers do), then the voltage impressed across the load would in fact change with frequency, even if the amplifier output voltage Vo remained invariant with frequency. Obviously, if the output impedance of the amplifier (Zout) were *zero*, then the amplifier would operate as an ideal voltage source, and the voltage across the varying load impedance would remain constant with frequency. In that case, DF would be infinite, and the amplifier output would maintain precise control of the voltage impressed across the load, no matter how much the load impedance varied.

$$DF = Z_{LOAD} / Z_{OUT} \qquad \text{EQ. 20.3}$$

In general, DFs in excess of 100 (with an 8-ohm load) are desirable for high-fidelity audio amplifiers. This would imply a maximum amplifier output impedance of no higher than 80mΩ to achieve this minimum DF. High negative feedback factors, in combination with high open-loop gain, as judiciously applied in Class-D amplifier design, have the benefit of reducing the output impedance of the closed-loop amplifier and allowing the amplifier to more closely approach the performance of an ideal voltage source, with high DF.

It is worthwhile, in this context, to also note that an open-loop amplifier, or a closed-loop amplifier that takes feedback *before* the output L-C filter, can never have an output impedance *less* than the DC Resistance (DCR) of the L-C filter inductor. If the filter inductor used in the reference design is employed, for example (Chapter 9), its DCR is 60.75mΩ. This would give a theoretical maximum DF of 8 ohms/60.75mΩ = 131.7.

Adding negative feedback *after* this inductor will improve this DF.

Any intervening impedance between the amplifier output and the load, including the return current paths, will degrade DF. This includes the impedance of connector contacts, PCB traces, speaker wiring, and speaker binding posts. This is why it is critical to properly size these interconnections, not only to achieve low impedance for current carrying ability, but also to reduce power loss and heat generation, and to maintain high DF.

Crosstalk Measurements

Crosstalk has been discussed extensively with regard to multi-channel amplifier designs (Chapter 19). It is included in this discussion of measurements relevant to Class-D amplifiers because the switching nature of Class-D amplifiers affects crosstalk in ways that are novel, compared to non-switching (linear) amplifier topologies, such as Class-AB.

Crosstalk is the phenomenon by which one amplifier channel (termed the aggressor or transmitter) feeds into, or otherwise disturbs, the signal in a second channel (termed the victim or receiver). In multi-channel amplifier designs, there may be more than one aggressor channel, which complicates the crosstalk considerations. Crosstalk usually needs to be considered in both directions (for example, for a Left/Right pair of stereo channels: Left feeding into Right, and Right feeding into Left). Crosstalk is usually expressed in dB, as the ratio of the measured signal voltage in the victim channel, to the larger signal voltage in the aggressor channel. For example, suppose the Left channel is producing 100W into an 8-ohm load (28.28VRMS), and a residual (crosstalk) signal of 28.28mVRMS is measured in the IDLE Right channel; the crosstalk will be:

$$= 20 \log (28.28\text{mVRMS}/28.28\text{VRMS}) = -60\text{dB}$$

Typically, specifications for Left/Right crosstalk in a stereo channel pair are less stringent than for other channel types, such as fully independent channels, because it is expected that the Left/Right channels will handle related signals, where some crosstalk of these signals will be less objectionable than if the channels were handling completely independent signals. For example, if a multi-channel amplifier had four channels configured as two independent stereo pairs, where each stereo pair supplied audio to a different room in a home, and each room had independent audio content, then crosstalk between "rooms" in this case would be more objectionable than Left/Right crosstalk within a stereo channel pair (i.e. within each room). This difference is reflected in the specifications. Typically, a minimum of −60dB of crosstalk within a stereo pair is desired, whereas diminished crosstalk of better than −80dB would be desired for independent rooms. Achieving this low level of crosstalk in Class-D design is no trivial matter. Crosstalk is typically specified at a test frequency of 1KHz, but it is prudent to check crosstalk over the full audio frequency range, because crosstalk will vary with frequency. This is due to the fact that different coupling mechanisms responsible for crosstalk may operate more or less efficiently as frequency changes. For example, capacitive coupling will cause crosstalk to increase with increasing test frequency.

Causes of Crosstalk

Crosstalk between amplifier channels can be caused by many mechanisms. These include:

1) Common impedance routing and layout
2) Capacitive coupling effects
3) Inductive coupling effects
4) Switching noise pickup on critical modulator signals.

Setting up Crosstalk Measurements

A typical setup for measuring crosstalk consists of configuring the aggressor channel to sweep from low to high output voltage amplitude at various fixed frequencies, while loaded with the specified loads. Simultaneously, the output voltage of the victim channel will be measured over this sweep. The victim channel will be at idle (producing no audio output), and its input should be properly terminated for lowest noise. This sweep can uncover threshold effects, for example if the signal in the victim channel displays a sudden change in signal level instead of a smooth linear increase in signal level, proportional to the output signal level of the aggressor channel. If the signal level rises with test frequency, this can indicate capacitive coupling as the crosstalk mechanism.

Capacitive and inductive effects are both exacerbated by circuitry and traces of neighboring amplifier channels placed or routed in close proximity to each other. One way to differentiate between these is to operate the aggressor channel at the same voltage levels but with no-load (no load current) and then, with lowest impedance load (maximum output current). Inductive crosstalk mechanisms will yield increased crosstalk with increased load current, whereas capacitive crosstalk effects will remain relatively unchanged with changes in current, but will increase proportional to voltage level and frequency.

Another common mechanism of crosstalk in multi-channel, fixed-frequency, Class-D amplifier designs is switching noise on the critical (shared) TRIANGULAR RAMP signal. Figure 20.2 illustrates this mechanism. Noise on the TRIANGULAR RAMP signal, caused by switching transitions in the aggressor channel, can prematurely trigger the PWM comparator in the victim channel, causing switching jitter in the victim channel. Note that this will be most noticeable when the victim channel is at idle, where 50% duty cycle PWM switching will be occurring near the zero-crossing of the ramp signal. There will be no audio present in the victim channel to mask the crosstalk, and it will sound like very low-fidelity feedthrough of the audio signal from the aggressor channel. The solution for this type of crosstalk is careful routing of the (shared) triangular PWM ramp, or separately buffering this ramp signal to each amplifier channel, or generating independent triangular PWM ramp signals for each channel. *Note:* This latter approach can still employ a common clock signal to generate these separate ramp signals.

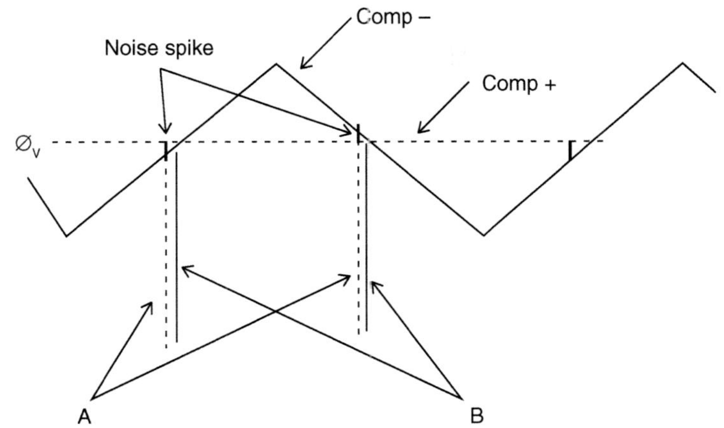

Figure 20.2 Triangular Ramp Voltage Signal with Noise Spikes

Figure 20.2 shows the two signals feeding the PWM comparator. The triangular ramp signal connects to the inverting comparator input, while the output of the error amplifier (E/A) connects to the non-inverting comparator input, and is depicted at being near zero volts due to the IDLE (no audio present) condition.

Time instance "B" is the nominal zero-crossing PWM switching transition times for the PWM comparator (in the absence of noise).

Time instance "A" indicates the premature switching times for the PWM comparator, triggered by a switching noise artifact in the ramp signal.

Bibliography

Analog Devices. (2009). Op amp distortion: HD, THD, THD + N, IMD, SFDR, MTPR. Analog Devices Tutorial, MT-053. Wilmington, MA: Analog Devices. www.analog.com

Audio Precision. (n.d.). Audio power amplifier testing (PowerPoint). Beaverton, OR: Audio Precision. www.ap.com

Cherry, E. M. (1979). Comments on "The theory of transient intermodulation distortion". *IEEE Transactions on Acoustics, Speech, and Signal Processing, ASSP-27*(6), 653–654.

Cordell, B. (2019). *Designing Audio Power Amplifiers* (2nd ed.). Abingdon: Routledge.

Elliott, R. (2019). Intermodulation distortion. Thornleigh, Australia: Elliott Sound Products (ESP).

Foley, D. (2018, February 20). Advanced distortion analysis methods. *Loudspeaker Industry Sourcebook.*

Ge, T., He, H., Zhou, J., Kang, Y., & Chang, J. S. (2016). An investigation of THD of a BTL Class D amplifier. In: *2016 IEEE International Symposium on Circuits and Systems (ISCAS)* (pp. 470–473). New York: IEEE Press.

Hofer, B. (2003, October). Measuring switch-mode power amplifiers. Audio Precision white paper. Beaverton, OR: Audio Precision.

Otala, M., & Leinonen, E. (1977). The theory of transient intermodulation distortion. *IEEE Transactions on Acoustics, Speech, and Signal Processing, ASSP-25*(1), 2–8.

Palmer, R. (2019). Guidelines for measuring audio power amplifier performance.

Texas Instruments Application Report SLOA068A. Originally published October 2001, revised September 2019. Dallas, TX: Texas Instruments.

Van der Hulst, P., Veltman, A., & Groenenberg, R. (2002). An asynchronous switching high-end power amplifier. Paper presented at the 112th AES Convention, Munich, Germany, May 10–13, 2002. New York: Audio Engineering Society.

Chapter Twenty One

Alternate Modulation Schemes for Class-D Amplifiers

Modulation Schemes

The modulator section of a Class-D amplifier is responsible for generating the pulse-stream that drives the power switching stage. There are many types of modulators employed in Class-D designs. These include fixed-frequency (clocked) designs, as used in the reference amplifier, as well as self-oscillating designs, which can be free-running (with no external clock signal) or clocked (synchronized) to an external clock signal.

Fixed-Frequency (Clocked) Modulators

The reference amplifier uses a fixed-frequency triangle waveform (linearly ramping up and down with identical rising and falling slopes), which is fed to one input of the PWM comparator. This waveform could alternatively be a sawtooth waveform (ramping up linearly, and falling instantaneously). The details of generating the reference ramp waveform, and the relative merits of triangle waveform over sawtooth waveform, are covered in Chapter 4. Here we will note that the triangle reference waveform results in two sampling "decisions" per period of the switching frequency, whereas the sawtooth waveform allows only one. The second comparator transition is fixed in time, and does not vary with the sampled error waveform. The comparator compares this reference waveform to the error amplifier output signal (when negative feedback is used) or it can be compared to the input signal in an open-loop design (i.e. when no feedback is used). In the

DOI: 10.4324/9781003327943-21

reference amplifier, feedback is used, and it is the error amplifier output that is compared to the triangle reference waveform. By inspection, it can be seen that if the error amplifier output voltage waveform is more positive than the reference triangle wave, the output of the PWM comparator will be high, and if it is more negative than the triangle reference wave, the output of the PWM comparator will be low. The comparator output therefore has only two states, and it is the job of the power switching stage to level-translate those two comparator states to the level of the power supply rail voltages. Therefore, the error amplifier output voltage is translated into a duty cycle ranging from 0% to 100% as it spans the voltage range of the triangle waveform, from its minimum to maximum peak voltages. The *gain* of the modulator-plus-power-switching-stage is then simply defined as the peak-to-peak rail voltages, or output voltage swing (peak-to-peak), divided by the peak-to-peak triangle wave voltage. In the case of the reference amplifier, the rails are +/−48VDC (or 96Vpp). The reference triangle waveform swings from −4V to +4V, or 8Vpp. Therefore, the *gain* of the modulator-plus-power-switching-stage is 96V/8V = 12V/V or 21.58dB.

Self-Oscillating Modulators

Self-oscillating modulators do not have ramp generation circuits. This is one advantage they have, that manifests itself in lower parts count and cost savings, as well as lower power consumption (i.e. power consumed by the ramp generation circuitry is saved). Instead, the PWM pulse train is generated by the self-oscillation of the modulator circuitry.

Broadly speaking, these types of self-oscillating topologies share another advantage when compared to fixed-frequency PWM designs, in that the control loop bandwidth (BW) in closed-loop designs is typically higher in self-oscillating designs. It is important to understand the basis for this difference.

In fixed-frequency designs (such as in the reference amplifier), when negative feedback is applied around the control loop, Nyquist stability criterion must be observed. This basically states that when the open-loop gain drops to unity (at the unity gain crossover frequency), the total phase shift of the loop must be less than 180 degrees, to avoid loop instability. *Note:* This assumes that there is already 180 degrees of phase shift due to negative feedback (with 360 degrees of phase shift total, causing instability and oscillation). The alternative way to express this is that open-loop gain must be less than unity, whenever total loop phase shift is 180 degrees. If this criterion is not met, positive feedback and oscillation will result. It is important to understand that in fixed-frequency designs, the unity gain crossover frequency is independent of the actual fixed switching frequency. Another stipulation for fixed-switching-frequency designs is that the Nyquist criterion states that the unity gain crossover frequency must be less than one-half the switching frequency. However, this is only a theoretical limit, and not suitable for conservative, stable real-world designs. Designers' opinions differ on how far to push this limit. One common rule of thumb states that the unity gain crossover frequency should not be pushed higher than one-quarter of the switching frequency, or at most, fswx/π for practical designs utilizing two-level modulation, and 2/π for designs using three-level PWM (Nielsen et al., 2009). Other references (Jiang, 2017; Huffenus et al., 2010) cite a limit imposed by the slope of the two signals fed to the PWM comparator. The first signal is the timing ramp signal (carrier) at the switching frequency, and the second is (usually) the output of the integrating error amplifier. The slope of this error amplifier output signal (seen at one PWM comparator input) must not exceed the slope of the timing ramp seen at the other input of the PWM comparator. This imposes a limit on the loop gain at the switching frequency, which will be significantly less than unity (0dB). In turn, this also puts a practical limitation on the closed-loop BW of fixed-frequency designs. For example, for the reference amplifier, running at 500KHz switching frequency, the closed-loop unity gain crossover frequency should not be pushed higher than 125KHz. This restriction on closed-loop BW will impact audio performance metrics, because it will determine how much feedback loop gain there will be at audio frequencies (for error correction). This is discussed in more detail in Chapter 12. For the purposes of this discussion, it will be understood that in a stable negative feedback design, the open-loop gain will fall in a controlled manner (typically between 6 and 12dB/octave) to the unity gain crossover frequency. So, knowing this, and working backwards from the unity gain crossover frequency, the open-loop gain at 20KHz can be approximated.

First, assuming the more conservative 6dB/octave (average) rolloff in open-loop gain: If the open loop gain is 0dB (unity gain) at 125KHz, for example, it will be +6dB @ 62.5KHz, +12dB @ 31.25KHz, and only +18dB @ 15.625KHz.

Next, assuming a very aggressive 12dB/octave rolloff in open-loop gain: If the open loop gain is 0dB (unity) at 125KHz, for example, it will be +12dB @ 62.5KHz, +24dB @ 31.25KHz, and +36dB @ 15.625KHz.

It should however be understood that the open-loop gain cannot fall at −12dB/octave over the whole frequency range from 20KHz to the unity gain crossover frequency. Phase shift is related to both gain slope and frequency span, so a gain slope of −12dB/octave occurring over a large frequency span of 3 octaves will accumulate a large phase shift, which may exceed the 180 degree stability criterion at the unity gain crossover frequency. For this reason, the gain slope can vary over the 3-octave frequency span, but will average to less than a −12dB/octave gain slope over this frequency span. This illustrates the BW restriction inherent in fixed-frequency modulator designs. The available feedback loop gain at 20KHz is restricted, and therefore the gain available for error correction at 20KHz is limited. This results in increasing THD and less control over frequency response, load effects (damping factor), power stage non-linearities (deadtime and propagation delays), and L-C filter non-linearity, with increasing audio frequency.

In some types of self-oscillating Class-D amplifiers, particularly globally modulated amplifiers that take feedback from the output, *after* the L-C filter, the closed loop is purpose-designed to oscillate ***at*** the closed-loop unity gain crossover frequency. In other words, feedback loop oscillation is intentional. This gives the self-oscillating topology a 4× increase in closed-loop BW over fixed-frequency topologies, using the switching-frequency-to-unity-gain-crossover-frequency rule of thumb cited above.

It is worthwhile to note here, that with the increasing adoption of new power switch technologies, such as SiC and especially GaN, switching frequencies for Class-D amplifiers can rise significantly. This means that for fixed-frequency designs, the closed-loop BW can increase as well. For example, if the switching frequency increases from 500KHz to 2MHz, using GaN power switches, the closed-loop unity gain crossover frequency can increase to 500KHz, rivaling the BW of self-oscillating topologies. As with most aspects of Class-D design, there are tradeoffs that need to be considered. The higher switching frequency yields smaller component values and sizes (particularly for the output filter inductor), but also yields higher switching losses. But here the GaN devices hold promise due to their fast switching speeds and lower parasitic capacitances and gate charge, which will reduce switching losses. Another approach to consider is multi-phase designs, as a way of increasing the effective switching frequency. In these designs, more than one Class-D power stage is utilized. These power stages can be Single-Ended = SE = Half-Bridge, or BTL. They are driven by different phases of the fixed-frequency clock. For example, if a 4-phase design were implemented, the phase difference between the switching frequency clock for each phase would be 90 degrees. The inductors in the L-C filter for each phase are connected at a common output node, where the filtered output currents from each Class-D phase are summed. This is shown in Figure 21.1.

An interesting benefit of this multi-phase arrangement is that each phase handles one-quarter of the output current, at the switching frequency. Therefore, the switching losses for each phase are one-quarter of what they would be for a single-phase design of the same output power. The effective switching frequency of the multi-phase design will be 4× the switching frequency of each phase, allowing a 4× increase in closed-loop BW compared to the single-phase design, and the total switching losses of the multi-phase design will be the same as for the single-phase design.

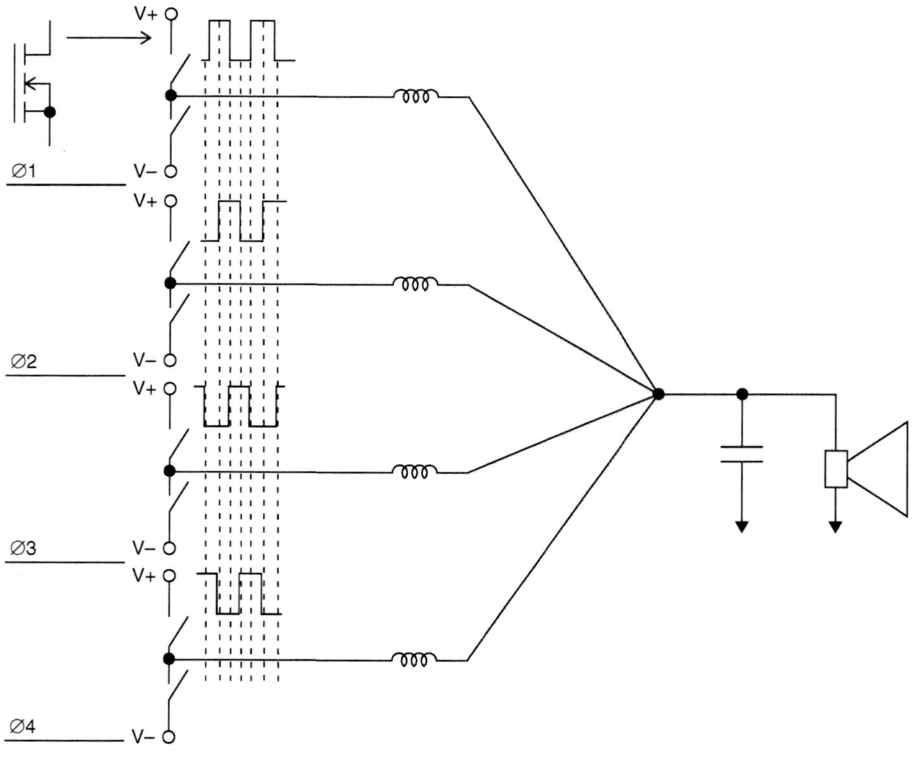

Figure 21.1 Schematic and Timing Diagram Illustrating a Four-Phase, Fixed-Frequency Class-D Power Stage

Multi-Level Switching Schemes

Multi-level switching schemes also benefit from this switching frequency multiplying effect, allowing increased loop BW. Infineon (2019) presents a five-level switching scheme, resulting in increased granularity in the output voltage from the Class-D switching stage. This results in lower ripple currents and decreased need for the output L-C filter.

Self-Oscillating Class-D Amplifier Topologies

Aspects of Self-Oscillating Modulators

As stated above, self-oscillating modulators have the advantage of not needing a clock oscillator or timing ramp generation circuits. This results in a lower parts count as well as cost and power savings. Instead, the PWM pulse train is generated by the self-oscillation of the modulator circuitry. The second main benefit is wider control loop BW, which allows for higher loop gain over the audio frequency range, affording greater error correction as compared to fixed-frequency PWM modulators.

There are, however, some disadvantages to self-oscillating designs that must be understood (*Note:* Not all types of self-oscillating modulators exhibit all the problems listed below):

- Wide switching frequency variation
- EMI signature and mitigation complexity
- Crosstalk between channels leading to idle tones
- Start-up issues
- Minimum and maximum duty cycle issues (modulation index limits).

Wide Switching Frequency Variation

In some types of self-oscillating modulators, there is a wide variation in switching frequency. Typically switching frequency varies with output voltage amplitude, with switching frequency being highest at idle (0V output) and lowest at the output voltage extremes. This is unfortunate because it means that the loop BW is decreasing as output voltage increases. This results in reduced audio performance specifications with increased output voltage, as well as increased output voltage ripple. It does however provide an efficiency advantage, because typically, as output voltage increases, so does load current, and switching loss in the power switches is proportional to load current switched, as well as switching frequency. (*Note:* Switched voltage amplitude in the power stage remains constant and is rail-to-rail.) Therefore, if switching frequency decreases as output voltage, power, and current increase, the power loss due to switching losses will be lower than it would have been if switching frequency had remained constant.

EMI Signature and Mitigation Complexity

Generally, having a fixed frequency is helpful for EMI mitigation. It makes the task of finding the EMI emitter easier, because the known switching frequency of all switching devices (and their harmonics) in a design is easily matched to the excessive EMI frequency peaks found by the EMI receiver. Once the guilty emitter is found, mitigation can be focused on a specific circuit or component. In a variable-frequency, self-oscillating design, however, the frequency peaks will shift dynamically, depending on a number of factors, including output voltage, and power, as well as power supply rail voltage variation. This sometimes makes the task of finding the specific emitting circuit or component more difficult. In addition, the mitigation task itself is more difficult than in a fixed-frequency design, because a fixed-frequency design can benefit from a "notch" filter, or another type of filter specifically designed for high attenuation at a specific switching frequency (high quality factor or "Q").

Figure 21.2 shows an example of a passive notch filter added to a fixed-frequency L-C filter output to attenuate switching frequency ripple, as implemented in

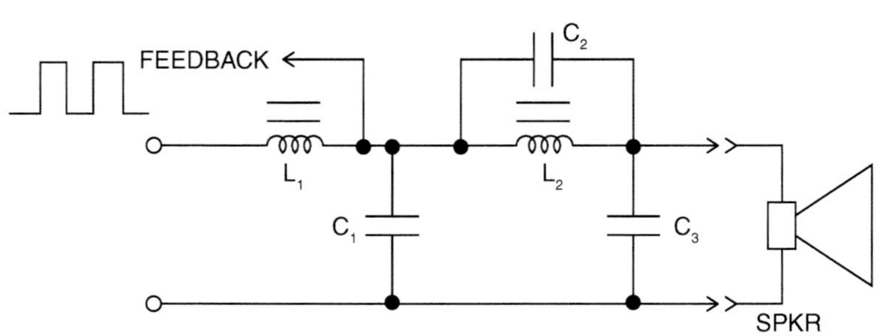

Figure 21.2 Passive Notch Filter that Can Be Added *after* the L-C Demodulation Filter

the reference design. The analysis of the passive notch filter of Figure 21.2 was discussed in Chapter 9.

It must be noted that there is also a reverse argument that is sometimes made about EMI mitigation. That argument is that when the switching frequency varies, the emitted energy is spread out over a wider frequency band, resulting in lower emissions at any specific frequency. This is an argument similar to "dithering". This "spread-spectrum" feature is used to advantage in switched-mode power supplies, to spread out the emissions energy over a wider frequency band, in order to reduce peak emissions.

Crosstalk between Channels, Leading to Idle Tones

In multi-channel implementations of self-oscillating (variable-frequency) designs, the switching frequencies of adjacent channels can be influenced by the switching noise of the adjacent channel, and if the *difference* in the self-oscillating frequencies is in the audible range, this can result in an audible "beat frequency" which is quite undesirable. This occurs at idle, when the adjacent channels are running at their maximum self-oscillation frequency, and producing no (or low-level) audio output, where any low-level noise signal will be audible over the otherwise low noise floor. In these multi-channel implementations of self-oscillating Class-D amplifiers, it is sometimes necessary to intentionally offset the idle frequencies of adjacent channels by more than 25KHz, so that this type of crosstalk is minimized.

Startup Issues

Another issue that must be addressed in self-oscillating designs is start-up. How does the self-oscillating condition start up, and what are its audible characteristics during start-up (pops, thumps, and clicks)?

Minimum and Maximum Duty Cycle Issues

As mentioned above, some types of self-oscillating topologies feature a large variation in switching frequency with input or output voltage. Maximum switching frequency occurs at idle (0V input) and minimum frequency occurs at the maximum voltage level condition. Putzeys (2009) refers to "modulation index", which is analogous to PWM duty cycle. At idle, where the duty cycle is 50%, the modulation index is 0.5. At the limits, the duty cycle is 0 (modulation index = 0) and 100% (modulation index = 1). In many self-oscillating designs, the switching frequency can drop to 0 at the modulation index extremes (0 and 1). This is very undesirable, and so some techniques must be used to limit the modulation extremes. Most of the references cite an 80% limit as being advisable (0.2 and 0.8 limits on modulation index). Unfortunately, this limits the utilization of the supply rails. The amplifier therefore will produce significantly less than a rail-to-rail output voltage to the speaker.

Types of Self-Oscillating Modulators

There are mainly two broad types of self-oscillation topologies. These are hysteresis-based and phase-shift-controlled (Putzeys, 2009).

Self-oscillating Class-D amplifiers oscillate at the frequency where the open-loop phase shift reaches 180 degrees (or 360 degrees if you include the 180 degrees inherent from negative feedback). As discussed above, this increased loop BW provides a valuable gain advantage in the audio BW, as compared to the fixed-frequency PWM topology (Nielsen et al., 2009). The two broad categories of self-oscillating modulators differ in the way the 180-degree phase shift is achieved.

Hysteresis Modulators

In hysteresis modulators, the phase shift is provided by a comparator with hysteresis. The oscillating control loop does not include the output L-C filter, and the actual hysteresis is applied either locally, around the comparator itself, or from the switching output of the power stage, taken before the L-C filter.

Phase-Shift Modulators

In the phase-shift modulator, the phase shift is provided by the total phase shift around the complete loop, *including* the output L-C filter, and also includes phase shift caused by switching time delays in the power stage.

One of the simplest, and most popular (in terms of units sold), hysteresis modulators is called the AIM (**A**stable **I**ntegrating **M**odulator) topology (Nielsen et al., 2009). It consists of an integrator in the forward path, with feedback taken *before* the output L-C filter, a power stage, and a comparator utilizing hysteresis.

This type of modulator is also known as a Sigma Delta Modulator (SDM), and can be further classified as either synchronous or asynchronous. The synchronous variety includes a flip flop with the comparator, which together comprise a quantizer that is clocked at a very high frequency. It suffers the disadvantage of in-band quantization noise. To reduce this quantization noise requires very high clock frequencies for the flip flop, which increase circuit complexity, cost, and power dissipation. The more popular variety is the asynchronous SDM, sometimes referred to as an analog SDM. In this variety, the hysteresis comparator alone is the quantizer and no clocked flip flop is needed. The quantization error of the asynchronous SDM is theoretically zero, because the switching does not occur in discrete time intervals, but instead occurs continuously, with infinite time resolution. The disadvantage of the asynchronous SDM is that its switching frequency varies with the amplitude of the input signal, decreasing as input signal increases.

It is worthwhile at this point to discuss the SDM in more detail, because of its great popularity in Class-D amplifier design.

Sigma Delta Modulators (Alternately Referred to as Delta Sigma Modulators)

An SDM usually consists of at least an integrator and a comparator, a DAC function (digital-to-analog converter), and a summer. The comparator is sometimes referred to as a 1-bit analog-to-digital converter (ADC) block. Typically, the power switching stage provides the two voltage states which are fed back and summed into the integrator. The two analog voltages provided by the power switching stage are analogous to the analog output voltages provided by the DAC block of the sigma delta converter (1-bit DAC). The main reasons for the popularity and success of the analog sigma delta converter are two features which it provides. These are oversampling and noise-shaping. Oversampling is a technique which improves conversion accuracy, which in Class-D amplifier design translates to lowering output noise and reducing distortion. Noise-shaping is a phenomenon which shifts quantization noise (present in any sampled system) up higher in frequency, ideally beyond the audible frequency range.

A basic delta sigma modulator is illustrated in Figure 21.3 (see also Yet Another Electronics Channel, 2020). It consists of a difference amplifier ("delta"), an integrator ("sigma"), and a comparator, often described as a 1-bit ADC. The textbook description of a delta sigma converter includes a clocked flip flop after the comparator (synchronous type mentioned above).

SDMs may be clocked (quantized) or continuous-time (non-clocked). Figure 21.3 illustrates a clocked SDM, whereas Figure 21.4 illustrates a continuous-time implementation. Theoretically, the continuous-time implementation could result in infinite switching frequency (and therefore unlimited oversampling), since the error signal hovers at the comparator switching threshold voltage. However, a small amount of hysteresis can be added to the comparator in this topology

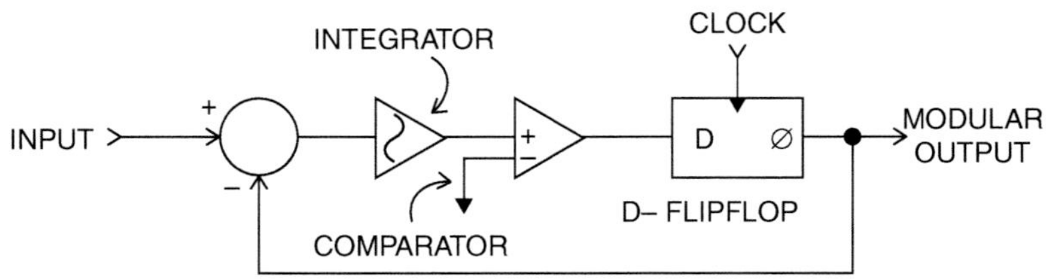

Figure 21.3 Block Diagram of a Clocked Sigma Delta Converter

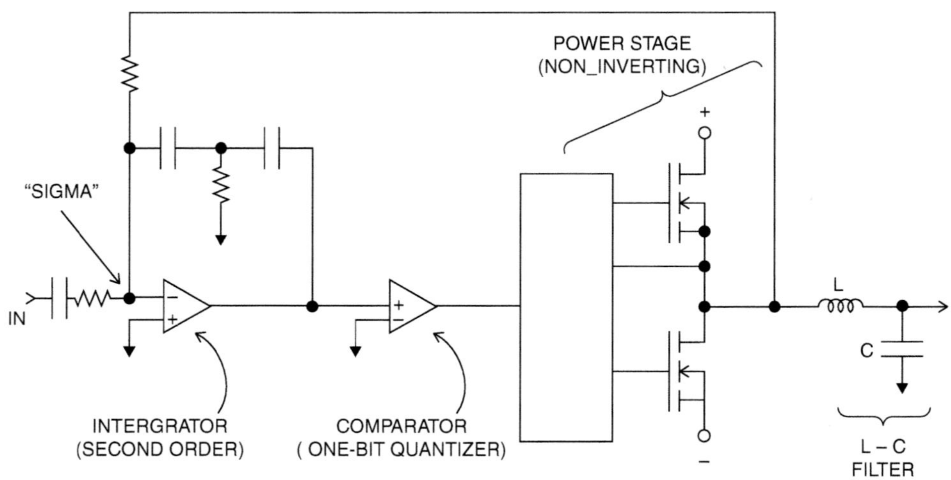

Figure 21.4 Second-Order, Analog Sigma Delta Converter

which will decrease the switching frequency into a range suitable for present-day Class-D modulators (300KHz–1MHz) (van der Hulst, Veltman, & Groenenberg, 2002). This hysteresis can be applied locally around the comparator itself, or by taking feedback from the switching node, which is switching rail-to-rail. In addition to providing the needed hysteresis, taking feedback from the switching node allows the timing and voltage errors of the Class-D switching stage to benefit from the negative feedback loop gain. It is also a very convenient connection point, because the switching node is available as a pin on Class-D integrated IC packages. This is another reason why this topology is so popular among integrated Class-D solutions.

Hysteresis controllers have been the workhorse of industrial electronics according to van der Hulst et al. (2002). However, they have not been as popular in audio electronics, and the reference postulates that the reason may be a perceived lack of accuracy or precision. In a hysteresis controller (or modulator, used interchangeably here), the "reference" switches between two discrete levels, and therefore there is an error that always exists between these two hysteretic switching levels. However, as the reference cites, if the error is ultrasonic, it can yield very low error and therefore low THD in the audio frequency range.

Since the hysteresis controller is an analog (continuous-time) circuit (as opposed to a clocked digital circuit), it does not suffer from quantization error, and no noise-shaping is necessary. van der Hulst et al. (2002) claims that because of this characteristic, lower-order feedback control loops, even first-order, are sufficient for good audio performance, as opposed to the higher-order feedback control necessary to provide noise-shaping in quantized control loops.

A good explanation of oversampling and noise-shaping can be found in Cordell (2019).

Oversampling

The Nyquist sampling rate needed for a 20KHz analog signal is two times the analog sampling frequency, or 40KHz. Oversampling refers to sampling at a rate higher than this Nyquist rate. This provides performance advantages. The term oversampling ratio (OSR) is the ratio of the sigma delta clock frequency to the Nyquist rate. One advantage afforded by oversampling is a reduction in in-band noise. This happens because the total noise is now spread out over a wider frequency range. For example, an OSR of two will spread the total sampling noise power over twice the frequency range, resulting in a halving of the in-band noise power (Cordell, 2019).

Noise-Shaping

In SDMs, oversampling works in conjunction with noise-shaping to provide further noise reduction. Whereas oversampling causes noise power to be spread out *evenly* over a given frequency band, noise-shaping refers to noise density differences across the same frequency range. The even spreading of noise by increasing OSR (due to oversampling) only results (typically) in a noise decrease of in-band noise equal to the square root of the OSR. Noise-shaping, as applied to SDMs, however, refers to a process whereby noise is *non-uniformly* pushed out of the audio frequency band and up to higher (ultrasonic) frequency bands (Meszlenyi, 2000). This is obviously beneficial for Class-D audio amplifier noise reduction. Noise-shaping utilizes negative feedback to push the noise power from lower frequencies to higher frequencies. In a negative feedback system, as used in a Class-D amplifier, loop gain (and therefore error correction) will be very high at low frequencies, and will decrease as frequency increases. The slope of this decreasing loop gain is dictated by the order of the system. A first-order integrator, for example, will have a gain decrease of 6dB per octave, whereas a second-order integrator will have a decrease of 12dB per octave. This means that quantization noise will be reduced by these same amounts. Now, since the total quantization noise power is fixed, if noise is reduced in one area of the frequency spectrum, it must be pushed out to another. In this case, the noise power removed from the low-frequency, in-band frequency range will be pushed out to ultrasonic range. In summary, high OSR will increase the available frequency range across which the (fixed amount of) quantization noise can be diluted, and noise-shaping (through negative feedback) will create a non-uniform distribution of this quantization noise. For example, for each doubling of the clock frequency (which doubles OSR), in-band quantization noise will decrease by 3dB, and implementing a second-order integrator in the negative feedback loop (which is commonly used in Class-D SDMs) will result in an additional 12 dB per octave of in-band noise reduction. This results in a net 15dB of in-band noise reduction due to the combined benefits of oversampling and noise-shaping (Meszlenyi, 2000).

Most modulators of this type take feedback from the power switching stage or "switching node" (i.e. *before* the output L-C filter). The advantages of this connection have been mentioned above, but a serious deficiency with this connection is that non-linearities in the L-C filter (discussed in Chapter 9) will not be corrected by the negative feedback loop. In this type of modulator, the components around the integrator (R's and C's) are adjusted to set the self-oscillating frequency at idle (Honda & Cerezo, 2003; Honda, Rodríguez, & Cerezo, 2007).

Phase-Shift Modulation

Another type of self-oscillating Class-D amplifier modulator topology is called phase-shift modulation. An excellent reference describing this is Putzeys (2005).

This type of modulator makes use of the phase shift inherent in the entire feedback loop of the amplifier (including the L-C filter), to provide the 180-degree phase shift required for oscillation. In this modulation scheme, the switching frequency of the amplifier *is* the feedback loop's unity gain crossover frequency. The desired frequency of oscillation is typically many times the corner frequency of the

L-C filter. At this high multiple of the corner frequency, the phase shift of the L-C filter is asymptotically approaching 180 degrees (but is still less than 180 degrees). The stability criterion states that the loop will oscillate at the frequency where the total phase shift of the loop is 180 degrees. Therefore, if the additional phase shift, which needs to be added to the loop in order to reach 180 degrees total, can be carefully controlled, the self-oscillation frequency can be purpose-designed. In the approach used in Putzeys (2005), this additional (small) phase shift is provided by the deadtime of the power switching stage. As discussed in previous chapters, this deadtime can be very well controlled. The benefits of this approach are that it allows feedback to be taken from the output of the amplifier (after the L-C filter) instead of before the L-C filter, and that the switching frequency is the same as the unity gain loop crossover frequency. The former benefit allows the feedback loop to correct for load-dependent non-linearities of the L-C filter, and the latter allows the loop gain to be higher over the audio BW as compared to a closed-loop fixed-frequency PWM amplifier operating at the same (idle) switching frequency. This, in turn, means greater error correction over the audio frequency BW, which results in lower THD and noise. This phase-shift oscillator also has much less of a switching frequency dependence on modulation index, compared with hysteresis modulators. In a hysteresis modulator, the minimum pulse width is half the pulse width at idle. This means that at the extremes of modulation index (duty cycle), the switching frequency drops dramatically. In contrast, the phase-shift modulator boasts much more stable switching frequency variation over a wide range of modulation index. The phase-shift modulator can also benefit from higher-order integrators in the feedback loop, which will increase loop gain at lower frequencies in the audio band, resulting in even better audio performance (principally lower THD+N).

Digital Modulators

An all-digital modulation scheme can be implemented. These can be divided into open-loop and closed-loop architectures. Open-loop architectures take advantage of the fact that digital audio sources and content proliferate. These are typically in pulse code modulation (PCM) format from an I2S bus (Inter-IC Sound developed by Philips Semiconductor). Digital circuitry converts the PCM audio data into PWM format, which directly drives the power switching stage of the Class-D amplifier.

Closed-loop all-digital implementations are more complicated. In order to close the negative feedback loop, the output signal of the Class-D amplifier (which is analog) must be digitized by an ADC. This converter must be very high speed, and with very low latency (delay), otherwise it will add phase shift in the feedback path, which could make the loop unstable. There are companies that have addressed this challenge and now offer all-digital closed-loop solutions (Schinkel et al., 2017; MPS, 2024).

More information on open- and closed-loop digital Class-D amplifiers can be found in Chapter 22.

Bibliography

Anderskouv, N., Nielsen, K., & Andersen, M. A. E. (1996). High-fidelity pulse-width modulation amplifiers based on novel double-loop feedback techniques. Paper presented at the 100th AES Convention, Copenhagen, Denmark, May 11–14, 1996.

Attwood, B. E. (1983). Design parameters important for optimization of very high-fidelity PWM (Class-D) audio amplifiers. *Journal of the Audio Engineering Society*, *31*(11), 842–853.

Bloechl, M., Bataineh, M., & Harrell, D. (2004). Class D switching power amplifiers: Theory, design, and performance. In: *IEEE SoutheastCon, 2004. Proceedings*, Greensboro, NC, March 26–29, 2004 (pp. 123–146). New York: IEEE.

Cordell, B. (2019). *Designing Audio Power Amplifiers* (2nd ed., Chapter 35). Abingdon: Routledge.

Dahl, N. J., Iversen, N. E., Knott, A., & Andersen, M. A. E. (2016). Comparison of simple self-oscillating PWM modulators. In: *Proceedings of the 140th Audio Engineering Society Convention*, Paris, France, June 4–7, 2016 (Article 9562). New York: Audio Engineering Society.

Dahl, N., Iversen, N., Knott, A., & Andersen, M. A. E. (2018). Optimal control of a high-frequency Class-D amplifier. *Journal of the Audio Engineering Society*, *66*(1–2), 34–43.

Duraij, M., Iversen, N. E., Petersen, L. P., & Boström, P. (2015). Self-oscillating 150 W switch-mode amplifier equipped with eGaN-FETs. In: *Proceedings of the 139th AES Convention*, New York, NY, October 29–November 1, 2015 (Article 9378). New York: Audio Engineering Society.

Gaalaas, E. (2006). Class D audio amplifiers: What, why, and how. *Analog Dialogue, 40*(6), 1–7.

Honda, J., & Cerezo, J. (2003, October 8). Class-D audio amplifier design tutorial. El Segundo, CA: International Rectifier.

Honda, J., Rodríguez, M., & Cerezo, J. (2007, July 27). Iraudamp5. El Segundo, CA: Infineon (International Rectifier). (Page 27 discusses difference frequencies between channels to reduce crosstalk. Page 26 calls out sigma delta modulation.)

Huffenus, A., Pillonnet, G., Abouchi, N., & Goutti, F. (2010). A comparison of phase-shift self-oscillating and carrier-based PWM modulation for embedded audio amplifiers. In: *Proceedings of the 128th AES Convention*, London, UK, May 22–25, 2010 (Article 8127). New York: Audio Engineering Society.

Infineon. (2019). MERUS^{TM} multilevel class D audio amplifier supports ultra-compact and low-power applications. Infineon white paper, 03-2019. Neubiberg, Germany: Infineon Technologies AG.

Jiang, X. (2017). Fundamentals of audio Class D amplifier design: A review of schemes and architectures. *IEEE Solid-State Circuits Magazine, 9*(3), 14–25.

Kang, Y., Ge, T., He, H., & Chang, J. S. (2016). A review of audio Class D amplifiers. In: *2016 International Symposium on Integrated Circuits (ISIC)*, Singapore, December 12–14, 2016 (pp. 1–4). New York: IEEE.

Kemp, P., Mouton, T., & Putzeys, B. (2011). High-order analog control of a clocked Class-D audio amplifier with global feedback using Z-domain methods. Paper presented at the 131st AES Convention, New York, NY, October 20–23, 2011.

Lehmann, C. (2009). A simple post-filter feedback topology for Class-D amplifiers. Paper presented at the AES 37th International Conference, Hillerød, Denmark, August 28–30, 2009.

Mei, S., Hu, Y., Xu, H., & Wen, H. (2022). The Class D audio power amplifier: A review. *Electronics, 11*(19), 3244. https://doi.org/10.3390/electronics11193244

Meszlenyi, I. (2000). Class D amplifier. US Patent 6320460 B1.

Moreno, S. S. (2005, June 4). Class-D audio amplifiers – theory and design (ed. & additional text by R. Elliott). Thornleigh, Australia: Elliott Sound Products (ESP).

MPS. (2024, February 21). Class-D audio with post-filter feedback using MPS Axign technology. Webinar. Kirkland, WA: Monolithic Power Systems, Inc. www.monolithicpower.com

Munk, S. M., & Andersen, K. S. (2003). State of the art digital pulse modulated amplifier system. Paper presented at the 23rd AES International Conference on Signal Processing in Audio Recording and Reproduction, Helsingør, Denmark, May 23–25, 2003.

Nguyen-Duy, K., Knott, A., & Andersen, M. A. E. (2014). Constant switching frequency self-oscillating controlled Class-D amplifiers. *Elektronika Ir Elektrotechnika, 20*(6), 84–88.

Nielsen, D., Knott, A., Pfaffinger, G., & Andersen, M. A. E. (2009). Investigation of switching frequency variations and EMI properties in self-oscillating Class-D amplifiers. In: *Proceedings of the 127th Audio Engineering Society Convention*, New York, NY, October 9–12, 2009 (Article 7907). New York: Audio Engineering Society.

Norsworthy, S. R., Shreier, S., & Temes G. C. (1996). *Delta-Sigma Data Converters – Theory, Design and Simulation*. Hoboken, NJ: IEEE/John Wiley & Sons.

Nyboe, F., Risbo, L., & Andreani, P. (2005). Time domain analysis of open loop distortion in Class-D amplifier output stages. Paper presented at the AES 27th International Conference, Hillerød, Denmark, September 2–4, 2005.

Poulsen, S., & Andersen, M. A. E. (2004a). Self-oscillating PWM modulators, a topological comparison. In: *Conference Record of the Twenty-Sixth International Power Modulator Symposium, 2004 and 2004 High-Voltage Workshop*, San Francisco, CA, May 23–26, 2004 (pp. 403–407). New York: IEEE.

Poulsen, S., & Andersen, M. A. E. (2004b). Simple PWM modulator topology with excellent dynamic behavior. In: *Nineteenth Annual IEEE Applied Power Electronics Conference and Exposition, 2004. APEC '04*, Anaheim, CA, February 22–26, 2004 (Vol. 1., pp. 486–492). New York: IEEE.

PowerPhysics. (2002). One-cycle sound. A PowerPhysics white paper. Newport Beach, CA: PowerPhysics Inc.

Putzeys, B. (2005). Simple self-oscillating Class D amplifier with full output filter control. Paper presented at the 118th AES Convention, Barcelona, Spain, May 28–31, 2005.

Putzeys, B. (2009). Globally modulated self-oscillating amplifier with improved linearity. In: *Proceedings of the 37th AES International Conference on Class D Audio Amplification*, Hillerød, Denmark, August 28–30, 2009 (Article 4). New York: Audio Engineering Society.

Putzeys, B. (2011, February 25). Ncore technology white paper. Groningen, Netherlands: Hypex Electronics BV.

Risbo, L. (2005). Discrete-time modeling of continuous-time pulse width modulator loops. Paper presented at the 27th AES International Conference: Efficient Audio Power Amplification, Hillerød, Denmark, September 2–4, 2005.

Schinkel, D., Groothedde, W., Mostert, F., Koerts, M.-J., van Iersel, E., Groeneveld, D., & Breems, L. (2017). A multiphase Class-D automotive audio amplifier with integrated low-latency ADCs for digitized feedback after the output filter. *IEEE Journal of Solid-State Circuits*, *52*(12), 3181–3193.

Shu, W., & Chang, J. S. (2008). THD of closed-loop analog PWM Class-D amplifiers. *IEEE Transactions on Circuits and Systems I: Regular Papers*, *55*(6), 1769–1777.

van der Hulst, P., Veltman, A., & Groenenberg, R. (2002). An asynchronous switching high-end power amplifier. Paper presented at the 112th AES Convention, Munich, Germany, May 10–13, 2002. New York: Audio Engineering Society.

Vanderkooy, J. (1985). Comments on "Design parameters important for the optimization of very-high-fidelity PWM (Class-D) audio amplifiers". Letter to the Editor. *Journal of the Audio Engineering Society*, *33*(10), 809–810.

Varona, J. (n.d.). Power digital-to-analog conversion using sigma-delta and pulse width modulations. ECE University of Toronto, ECE1371 Analog Electronics II.

Yet Another Electronics Channel. (2020). #21, Self-oscillating Class-D. *YouTube*. https://www.youtube.com/watch?v=E2LPpjhGIYk

Chapter Twenty Two

Digital Applications, Interfaces, and Digital Signal Processing

Digital versus Analog Circuitry

A good starting point for a discussion of digital circuit applications in Class-D audio amplifier design is to address the fact that, in some circles, Class-D amplifiers are referred to as "digital amps". This is at best misleading, and is dealt with quite well by some of the references cited at the end of this chapter (e.g. Putzeys, 2003), but is also worth mentioning here for clarification. The name "digital amp" is a misnomer for the great majority of Class-D amplifiers. The term digital should be reserved for circuits where only the logic state (high or low) and timing matter. In such circuits, the absolute *voltage* associated with the high or low logic state is of little importance, and contains no data. For example, a microprocessor or Digital Signal Processing (DSP) IC processes data bits as logic states: high or low (1's or 0's). The digital designer, using these ICs, does not concern himself or herself with the actual voltages associated with these logic states, as long as they meet the voltage thresholds for that logic state and do not violate noise margins. The information (data) processed by this digital circuitry is completely contained in the logic state, and not in the absolute voltage associated with that logic state. Analog circuitry is very different, in that the actual circuit voltages *do* matter, and do contain the information (and noise) processed by the analog circuits.

With these "definitions" in mind, the circuitry of a Class-D amplifier can be reviewed and parsed into analog and digital circuits. Taking the reference amplifier as an example, the input circuitry, up to the PWM comparator, is purely

 DOI: 10.4324/9781003327943-22

analog. The audio signal input is an analog voltage waveform containing all the desired audio information plus undesired noise, in a time-varying analog voltage waveform. From the output of the PWM comparator to the output of the gate drivers (driving the gates of the upper and lower half-bridge mosfets), an argument could be made to consider that circuitry to be digital, because the information is contained in the logic state (high or low) of those signals, and their timing relative to each other. The actual voltage at the input pins to the gate driver is not overly critical, as long as it meets the logic high/low threshold requirements, and timing requirements, such as slew rate and minimum pulse width. The output voltages of the gate driver, driving the gates of the mosfets, must have a logic low state that switches the mosfet fully off, and a logic high state that produces complete enhancement of the mosfet gate, for low R_{DS}ON, in the ON (conducting) state. Likewise, the timing of these drive signals must be precise, to prevent overlap and cross-conduction. This *digital* characterization degrades, however, when we start to consider the slew rate of the gate-drive signal, and the effects of the Miller plateau, gate charge, etc. These are all *analog* considerations. Next, when we consider the output stage and the switching node, it is switching rail-to-rail, which seems like a digital phenomenon, but the output of this switching stage is completely dependent on the voltage(s) (including ripple and noise) of these rail(s). The next stage in filtered Class-D amplifiers (as opposed to filterless) is the L-C demodulation filter. Here, the switched power rail voltages mentioned in the previous stage are applied to the L-C filter and become the demodulated (integrated or averaged) amplifier output signal, and thereby determine the *analog* output voltage of the amplifier, including its audio performance characteristics, such as gain and noise performance. Therefore, the L-C filter stage is a purely analog stage.

In summary, the great majority of Class-D amplifier designs are most definitely analog circuits, but incorporate digital circuitry where the benefits of digital circuitry can be exploited (these being high speed and frequency, noise immunity, low crosstalk, and low power dissipation).

With this understanding of analog and digital circuitry, we can explore other approaches to Class-D amplifier design that *increase* the involvement of digital circuitry and digital techniques in the design.

Field-Programmable Gate Arrays (FPGAs)

Field-Programmable Gate Arrays (FPGAs) can enhance the performance of a Class-D amplifier, by performing many of the housekeeping roles. Some examples of tasks that can be delegated to an FPGA are:

1) **Enable/Disable control:** The FPGA can use the outputs of UVLO and housekeeping circuits (voltage monitors and detectors) to determine when it is safe to enable the power stages of the amplifier. They can also include delays where appropriate.
2) **"Pop-less" PWM start-up and shutdown:** There are specific start-up sequences of the PWM timing for the Class-D power stage that will result in minimum start-up artifact (audible as a "pop" or "click"). An FPGA can control the PWM start-up pulse train and then relinquish control to the feedback error amplifier and PWM comparator after a specific time interval. It can also control the PWM stream during shutdown, if an early warning signal is used to alert the FPGA that power loss (rail collapse) is imminent, for example. Some of these PWM start-up and shutdown sequences are intellectual property (IP) protected by patents.
3) **Fault processing (from analog or digital fault detection circuits):** An FPGA can monitor the various fault detection circuits and take the desired action (such as limiting duty cycle, reducing signal path gain, disabling the power stage, or other). The FPGA can also re-enable the faulted channel after a specific time interval, if the fault has cleared, for example.
4) **Duty Cycle control:** The FPGA can take over control of the PWM stream from the error amplifier and PWM comparator in specific situations. These can be PWM start-up and shutdown, as mentioned above, but also to enforce minimum and maximum duty cycle limits. For example, some variable-frequency, self-oscillating designs can vary frequency over a very wide range. At excessively low switching frequency, or high output amplitude, the duty cycle extremes can have negative effects. These can be excessive output voltage ripple, or excessive ON-time for the upper mosfet in the half- or full-bridge power circuit, or excitation of the L-C filter resonance. If the gate-drive bias supply for these upper mosfets is recharged by a conventional bootstrap circuit, the boot capacitor voltage will droop below the UVLO threshold for the upper mosfet driver, and the upper mosfet will switch off. This will cause the

output voltage of the amp to droop. The feedback network will sense that the output is too low and will attempt to keep the upper mosfet ON even longer. This creates a positive feedback scenario where the output will remain at zero voltage until the input signal reverses polarity, calling for the lower mosfet to switch on, finally recharging the bootstrap capacitor. This behavior can result in severe audible output signal dropouts. By enforcing minimum and maximum duty cycle limits, the FPGA eliminates these dropouts, and causes a clipped output voltage response instead, which is much preferable to audio signal dropout.

5) **Clipping Detection:** If the allowable (unclipped) range of PWM duty cycle is known (especially for a fixed-frequency design), the FPGA can indicate *clipping* if the duty cycle extends beyond these extremes. In addition, the FPGA can take action based on this, such as forcing a reduction in signal path gain, to bring the amplifier out of clipping.

DSP

DSP (Digital Signal Processor, or Processing) refers to ICs that manipulate digital data, to perform roles as elementary as audio signal routing and volume control, or as sophisticated as mixing, audio effects such as spatial audio, echo cancellation, and other sound-processing effects. The implementation of these devices is beyond the scope of this book, but they are mentioned here because their role is increasing. In the future, DSP features will find increased integration in Integrated Circuit (IC) solutions for Class-D amplifiers.

DSP is also used in the feedforward and feedback paths of Class-D amplifiers, and is discussed in the following sections.

Digital-Input Class-D Amplifiers

Many currently available Class-D amplifier solutions feature ICs that process digital audio input signals. Various digital formats can be accommodated, such as PCM (Pulse Code Modulation), I2S (Inter-Integrated Circuit Sound), TDM (Time Division Multiplexing), PDM (Pulse Density Modulation), PWM, etc. These types of amplifiers process the digital data in these various standard formats, and ultimately produce a PWM or PDM data stream that drives the switching power stage. Many of these Class-D designs are open-loop designs (lacking negative feedback), because the output signals to the speakers are analog, and the input signals to which they would need to be compared (by the feedback error amplifier, for error correction) are digital. These open-loop amplifiers cannot rely on negative feedback to correct the many sources of error that cause distortion in a Class-D amplifier. To make up for this deficiency, these open-loop amplifiers sometimes try to correct these errors in other ways. For example, it has been discussed how Class-D switching stages have zero power supply rejection. Any voltage perturbation on the power supply rails is applied to the L-C filter and appears at the output. *Note:* These perturbations will be attenuated by the L-C filter only if the perturbation frequency is above the cutoff frequency (F_0) of the filter (usually greater than 50KHz). However, any power supply ripple or noise in the audio frequency band (20Hz–20KHz) will be passed through to the output unattenuated.

Likewise, any timing errors in the Class-D switching stage, introduced by deadtime, finite slew rate, and propagation delays, will produce distortion in the analog output of the amplifier. An approach taken by fully digital open-loop Class-D amplifiers is to correct for these errors in the forward signal path. This involves *knowing* what these errors are a priori, and building in correction algorithms in the DSP block. Using these algorithms, the DSP block makes the appropriate timing changes to the PWM data stream to compensate for these known errors.

In order to handle the PSRR problem, there are usually one or two approaches taken. The first is to use a tightly regulated power supply to generate the amplifier power rail. This power supply needs to have precise output load regulation, meaning that changes in output load current produce little output voltage change. This type of power supply also requires fast transient response, meaning that its response to load changes is extremely fast, preventing transient dips and overshoot in the output voltage in response to sudden load changes. The second approach is to implement a feedforward circuit (Putzeys, 2006b). This circuit modifies the PWM stream in direct response to changes in the power supply rail voltage. Since a disturbance (perturbation) of a known amplitude in the rail voltage will result in a disturbance of a known amplitude in the output voltage, it is possible to sense the amplitude of the power rail disturbance, and use an algorithm that modifies the PWM timing an exact amount in response to that disturbance. This will cancel the output perturbation.

Note the difference here, between this feedforward action and a feedback response. The feedback response relies on a small perturbation making it through to the output, where it is sensed by high loop-gain compensation circuitry, which seeks to reduce that error. This feedback loop is subject to the stability criterion mentioned in other chapters. In contrast, this feedforward response just described is *not* subject to the stability criterion, and instead relies on known correction algorithms to correct known errors, all in the forward signal processing path. The disadvantage to this approach is that the corrections are never perfect. The fixed amount of error correction applied may be slightly too much for some units, and slightly too little for others, based on the normal expected variations between units in high-volume production. In spite of these limitations for all-digital open-loop amplifiers, they do find application in low-power digital products, where ultra-low THD+N and ruler-flat frequency response are traded off, in favor of small size, low component count, and low cost.

"Fully Digital" Class-D Closed-Loop Amplifiers

This is a type of Class-D amplifier (Andersen et al., 2007; Bell, Jones, & Watts, 2007; Mouton & Putzeys, 2009) where the feedback incompatibility problem mentioned above is solved. The analog output signal of these amplifiers is converted to digital by high-speed, low-latency analog-to-digital-converters (ADCs) with extremely high resolution (Andersen et al., 2007). These then use DSP to compare the input and output digital data, and provide error correction to the data stream driving the output power switching devices. This type of Class-D amplifier topology deserves to be called "fully digital", even though it could be argued that the power switches still switch rail-to-rail, and the actual rail voltage levels and noise *do* affect the audio output performance (in terms of loop gain, THD, and noise performance). The enabling technology for this class of amplifiers is the low-latency ADC, and the correction algorithms implemented in the DSP block.

Bibliography

Andersen, J., Chieng, D., Harris, S., Klaas, J., Kost, M., & Taylor, S. (2007). All digital high resolution Class-D amplifier designs using power supply feed-forward and signal feedback. Paper presented at the AES 31st International Conference: New Directions in High Resolution Audio, London, UK, June 25–27, 2007.

Axign website. (n.d.). Enschede, Netherlands: Axign. https://www.axign.nl/technology/

Bell, C., Jones, D., & Watts, R. (2007). The application of direct digital feedback for amplifier system control. Paper presented at the 123rd AES Convention, New York, NY, October 5–8, 2007 (Paper 7226).

Esslinger, R., Gruhler, G., & Stewart, R. W. (2003). Feedback strategies in digitally controlled Class-D amplifiers. Paper presented at the AES 114th Convention, Amsterdam, Netherlands, March 22–25, 2003.

Mouton, T., & Putzeys, B. (2009). Digital control of a PWM switching amplifier with global feedback. Paper presented at the AES 37th International Conference, Hillerød, Denmark, August 28–30, 2009.

MPS. (2024, February 21). Class-D audio with post-filter feedback using MPS Axign technology. Webinar. Kirkland, WA: Monolithic Power Systems, Inc. www.monolithicpower.com

Putzeys, B. (2003, March). Digital audio's final frontier. *IEEE Spectrum Magazine.*

Putzeys, B. (2006a). All amplifiers are analogue, but some amplifiers are more analogue than others. Paper presented at the 120th AES Convention, Paris, France, May 20–23, 2006 (Paper 6690).

Putzeys, B. (2006b). Simple, ultralow-distortion digital pulse width modulator. Paper presented at the 120th AES Convention, Paris, France, May 20–23, 2006 (Paper 6694).

Schinkel, D., Groothedde, W., Mostert, F., Koerts, M.-J., van Iersel, E., Groeneveld, D., & Breems, L. (2017). A multiphase Class-D automotive audio amplifier with integrated low-latency ADCs for digitized feedback after the output filter. *IEEE Journal of Solid-State Circuits, 52*(12), 3181–3193.

Sozański, M. P. (2017). *Digital Signal Processing in Power Electronics Control Circuits* (2nd ed.). London: Springer.

Chapter Twenty Three

Integrated Circuit (IC) Approaches and Advances in Components and Materials

Integrated Circuit (IC) Solutions

Any thorough treatment of Class-D amplifier design must include a discussion of the currently available integrated circuit (IC) solutions. Of all the topics in this book, this is likely the most rapidly changing, as semiconductor manufacturers continue to design and develop new ICs with new levels of integration and features. Always check the manufacturer's website for the latest product information.

Historically, the adoption of paradigm-shifting technologies in electronic circuit design usually follows a pattern. The first circuits implementing the technology are constructed using the available building blocks. Arguably, these could be referred to as "proof-of-concept" designs. These designs may use devices which have very low levels of integration, perhaps performing just a single function. Then, if the new technology is found to provide benefits in terms of cost, performance, size, efficiency, etc., IC manufacturers will perform the ROI (return-on-investment) calculations to decide whether there is a business case for producing an integrated device, which addresses this market and includes a higher level of integration over the initial proof-of-concept designs. If the market grows, more manufacturers enter the arena, producing competitive devices, with greater levels of integration. The key feature here is integration. Integration offers many advantages over discrete designs, including lower component count, smaller size, and higher efficiency, due to reduced power loss, better noise control due to shorter interconnects, and better control over parasitics. Integration also eases the

DOI: 10.4324/9781003327943-23

circuit design task for the design engineer, which greatly facilitates adoption of the technology. So, there is a positive feedback effect here... The greater the adoption of the technology, due to the ease of design, the more the market grows, which then causes more development investment by the IC manufacturers, resulting in greater device offerings, with ever-increasing levels of integration.

It is straightforward to see this pattern in the evolution of Class-D amplifier design solutions. The first Class-D amplifiers used op-amps and comparators that were optimized for other applications. They used discrete transistor circuits for logic, level-shifting, and gate drivers, and of course the power switches were individually packaged mosfets. Integration increased with the adoption of IC op-amps and comparators with specifications suitable for Class-D requirements, such as low noise, wide bandwidth (BW), and low propagation delay. Then purpose-designed gate-driver ICs were developed. Originally these were only "low-side" types, but later, around 1991, International Rectifier introduced the first half-bridge driver IC with a floating high-side driver, with boot-strapped supply (Infineon Technologies (International Rectifier), 2019a; Nguyen, Fang, & New, n.d.). It was originally developed to address the off-line ACDC power supply market, but as a general half-bridge driver with high voltage capability, it was adopted for Class-D designs as well. The next level of integration was to integrate the gate drivers and the error amplifier and comparator, with or without protection circuitry, into a single IC. This approach still required discrete mosfets as power switches, and of course the separate inductor and capacitor for the demodulation filter. The next level of integration included the power switches in the IC package. At this level of integration, a Class-D amplifier could be designed with this single IC, the L-C filter, and a handful of discrete resistors and capacitors, and perhaps a heatsink, depending on output power level.

This chapter will present an overview of the present state-of-the-art offerings by the most well-known manufacturers of integrated Class-D amplifier solutions.

Note: When reviewing IC Class-D offerings, it is important to evaluate the following features, in order to make an intelligent selection:

1) Know beforehand what range of loudspeaker impedance you will need to drive, and to what power levels, and at what maximum THD+N level.
2) What are the maximum specified operational power supply rail voltages? This will determine the maximum output power available for driving a given loudspeaker impedance.

3) What input types are available? Single-ended analog, differential input analog, digital? And, if so, which data formats?
4) How many amplifiers are included in each package (and what topology; see also item (7) below)?
5) Is fixed-frequency or variable switching frequency topology used, and what is the range of switching frequency?
6) Is the device synchronizable to an external clock?
7) What is the power stage topology: half-bridge or full-bridge (BTL), parallel BTL (PBTL)?
8) Is filterless operation possible?
9) Is parallel BTL operation possible, and if so, can it be implemented pre-filter or post-filter?
10) Is negative feedback used? Is the feedback taken before or after the L-C filter? Can external feedback be added (Kisling & Beardsworth, 2018)?
11) What type of fault protections are included, and what is the response to each fault type? Typically, these protections include overcurrent and thermal, but may also include DC at output, and clipping. Will there be audio dropouts during protection events?
12) What is the quiescent power?
13) What is the operational power conversion efficiency?
14) Audio performance: typically, at least THD+N, frequency response, and SNR are spec'ed.
15) Does the device have an enable/disable and/or *mute* function?
16) What EMI mitigation features are included (such as spread spectrum)?
17) Package type and cooling features: top/bottom cooling, and compatibility with external heatsink, or cooling via PCB copper?
18) Will there be special signal processing features, such as DSP?

Texas Instruments (TI)

Texas Instruments has a very wide Class-D offering, including extremely low-power to high-power devices. Most devices are fixed (but selectable) frequency types, and "closed-loop" parts which take feedback from the switching node and do not include the L-C filter in the loop. This is a convenient topology for integrated designs, because the switching node is at a package pin, whereas the output of the L-C demodulation filter is not, and would therefore require a PCB trace

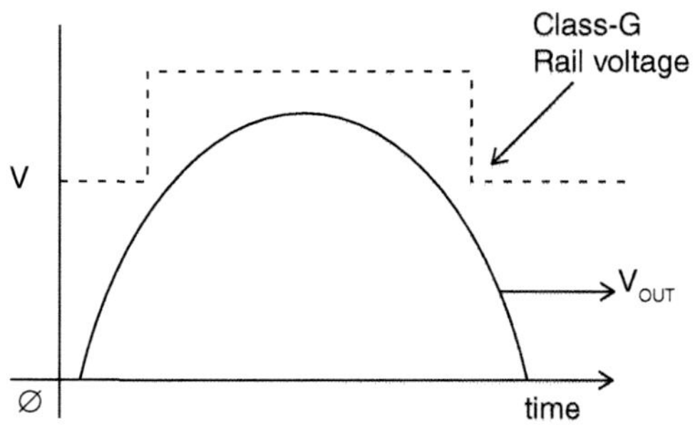

Figure 23.1 Boosted Class-G

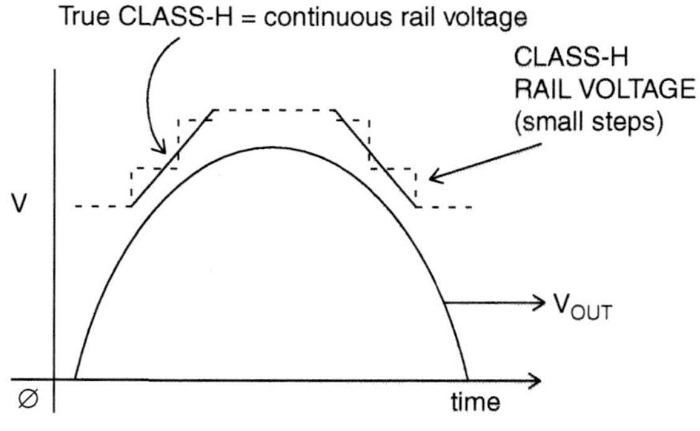

Figure 23.2 Boosted Class-H

connection. TI does, however, provide guidance for adding post-filter feedback (PFFB) in application notes (Kisling & Beardsworth, 2018). TI also offers analog, digital, and PWM input devices, as well as devices optimized with features for low quiescent current, mainly targeted for battery-powered applications. There is great versatility offered, including powerful DSP and interface features. Multiple power stage architectures are supported, including SE (single-ended, half-bridge), BTL (AD, BD, and ternary modulation), as well as parallel bridge-tied load (PBTL), applied either *before* (pre-filter PBTL) (Clifton, 2018) or *after* the inductors (post-filter PBTL). Extensive protection features are offered as well, including cycle-by-cycle current limiting, output DC protection, and thermal protection. TI also has Class-D product featuring "boosted Class-D" mode (Bryson, 2019). This allows the power supply (single rail) to boost its output voltage in coordination with the required output power level. Some devices feature Class-G boost (single-step power supply voltage change) and Class-H (multiple-step power supply output voltage change, tracking the envelope of the audio output voltage) (Bryson, 2019). The boost power switch is included in the integrated package, with the boost inductor external.

Texas Instruments (2016) is a datasheet for a representative high-power integrated Class-D amplifier solution. The Texas Instruments website (www.ti.com) also provides extensive application notes for Class-D applications.

Infineon

Infineon acquired the International Rectifier line of Class-D ICs. This has resulted in an extensive Class-D IC offering. The line includes controller ICs (which drive external mosfets), as well as devices with integrated power switches. Very high power designs are possible due to the high voltage capabilities of the devices (up to +/−200V) (Infineon Technologies (International Rectifier, 2019b). The innovative MERUS (Infineon, 2019) line touts multi-level switching capability, with the following claimed performance features:

1) Low quiescent current, enabling high efficiency, even at low output power
2) Compact size due to internal switches
3) Lower EMI, requiring smaller mitigation components
4) Digital and analog input versions
5) Fourth-order internal feedback loop
6) Three-level PWM for half-bridge and five-level PWM for full-bridge topologies.

MERUS multi-level audio ICs use four transistors per half-bridge instead of the typical two. Each of these four transistor switches is driven by a dedicated PWM drive signal. If a full-bridge topology is used, the PWM drive signals create a five-level switched output across a differentially connected speaker load. This has the benefit of greatly increasing the effective switching frequency, which allows for an increase in the feedback loop BW, as well as smaller voltage steps across the load. Thus, the switched output (before demodulation filtering) already more closely approximates its demodulated result. For low-power (less than 20W typical) applications, the L-C filter is therefore not needed (i.e. a filterless design is sufficient).

The Infineon website (www.infineon.com) also provides extensive application notes for Class-D applications.

Analog Devices

Analog Devices, which acquired Maxim, offers devices featuring differential analog inputs, digital inputs, and output power ranging from 165mW to 39W, with total supply rail voltage up to 40V. Features include external clock synchronization, low quiescent current, programmable volume control, spread-spectrum for reduced EMI, boosted Class-D, and Dynamic Speaker Management (DSM). Maxim Integrated (Analog Devices) (2020) is an example of an integrated solution, with boosted Class-D operation and DSM.

Monolithic Power Systems (MPS)

MPS, which acquired Axign, offers devices with analog inputs and PWM digital inputs, with output power ranging from 2.7 to 90W. MPS, through acquisition of Axign, now also offers fully digital Class-D solutions featuring post-filter digital feedback, using low-latency A/D conversion.

Their audio portfolio also includes digital controllers and processors, accepting TDM and I2S input formats, as well as analog input Class-D ICs with integrated power switches, operating in half-bridge or full-bridge BTL topologies. MPS (2013) is an example of a 2 × 30W integrated solution. MPS also offers Class-D power switching stages, featuring half- and full-bridge outputs and accepting PWM inputs. These incorporate various protection features and fault flags as well. MPS (2018) is an example of such an integrated Class-D power driver.

NXP

NXP offers devices ranging from 7 to 250W. They feature I^2C control and selectable AD or BD modulation, as well as low quiescent current, and spread spectrum for low EMI. Some devices feature the ability to readback load diagnostics and fault conditions over I^2C (NXP, 2013).

ST Microelectronics

ST Microelectronics features integrated Class-D offerings ranging from 2W to 2 × 160W (ST Microelectronics, 2020a). They are externally synchronizable, and can be configured for filtered or filterless operation. The analog input offerings can accept differential inputs. They also offer digital Class-D power stages at various power levels. These feature PWM inputs and various protection features (ST Microelectronics, 2020b).

Cirrus Logic

Cirrus Logic offers 5.3W and 6.8W integrated Class-D amplifier solutions with DSP. These feature low quiescent current, I2S or TDM digital inputs, and Class-H, envelope-tracking, rail-boost circuitry (Cirrus Logic, 2018).

ICEpower

ICEPower offers Class-D audio chipsets as well as module (turnkey) solutions (see the "Class-D Modules" paragraph below). The portfolio covers an extremely wide power range, up to 4KW (ICEpower, 2022), due to supply voltage rail capability up to 220V. The Class-D amplifier module solutions are offered with and without ACDC power supplies which operate over the full universal AC mains voltage range.

Class-D Modules

Another type of integrated Class-D solution is the Class-D *module*. These are sometimes called "turnkey solutions". These are complete Class-D audio power amplifier solutions, featuring input circuitry, modulators, power stages, L-C filters, connectors, and heatsinking (if needed), on a PCB. They may or may not also include an ACDC power supply to power the AMP, comprising a complete turnkey solution. These are made by many audio companies, including ICEpower (ICEpower, 2022), Hypex, Pascal, and Purifi Audio.

Advances in Components and Materials

Any discussion of Class-D audio amplification would be incomplete without a discussion of advances in component and material technologies. Indeed, the steady improvement in Class-D performance, and its widening acceptance and adoption, have been made possible by advances in electronic components and materials.

One of the major advances that has made Class-D amplification viable, as a technology that can provide audiophile performance with outstanding power efficiency, is the development of the silicon power mosfet. The superior switching capabilities of the mosfet compared to its predecessor, the bipolar power transistor, allowed switching frequencies to increase dramatically, from the low 100KHz–200KHz range for bipolar devices, up to about 800KHz for mosfet-based designs. Presently, mosfets are used almost exclusively as the power switch for Class-D audio amplifier designs. In the future, wide-bandgap (WBG) devices, such as Silicon Carbide (SiC) and Gallium Nitride (GaN) in particular, will make possible the next big leap in Class-D amplifier performance. Technological advances, however, are not restricted to power switches. There have been critical advances in gate-driver circuits, isolation technologies (for level-shifting), high-speed comparators, high-speed and low-noise op-amps, current-sensing devices, inductor and transformer core materials and magnet wire, capacitor technologies, as well as PCB materials and thermal management materials.

Power Switches

Silicon Mosfets

Improvements in silicon mosfets relevant to Class-D operation include:

1) Improved figure of merit (FOM), defined as the product of $R_{DS}ON$ times gate charge (Q_G)
2) Enhanced body diodes, including improved reverse recovery characteristics such as shorter reverse recovery time (t_{RR}), lower reverse recovery charge (Q_{RR}), and softer reverse recovery characteristics (less snappy for less noise and EMI)
3) Improved packaging, including top- and bottom-side cooled SMD packaging, and low package inductance
4) Higher dv/dt capabilities for both mosfet and body diode
5) Lower capacitances (C_{OSS}), allowing higher-frequency operation with less power loss.

GaN

GaN devices have many advantages, which make them well suited for Class-D operation and superior to silicon mosfets in many respects. Arguably the most significant of these benefits is the absence of the p-n body diode in GaN devices. GaN devices have the ability to conduct current in the reverse direction from Source to Drain (third-quadrant operation) (Sun, 2019) without a body diode. This occurs whenever the drain is forced to a lower voltage than the source, even at 0V gate bias (V_{GS}). This operating mode is essential in Class-D designs, so that the switch-node voltage remains clamped to the power supply rails during the switching deadtimes. The p-n junction body diode in silicon mosfets is the principal source of switching noise in Class-D designs in which they are employed. The absence of this body diode in GaN devices allows for pristine switching waveforms in Class-D designs, when careful layout rules are followed. It must be noted, however, that the V_{SD} voltage, when current flows from source to drain, is significantly higher than the (approx.) 1.2V of the silicon mosfet. Typical voltages are in the 2–3V level, and increase with current, and decrease with temperature. As will be understood from other chapters in this book, this higher V_{SD} voltage is only present during switching deadtime, when V_{GS} = 0V. However, it will increase the open-loop distortion of the power stage. This can be mitigated to some extent by minimizing deadtime, which the faster switching speeds of GaN facilitate. It can also be helped by adding Schottky diodes in shunt across each GaN device in the power stage, or other changes to the bootstrap circuitry (discussed in the gate drivers section below). Adding clamp diodes, however, decreases the attractiveness of the GaN solution, because it complicates the layout of the power stage, which should be kept as small and tight as possible to minimize parasitics and ringing, especially since switching frequency and switching edge speeds have been increased.

GaN devices have tighter restrictions on gate bias voltage than silicon mosfets. Whereas silicon mosfets typically tolerate +/−20V for V_{GS}, most GaN devices utilize a gate voltage of 0V to +5V, with an absolute max of between 6V and 7V depending on the manufacturer. Different manufacturers may advocate the use of a small negative (OFF) gate bias, but be aware that this negative gate bias may increase the V_{SD} voltage discussed above. Careful control of gate voltage is therefore mandatory in gate drivers employed in GaN designs. To address this gate voltage sensitivity, some manufacturers integrate gate drivers into their GaN power switches. Other gate drivers targeted for GaN include voltage regulators for high-side and low-side gate switches, to provide better control over gate voltage. GaN devices are offered by many manufacturers with V_{DS} ratings of 100V and 650V. The selection of devices with V_{DS} ratings between 100 and 650V is much more limited. GaN devices are primarily offered in surface-mounted (SMD) packages, allowing bottom-side (through-the-PCB) cooling, and sometimes top-side cooling package configurations as well. These cooling techniques are well suited to compact Class-D amplifier designs.

SiC

SiC is another majority carrier semiconductor material, but whereas GaN power devices do not have a body diode, SiC power devices *do* have a body diode, but it is a Schottky device, with (theoretically) zero reverse recovery time, because there is no minority carrier current flow during reverse recovery. In reality, SiC device datasheets do specify some small recovery times, principally due to junction capacitance and the small electron drift (recovery) current (ROHM Semiconductor, 2017). However, these reverse recovery times for SiC are much smaller than for silicon mosfets with p-n body diodes.

At present SiC is targeted towards very high-power (>1KW) and high-voltage (650V/1200V) applications. These include electrical vehicle drive systems (inverters), and high-power battery charging systems and solar power inverters. Also, because of these targeted markets, the large through-hole packages available for SiC devices are, for the most part, not amenable to small, compact Class-D amplifier designs. So, though SiC is a viable switch technology for Class-D power switching stages, GaN will supersede SiC for Class-D usage, at least for the foreseeable future.

Gate Drivers Optimized for GaN

Gate drivers optimized for GaN devices should have the ability to operate at higher frequencies (compared to conventional gate drivers). This means at least 2MHz operation, with extremely high precision in selectable or adjustable deadtime. Because GaN devices are less forgiving with regard to gate overvoltage than are silicon mosfet power transistors, gate drivers should have provisions for precise limiting of gate-drive output voltage or Vcc voltage. In particular, high-side driver bias must be dealt with carefully. It is quite easy for the high-side bias voltage to vary greatly if it is derived from a diode-bootstrap circuit, which is recharged when the lower power switch is ON (as is typically employed in half-bridge drivers used in Class-D amplifier applications). Depending on the current direction in the lower switch, and inductances in the power loop, overvoltages in the high-side bootstrap supply can easily result. A purpose-designed GaN gate-driver circuit must take all this into account, and prevent this overvoltage from occurring. One method that is used is to include a fast switch in the bootstrap circuitry that only connects the bootstrap diode when the lower GaN switch is driven ON. This works as follows: When the lower GaN switch is OFF, and current is flowing source-to-drain (i.e. third-quadrant operation; Sun, 2019), the switch-node voltage's excursion below the negative rail is only limited by the third-quadrant characteristics of the GaN switch. It behaves as a symmetrical device, and exhibits "diode-like" behavior in this mode, but the source–drain voltage can be significantly higher than a p-n junction diode. This extra voltage could result in overcharging of the bootstrap supply for the upper switch. Including a high-speed switch in the bootstrap circuit *disconnects* the bootstrap diode when the lower switch is off (during deadtime), while this excessive source–drain voltage is present at the switch-node. At the end of the deadtime, when the lower GaN switch is turned on, its V_{DS} voltage will be very low, and the high-speed bootstrap switch turns on, allowing the bootstrap diode to recharge C_{BOOT} to the safe, nominal level. Drivers like TI's LMG1210 (Texas Instruments, 2019) include this bootstrap switch. Also, because of the lower gate voltage operation of GaN, newer gate drivers, optimized for GaN, will have options featuring lower UVLO thresholds for the upper and lower gate drivers.

Other important specifications for these newer drivers will be very low propagation delays and propagation delay-matching between upper and lower drivers, as well as high dv/dt rating and common-mode transient immunity (CMTI). Both of these latter specifications concern how fast the switch-node and associated upper gate-driver circuit nodes can slew, without incurring logic errors. This is an indication of the robustness of the level-shifting circuitry used in the driver.

Comparators

As discussed in other chapters in this book, there are particular performance demands for the PWM comparator. It must have very high switching speed, and low propagation delays. It would be quite beneficial if the common-mode input range for the PWM comparator were increased. For Class-D amplifier designs incorporating a triangle RAMP, as does the reference design, increasing the peak-to-peak amplitude of the triangle ramp (all things being equal) will decrease the noise floor of the amplifier. This is because *fixed* noise amplitude on the triangle ramp gets amplified by the gain of the modulator (V_{RAILS}/V_{RAMP}). However, if the peak-to-peak amplitude of the ramp is increased, the common-mode input range of the PWM comparator must be increased to accommodate it. Typically, the best high-speed comparators have rail voltages in the +/−5V to +/−6V range, and some have rail-to-rail input voltage capability. This means they would be limited to a triangle ramp amplitude of this same range.

Other important characteristics for the PWM comparator are the ability to handle a bipolar input signal (i.e. a ground-referenced signal swinging from positive to negative voltage), while also having an output stage capable of high-speed unipolar swing. This allows easy interfacing of the "digital" comparator output signal to level-shifters and gate drivers, with minimum propagation delay.

Op-Amps

The critical specifications for op-amps used in Class-D application are low-noise and wide BW (15MHz–100MHz typical). Other features which can be optimized are wide input common-mode voltage range, input overdrive protection, as well as protections against output phase reversal, and wide (rail-to-rail) output voltage swing. Minimum quiescent current is desirable (as always to maintain high overall amplifier efficiency, especially at low output power levels).

Current Sensors

The requirements for current sensors targeted for Class-D application are wide BW and minimum propagation delay. They must handle bidirectional currents, with high accuracy. Hall-effect devices deliver these attributes, while providing isolation. The isolation allows for the current-sensing path to be at the potential of the amplifier output voltage, while the output section of the current sensor can be freely referenced as desired (usually to circuit ground), so that it can interface with fault detection circuits. Hall-effect sensors also increase overall amplifier efficiency because they eliminate a resistive shunt, which would otherwise contribute to power dissipation losses.

Inductor Core Materials

The Class-D inductor used in the L-C demodulation filter is a critical component of Class-D design, and also one of the main power-dissipating components, usually second only to the power switches. Therefore, optimization of this component, in terms of performance and power loss, is critical for an optimum Class-D design. Presently, ferrite core materials exhibit the lowest core loss, but being gapped structures, these also exhibit a sharp saturation characteristic. This means they are not very tolerant of excessive current levels, which places greater demands on the overcurrent protection circuitry to achieve a reliable design. In contrast, distributed-air-gap core materials provide a much smoother saturation characteristic, indicated by a gradual rolloff of inductance with current, making them more tolerant of transient overcurrent events. Recent progress in core materials involves lower core loss distributed-air-gap materials, in new shapes, which can result in more efficient and compact Class-D designs. Progress in ferrite materials also involves new ferrites with lower core loss, and optimization for higher-frequency operation, with higher flux density swings. In addition, new core shapes are always being developed, with both through-hole and surface-mounted options.

Magnet Wire

Advances in magnet wire include flat wire technologies (instead of round wire) and litz wire, to allow for more compact windings and smaller inductors. They also include improved coatings for higher temperature and breakdown voltage.

Capacitor Technologies

The trend in multi-layer ceramic capacitors (MLCCs) is to provide greater capacitance in smaller package sizes. Ceramic capacitors boast very much lower ESR and ESL, and different geometries are available to improve ESL. Reverse geometry packages, for example, with the terminals on the long package dimension, help to reduce inductance. The most popular dielectrics remain X7R, X5R, and C0G/NP0 when the most stable characteristics are required.

Polymer electrolytic capacitors are becoming more popular, especially as tantalum replacements and enhancements of Aluminum Electrolytic (AE) types. They feature lower ESR and higher ripple current ratings than AE types, but are not yet available in voltage ranges as high as AE capacitors. They are becoming more popular in low-voltage DCDC converter circuits.

PCB Materials and Construction

The trend in Class-D amplifier thermal management, especially for low- to moderate-power designs, is to use the PCB for heat conduction and cooling. To this end, FR-4-based PCBs are still popular. Adding copper areas and thermal vias are the most economical and practical ways to increase heat conduction to the PCB from power semiconductor packages. PCBs with metal substrates are available, and have excellent heat conduction characteristics, but these are typically too costly to be considered for mainstream Class-D amplifier applications.

Thermal Management Materials: Gap Pads

Due to the high, and increasing, efficiency of Class-D audio power amplifiers, it has become possible to eliminate discrete heatsinks in many low- to moderate-power applications, and instead rely on the PCB as the primary "heatsink." Power loss, generated by heat in PCB-mounted power semiconductors, is conducted from the packages of these devices into the PCB, through wide copper areas, the liberal use of thermal vias, and multiple copper layers. In addition, compliant gap pads are often used. These are thermally conductive, electrically insulating materials that are conformable to the surface contours of components mounted to PCBs. By conforming to these surfaces, they can maintain contact with both the semiconductor package top, as well as the surrounding PCB copper, and can conduct heat from both of these surfaces, to a chassis or other heat-radiating surface. Pressure is applied to compress the gap pad, and thermal performance of these pads is specified under different degrees of compression. They are offered in many different thicknesses, thermal conductivities, and grades of "softness".

Integrated Class-D (IC) Solutions

As mentioned earlier in this chapter, progress in this area includes devices featuring multi-level switching topologies, and fully digital implementations, including feedback after the L-C filter.

Bibliography

Bryson, S. (2019, April). Benefits of Class-G and Class-H boost in audio amplifiers. Texas Instruments Application Report ALAA888. Dallas, TX: Texas Instruments.

Cirrus Logic. (2018, November). Audio product bulletin, CS35L41. Austin, TX: Cirrus Logic.

Clifton, R. (2018, September). Pre-filter parallel bridge-tied load (PBTL). Texas Instruments Application Report SLAA822. Dallas, TX: Texas Instruments.

Honda, J., & Cerezo, J. (2003, October 8). Class-D audio amplifier design tutorial. El Segundo, CA: International Rectifier.

ICEpower. (2022). Datasheet, 2000ASHV, Version 1.1. Søborg, Denmark: ICEpower.

Infineon. (2019). MERUS™ multilevel class D audio amplifier supports ultra-compact and low-power applications. Infineon white paper, 03-2019. Neubiberg, Germany: Infineon Technologies AG.

Infineon Technologies (International Rectifier). (2019a, June 3). Datasheet, IR2110, IR2113, No. PD60147 rev. V. El Segundo, CA: International Rectifier.

Infineon Technologies (International Rectifier). (2019b, June 4). Datasheet, IRS2461S. Rev. 2.3. El Segundo, CA: International Rectifier.

Kang, Y., Ge, T., He, H., & Chang, J. S. (2016). A review of audio Class D amplifiers. In: *2016 International Symposium on Integrated Circuits (ISIC)*, Singapore, December 12–14, 2016 (pp. 1–4). New York: IEEE.

Kisling, D., & Beardsworth, M. (2018, March). TPA324x and TPA325x Post-filter feedback. Texas Instruments Application Report SLAA788A. Dallas, TX: Texas Instruments.

Maxim Integrated (Analog Devices). (2020, July). Datasheet, MAX98390, Rev. 3. San Jose, CA: Maxim Integrated Products, Inc.

Mei, S., Hu, Y., Xu, H., & Wen, H. (2022). The Class-D audio power amplifier: A review. *Electronics*, *11*(19), 3244. (*Note:* This reference has an extensive reference section.)

Monolithic Power Systems (MPS). (2013, October 14). Datasheet, MP7748S. Rev. 1.0. Kirkland, WA: MPS.

Monolithic Power Systems (MPS). (2018, October 24). Datasheet, MP8049S. Rev. 1.01. Kirkland, WA: MPS.

Nguyen, J., Fang, M., & New, D. (n.d.). IRS211(0,3) and IR211(0,3) comparison.

Infineon Technologies (International Rectifier) Application Note AN1100. El Segundo, CA: International Rectifier.

NXP. (2013, May 2). Datasheet, TDF8599A. Rev. 3. Eindhoven, Netherlands: NXP Semiconductors.

ROHM Semiconductor. (2017). Effective use of power devices: Silicon carbide power devices: Understanding & application examples utilizing the merits. Santa Clara, CA: Rohm Semiconductor Co., Ltd.

ST Microelectronics. (2020a, June). Datasheet, TDA7498E, DS8807. Rev. 4. Geneva, Switzerland: ST Microelectronics.

ST Microelectronics. (2020b, November 16). Datasheet, ST516B, DocID 13183. Rev. 7. Geneva, Switzerland: ST Microelectronics.

Strydom, J. (2013). Dead-time optimization for maximum efficiency. Efficient Power Conversion (EPC) white paper: WP012. El Segundo, CA: EPC.

Sun, B. (2019, February). Does GaN have a body diode? Understanding the third quadrant operation of GaN. Texas Instruments Application Report SNOAA36. Dallas, TX: Texas Instruments.

Texas Instruments. (2016). Datasheet, TPA3255 315-W stereo, 600-W mono PurePath™ ultra-HD analog-input, SLASEA8A. Originally published February 2016, revised October 2016. Dallas, TX: Texas Instruments.

Texas Instruments. (2019). Datasheet, LMG1210, SNOSD12D. Originally published November 2018, revised January 2019. Dallas, TX: Texas Instruments.

Index

AC-coupled 53, 59, 72, 138, 139–140, 256, 392, 393
ACDC 342
ADC 415, 420; low-latency 429
AD Modulation 138, 142, 142–144, 300, 301, 352
AES17 391
air core 158, 194, 327
air gap 194, 195, 381, 443
aluminum nitride 290
Analog Devices, Inc 436
anti-parallel 19
asymmetric drive 110
Attwood, Brian 20, 197, 237
audio precision (AP) 23, 389, 393
AUX-0025 23
A-weighting 23, 391Ayrton Perry 194

Baker clamp 19
balanced 58–60, 112, 149, 212, 257, 354
bandwidth (BW) 3, 14, 18, 23, 50, 51, 74, 143, 150, 176, 219, 282, 315, 387, 391, 407, 432
Bang and Olufsen 21
BD modulation 138, 142–147, 300, 302, 434, 436
beat frequency 413
belly band 381
Bergquist 333
beta 18, 85, 253, 292
B-field 296, 327, 381
bias: current 9, 51, 81, 250, 251; supply 114
bipolar: drive 121; input signal 442; supplies 27, 53, 126, 140, 357, 372; transistor 18, 19, 20, 76, 78, 92, 110, 125, 438
Bocherot Cell 154
Bode plot 150–54, 157–160, 222–223, 235, 236
body diode 188–189
boosted Class-D 434, 436
bootstrap: capacitor ; circuitry 104, 115, 196, 325, 326, 426, 441; diode 325, 327, 441; rectifier 380; resistor 327
BTL 17, 125, 177–180, 306, 308, 350–352, 372, 390, 409, 433, 434, 436
buck converter 355–356
buffer 80, 387
Butterworth response 149, 151

capacitor (capacitive): aluminum electrolytic 140, 192, 297, 314, 335, 392, 443; ceramic: X7R 175, 191, 336, 392, 443; X5R 174–175, 191, 336, 392, 443; NP0/C0G 172, 175, 191–192, 336, 443; coupling 54, 77, 108, 164, 329,

377, 380, 402; ESL 54, 150, 191, 192, 297, 303, 305, 309, 314, 389, 443; ESR 150, 173–174, 191–193, 389, 443–444; feedthrough 305; film 172–176, 192, 194, 303, 335, 392; polymer 443; shunt 150, 159–160, 303, 389; temperature coefficient 191, 192; voltage coefficient of 174–175, 191, 192, 392
CCIF 395
CCIR-468 391
Cirrus Logic 437
Class-A 5
Class-AB 6, 8, 10
Class-B 7, 9
Class-D: boosted 434, 436; filterless 14, 17, 143, 176–181, 300, 389, 390, 425, 433; module 437
Class-G 434
Class-H 434
clipping: detection 68
closed loop 53, 102, 145, 150, 154, 173, 184, 212, 219, 221, 224, 225, 237, 285, 395, 399, 400, 407–410, 419, 420, 429, 433
CMOS (Complementary Metal Oxide Semiconductor) 105, 111
CMRR (Common Mode Rejection Ratio) 50, 52, 58
common impedance 313, 320–322, 324, 337, 378, 382, 402
common mode (CM): common mode rejection (CMR) 291; Common Mode Rejection Ratio (CMRR) 50, 52, 58; common mode transient immunity (CMTI): 87, 441
comparator 17, 441
compensation 219
compliance voltage 80
composite core 327
conduction (cooling) 444
continuous mode 355, 356, 358–361
convection cooled 255, 330, 353
corner frequency 57, 149, 191, 234, 256, 267, 369, 418–419
continuous conduction mode (CCM) 355–357
continuous mode 355, 356, 358–361
constant voltage (amplifiers) 365
critically damped 149, 150
crest factor 294, 343
crowbar 269
cross conduction: 15, 92, 93, 100–102, 110, 127, 131, 133, 185, 212, 275, 276, 425
crossover distortion 8, 10, 51, 395
crossover frequency 151, 222, 225, 230, 237, 239, 240, 407–409, 418–419
crosstalk 54, 378, 384, 401, 402; aggressor 386, 401; victim 327, 378–381, 383, 386, 401, 403
crystal oscillator 77, 172, 335, 336, 361
current containment 316–318, 320, 337, 361
current mirror 76, 77, 80
current sink 131
current source 76, 80, 251, 252, 292

DAC 415
damping factor (DF) 173, 232, 399, 401, 409
dB (decibel) 52, 239, 343, 401, 418
DC-coupled 59
DCDC 354
DC-offset 26, 27, 53, 72, 82, 128; fault 71, 262
deadtime: 3, 15, 92, 100, 101, 102, 110, 117, 131, 132, 185–190, 196, 201, 204
delta sigma modulator (DSM) 415
demodulation filter 26, 35, 75, 127, 129, 136–138, 153, 173, 176, 189, 192, 194, 211, 219, 271, 303, 327, 389, 395, 398, 399, 412, 425, 432, 433
DFD 395
differential mode (DM) 52, 142, 145, 299
digital: amp 17, 420, 427, 429; audio sources 419; circuitry 4, 28, 50, 354,

419, 424–425; control 98; feedback 237; inputs 435, 436, 437; isolator 87; logic 27; modulator 419
diode: body 109, 117, 134, 136, 137, 187–190, 195, 196, 204–213, 303, 318; clamp 56, 113; current steering 78; Schottky 85, 86, 210, 213, 214, 303, 357, 358, 439
direct drive 370, 371, 374
discontinuous (conduction) mode 355–357
distributed gap material 156, 195, 327, 382, 443
dithering 413
DSP 285, 424, 427, 428, 433, 434, 437

eddy current 194
E-field 298, 325, 380
EMC (Electromagnetic Compatibility) 164
ESD 25, 55, 57, 309, 311

fast Fourier transform (FFT) 391, 393
feed forward 427–429
ferrite: bead 55, 57, 306, 387; core 156, 160, 194, 306, 381, 395, 443; gap 270, 382; PQ 156, 381, 382; RM 156, 381, 382
filterless Class-D 14, 17, 143, 154, 176–181, 300, 389, 390, 425, 433, 435, 436
fixed-frequency 78, 171, 403, 411
flip flop 414, 415
flux band 381
forced PWM (FPWM) 355
forward recovery time 117, 211
FPGA (Field Programable Gate Array) 50, 72, 98, 426, 427
FR4: 298, 334, 335
frequency response 140, 151, 176, 226, 232, 306, 368–370, 391, 399, 409, 429, 433
fringing flux 158, 194, 381, 382
full-bridge 138, 139, 145, 147, 300, 352

gain margin 228, 239, 240
GaN (gallium nitride) 19, 438
gap pad 330–335, 444
gapped core 156, 158, 160, 194, 270, 327, 381, 382, 443
gate charge 88, 216
gate driver 99–102, 108–110, 384, 439, 440–442
ground loop 58

half-bridge 27, 100, 114, 290, 337
heatsink 289, 298, 330–335, 385, 432, 433, 444
Hi-Z 366
housekeeping 242
hysteresis 267, 287–288; controller 417; input threshold 105–107, 119; modulator 414, 419; power loss 382

I2S 419, 427, 436, 437
ICEpower 21, 437
idle tones 411, 413
IGBT 110
IMD (Intermodulation Distortion) 145, 393, 394, 398; SMPTE IMD 394; Transient 396; twin tone 395
inductive coupling 329, 402
inductor: DCR 57, 150, 160, 280, 400
Infineon 435
input overdrive tolerance 62, 64
insertion loss 368
integrator 221
intellectual property (IP) 426
inverter 143
inverting buck-boost 28, 358, 359
IPC 327

Jitter 47, 76, 195, 196, 384, 403
junction temperature 102, 190, 281, 282, 287, 289, 330, 331, 333

Karsten Nielsen 21

latch 27, 28, 36, 37, 249, 267, 269, 270, 364
layout 312

L-C demodulation filter 148
LDO (Low DropOut) 28, 39, 48, 108, 340, 354, 361–363
LED (Light Emitting Diode) 32, 274, 336
level shifter 14, 16
linearity 76, 80, 138, 156, 184, 194, 196, 240, 394, 395, 409
Litz wire 443
load regulation 220, 368, 369, 378, 428
logic gate 47, 66, 67, 85, 249, 250
loop gain 3, 143, 145, 221, 224–226, 230–231, 234, 237–239, 320, 379, 408, 409, 411, 418, 419
LTspice 111

Magnetic: blow-outs 260; core 156, 163, 327; coupling 108; field 176, 194, 260, 327, 381; flux 156, 158, 327, 381; flux density 381; flux leakage 382; gap 257; hysteresis 156; leakage field 381; losses ; non-linearity 156; path 165; permeability 156, 158, 160, 161, 163, 165, 172, 194, 285, 369, 382; Saturation 156, 160, 161, 165, 170, 194, 270, 272, 283, 306, 327, 368, 369, 382, 395, 443
magnetizing current 369
Magnetizing force 165
Micrometals (Inc.) 35, 46, 163, 165, 168, 194
microprocessor 424
microstrip 329
miller plateau (interval, region) 89–91, 107, 112, 216, 425
minority carrier 18, 19, 208, 440
MLCC 191
modulation index 138–139, 226, 411, 413, 414, 419
Monolithic Power Systems, Inc. (MPS) 436
MOV (Metal Oxide Varistor) 48
MTTF 255, 353
multifunction gate 65, 67, 246, 284

N-channel 19, 77, 88, 93, 98, 189
natural sampling 75
NEC 21
noise: density 417; power 417, 418; shaping 418
notch filter 26, 149, 171–173, 391, 412, 413
NXP Semiconductors 436
NTC (Negative Temperature Coefficient) 287
Nyquist 407, 408, 417

OEM 343
Onkyo 21
op amp (operational amplifier) 25, 26, 50–59, 64, 225, 226, 314, 384
open loop 96, 138, 145, 183, 184, 196, 219–221, 225, 226, 237, 379, 392, 399, 406, 408, 419, 427–429
oscilloscope 68, 69, 185, 206, 209, 238–240, 276, 277, 390
output impedance 64, 173, 232, 399
overcurrent 270
overdamped 399
oversampling 391 415–418; ratio (OSR) 417
over temperature 27, 37, 286, 336
overvoltage 27, 28, 55, 57, 85, 116, 118, 181, 209, 249, 257, 269, 280, 440–441

P-channel 19, 47, 93, 116, 118
Panasonic 21
PBTL (Parallel Bridge Tied Load) 17, 125, 433, 434
PCB (Printed Circuit Board) 298, 312
PCM 419, 427
PDM 427
Peavey 20
PFFB (Post Filter Feedback) 434
phase: lag 222; Lead 226, 228; margin 173, 183–184, 230, 234, 239, 240; Reversal 50, 53, 54, 64, 263, 442; Shift 102, 149, 151, 173, 221, 225, 226, 232,

234, 399, 407–409, 414, 419, 420; Shift Modulator 418
phasing 60, 351
piezoelectric 175, 191
p-n junction 19, 117, 187–189, 208, 439, 441
Point-of-Load (POL) 28, 353
polarity protection diode 358
pole 58, 79, 161, 219, 221, 225, 256, 264, 369, 399
pole frequency 58
powdered iron 156, 158, 160, 163, 195, 270, 306, 327
power density 1, 2, 281
propagation delay 49, 62, 63, 65, 84, 87, 102, 110, 161, 196, 220, 246, 409, 428, 432, 441, 442
PSRR (Power Supply Rejection Ratio) 52, 58, 138, 145, 192, 220, 361, 377, 390
Putzeys, Bruno 50, 314, 413, 414, 418, 419, 424, 428, 429

Q 149, 412
quality factor 149, 412
quantization noise 416, 418
quantizer 414–415, 416

rail pumping 28, 138–140, 263, 346, 347, 350, 351, 371, 372, 374, 379
rail-to-rail: input (RRI) 50, 63; input output (RRIO) 52; output (RRO) 50, 52, 53, 57
RCWV 104
Reeves, Alec 20
regenerative drive 269
relay: coil 259, 261; contact 242, 259, 260, 261, 262; electromechanical 244; latching 261, 262
residual output voltage ripple 15, 104, 129, 130, 148, 151, 153, 154, 160, 161, 172, 174, 181
resistor: current sensing 273–275; reverse geometry 281, 443; thick film 194; thin film 194
resonant: frequency 150, 161, 162, 225; switching 185, 201–203
reverse recovery: characteristic 19, 115, 134, 188, 208, 209, 211, 303, 438; charge (Qrr) 19, 209–212; current 189–190, 209, 297, 303; interval 137, 188, 205; losses 199, 212; time (Trr) 19, 78, 117, 135, 189, 207, 209–211, 438
RFI 26
right hand rule 381
ringing 20, 96, 117, 121, 196, 209, 212, 213, 279, 281, 304, 312, 355, 372
Risbo, Lars 21, 183

safe operating area (SOA) 19, 29
Samsung 21
Sanyo 21
saturation: flux density 382; hard 160, 161, 396; soft 160, 356
sawtooth ramp 74, 75, 355, 406
Schmitt trigger 105, 249
SCR 269
self-bias 98, 99
self-inductance 281
self-oscillating 102, 407, 409, 411–413, 418, 426
self resonant frequency (SRF) 285, 303, 314, 315
servo 26, 71, 79, 288, 363
shunt reference 47, 250, 252, 291–292
SiC (Silicon Carbide) 19, 438
sigma delta modulation (SDM) (see delta sigma modulator) 391, 414–417; analog 415, 416; asynchronous 414, 415; synchronous 414
Sinclair 20
single ended (SE) 409, 433, 434
Skyworks Solutions Inc 90, 95, 105, 108, 245

slew rate (SR) 20, 51, 84, 108, 129
small signal 49, 77, 289, 354, 383
SMD 19, 175, 192, 281, 287, 289–290, 303, 328, 330, 336, 438, 440
SNR (Signal to Noise Ratio) 23, 63, 76, 329, 353, 390, 433
snubber 304
Sony 21
speaker (electrodynamic) 176, 177, 257, 400
split rail 47, 62, 63, 69, 84, 350, 351, 372, 373, 390
spread spectrum 304, 413, 433, 436
square wave 71, 129, 132, 154, 162, 222, 277, 281, 300
stability (loop) 50, 53, 150, 183, 219–221, 224–226, 238, 407, 409, 419, 429
star ground 322–324, 383
start-end gap 158, 160, 164, 170
start-up: pop-less 98, 103, 138, 140, 269, 413, 426
ST Microelectronics 436
stitching vias 328
stripline 329
summer 415
summing junction 221, 226
switching loss 147, 185, 202, 206, 214–217, 370, 375, 412
switching node 117, 126–133, 142, 187, 188, 204, 298, 304, 325, 380, 416
symmetry (ramp) 76
synchronous counter 196
synchronous rectification 325, 355–356, 358, 379

TA1101 21
Tact Audio 21
Tact Millenium 21
TDM 427, 436, 437
ternary modulation 434
Texas Instruments (TI) 433
Thandar, Thurlby 20
THD (Total Harmonic Distortion) 3, 50, 76, 96
THD+N (Total Harmonic Distortion plus Noise) 390
thermal: bridging device 290; mass 193, 286, 287, 289–290, 328, 385; vias 335, 444
thermistor (see NTC)
thermostat 289
third quadrant operation 19, 439, 441
TIM (Transient Intermodulation Distortion) 52, 396
TIM (Thermal Interface Material) 330, 332, 335
TO-220 289, 335
TO-247 335
TO-3P 286
toroid (toroidal) 35, 46, 156, 158, 160, 161, 163–167, 172, 223, 283, 327, 382
Toshiba 21, 47
transformer: current 285; driver 120; gate-drive 19, 112, 119, 120–123; isolation 119, 238, 240, 370; load regulation 369; output 368; power 293; step-down 18, 367, 368, 371; step-up 368–370; turns ratio 369; winding 112, 369, 370
transient response 428
transient voltage suppressor (TVS) 56, 195, 374
transient voltage undershoot 101
tranzorb 55
triangular ramp (generator) 16, 74, 75, 77, 143, 144, 168, 195, 200, 222, 226, 320, 384, 386, 403, 404
Tripath 21
Tripathi, Adya 21
TTL 85, 86, 105

UL62368 386
Ulrick, John 20
unbalanced 58
underdamped 181

unity gain: crossover frequency 102, 151, 173, 222, 225, 226, 228, 230, 237, 239, 240, 407–409, 418–419; stability 50, 53
UVLO (Under Voltage LockOut) 97–100, 242–250, 259

VFETS 20
victim (see crosstalk)
voltage feedback 27, 337
volt-second product 112, 113

WBG (Wide BandGap) 5, 19, 101, 108, 110, 210, 438

XNOR 32, 67

zero frequency 221, 225
zero voltage switching 132, 138
Zener 64
Zobel network 26, 154–157, 159, 181, 224–228, 230–234, 283, 380